REMEMBERING
the SPACE AGE

ISBN 978-0-16-081723-6

For sale by the Superintendent of Documents, U.S. Government Printing Office
Internet: bookstore.gpo.gov   Phone: toll free (866) 512-1800;   DC area (202) 512-1800
Fax: (202) 512-2104 Mail: Stop IDCC, Washington, DC 20402-0001

ISBN 978-0-16-081723-6

REMEMBERING the SPACE AGE

Steven J. Dick

Editor

National Aeronautics and Space Administration
Office of External Relations
History Division
Washington, DC

2008

NASA SP-2008-4703

Library of Congress Cataloging-in-Publication Data

Remembering the Space Age / Steven J. Dick, editor.
    p. cm.
  Includes bibliographical references.
  1. Astronautics--History--20th century. I. Dick, Steven J.
  TL788.5.R46 2008
  629.4'109045--dc22

                                   2008019448

CONTENTS

PART I. NATIONAL AND GLOBAL DIMENSIONS OF THE SPACE AGE

PART II. REMEMBRANCE AND CULTURAL REPRESENTATION OF THE SPACE AGE

PART III. REFLECTIONS ON THE SPACE AGE

Acknowledgments

I wish to thank the members of the conference organizing committee, including Roger Launius (National Air and Space Museum), Linda Billings (SETI Institute), Asif Siddiqi (Fordham University), Slava Gerovitch (MIT), Bill Barry (NASA Headquarters Office of External Relations), and, on the staff of the NASA History Division, Stephen Garber and Glen Asner. I also want to thank Michael Neufeld, who took over from Roger Launius as the Chair of the National Air and Space Museum Division of Space History during our planning and gave us his full support. My thanks to Nadine Andreassen in the NASA History Division for her usual good work in planning the logistics for the meeting, as well as Kathy Regul and Ron Mochinski for their logistical work. Finally, thanks to the Communications Support Services Center team at NASA Headquarters for their crucial role in the production of this book.

Introduction

Fifty years ago, with the launch of Sputnik I on October 4, 1957 and the flurry of activity that followed, events were building toward what some historians now recognize as a watershed in history—the beginning of the Space Age. Like all "Ages," however, the Space Age is not a simple, straightforward, or even secure concept. It means different things to different people, and, space buffs notwithstanding, some would even argue that it has not been a defining characteristic of culture over the last 50 years and therefore does not deserve such a grandiose moniker. Others would find that to be an astonishing viewpoint, and argue that the Space Age was a saltation in history comparable to amphibians transitioning from ocean to land.[1]

There is no doubt that the last 50 years have witnessed numerous accomplishments in what has often been termed "the new ocean" of space, harking back to a long tradition of exploration. Earth is now circled by thousands of satellites, looking both upward into space at distant galaxies and downward toward Earth for reconnaissance, weather, communications, navigation, and remote sensing. Robotic space probes have explored most of the solar system, returning astonishing images of alien worlds. Space telescopes have probed the depths of the universe at many wavelengths. In the dramatic arena of human spaceflight, 12 men have walked on the surface of the Moon, the Space Shuttle has had 119 flights, and the International Space Station (ISS), a cooperative effort of 16 nations, is almost "core complete." In addition to Russia, which put the first human into space in April 1961, China has now joined the human spaceflight club with two Shenzhou flights, and Europe is contemplating its entry into the field.

---

1. Walter McDougall (see chapter 18 of this volume) opens his Pulitzer-Prize winning book . . . *the Heavens and the Earth* with such a scenario. See also Walter A. McDougall, "Technocracy and Statecraft in the Space Age: Toward the History of A Saltation," *American Historical Review* 87(1982), 1025. By the 40th anniversary of Sputnik in 1997 McDougall had revised his thesis to say "I no longer think that *saltation* was the right label for the chain of events kicked off by Sputnik." But he thought in the long term, when a new launch technology had replaced the "clumsy chemical rocket," *saltation* might still prove an apt term. Walter A. McDougall, "Was Sputnik Really a Saltation?" in *Reconsidering Sputnik: Forty Years Since the Soviet Satellite,* ed. Roger D. Launius, John M. Logsdon, and Robert W. Smith (Harwood Academic Publishers, 2000), pp. xv–xx. In chapter 4 of this volume, Robert MacGregor also challenges the view of Sputnik as a technological saltation, arguing that technocratic ideas of the relation of science to the state were already well established by this time. In particular he points to the parallels between the Atomic Energy Commission and NASA, and further argues that "NASA's rise in the 1960s as an engine of American international prestige was rooted in atomic diplomacy, and that certain debates in Congress about the new Agency were largely approached from within a framework of atomic energy, thereby limiting the range of discourse and influencing the shape of the new Agency."

After 50 years of robotic and human spaceflight, and as serious plans are being implemented to return humans to the Moon and continue on to Mars, it is a good time to step back and ask questions that those in the heat of battle have had but little time to ask. What has the Space Age meant? What if the Space Age had never occurred? Has it been, and is it still, important for a creative society to explore space? How do we, and how should we, remember the Space Age?

It is with such questions in mind that the NASA History Division and the National Air and Space Museum Space History Division convened a conference on October 22-23, 2007, to contemplate some of the large questions associated with space exploration over the last half century. The conference was designed to discuss not so much the details of what has happened in space over the last 50 years, nor even so much the impact of what has happened, but rather its meaning in the broadest sense of the term.[2] In doing so, the organizers made a conscious attempt to draw in scholars outside the usual circle of space history. This was not an easy task; we found that, with few exceptions, historians had not contemplated the meaning of the Space Age in the context of world history, even though the Space Age has given rise to an embryonic movement known as "big history" encompassing the last 13.7 billion years since the Big Bang.[3] We therefore turned to "big picture" historians, among whom is John R. McNeill, who had recently coauthored *The Human Web: A Bird's-eye View of Human History* with his father, William H. McNeill, another big picture historian.[4] With the idea that space is the ultimate "bird's-eye view" and that it has enlarged and enhanced the human web, we invited the younger McNeill to deliver our opening keynote lecture. Readers will find his provocative thoughts in chapter 1.

The conference encompassed two main themes, reflected in the first two sections of this book. The first, "national and global dimensions of the Space Age," was meant to examine the place of space exploration in human history. Here the guiding questions were as follows: Has the Space Age fostered a new global identity, or has it reinforced distinct national identities? How does space history connect with national histories and with the histories of transnational or global phenomena such as the Cold War, the rise of global markets, or global satellite communications? One might argue there is a fundamental

---

2. On the question of societal impact, see Steven J. Dick and Roger D. Launius, eds., *Societal Impact of Spaceflight* (NASA SP-2007-4801: Washington, 2007).

3. On "big history," see David Christian's "The Case for 'Big History'" in *The Journal of World History*, 2, No. 2 (Fall 1991): 223-238 (*http://www.fss.uu.nl/wetfil/96-97/big.htm*); David Christian, *'Maps of Time': An Introduction to 'Big History'*, Berkeley, CA: University of California Press, 2004; as well as Fred Spier, *The Structure of Big History: From the Big Bang Until Today*, Amsterdam: Amsterdam University Press, 1996, and Marnie Hughes-Warrington, "Big History," *Historically Speaking*, November, 2002, pp. 16-17, 20 (available at *http://www.bu.edu/historic/hs/november02.html#hughes-warrington*).

4. J. R. McNeill and William H. McNeill, *The Human Web: A Bird's-eye View of Human History* (W.W. Norton, New York, 2003).

tension between national goals of preeminence in space, and the global identity that the Space Age fosters in a variety of ways—through communications satellites, through the global awareness brought by photos of the whole Earth and Earthrise from the Moon, and through international cooperation in huge human spaceflight endeavors like the ISS. Scholars from several disciplines and backgrounds examined these questions from a variety of perspectives, and the debate spilled over into the entire conference. Speaking in his article of applications satellites such as the Global Positioning System (GPS) and Iridium, Martin Collins concludes that "Spaceflight—especially those near-Earth applications cited here—has been a major element in creating the incarnation of the global we have experienced over the last 40 years. It has provided images and practices that have made the category of the global natural and insistent, even when different actors give it different meanings. It has been a primary site in which prior categories of the modern—the nation state, the military, civil society, capitalism—have been refashioned and given new meanings." Others in this volume differ with that interpretation.

The second theme, "remembrance and cultural representation of the Space Age," posed questions about how the historical record of the Space Age has been collected, preserved, displayed, and interpreted around the world, especially in the United States, Russia, the European Union, Canada, and China. How do the "official" versions of events square with the document trail and with eyewitness accounts? How has the Space Age been represented in the arts, the media, the movies, in propaganda discourse, and so on? What does space exploration tell us about culture and the space endeavor's relation to culture? Such questions are not confined to the realm of space. They permeate all of history, and especially global events affecting the masses. In the context of the 20th century out of which the Cold War and Sputnik emerged, for example, one thinks in the fictional realm of Herman Wouk's novel *War and Remembrance*, or Emily Rosenberg's non-fiction work *A Date Which Will Live: Pearl Harbor in American Memory*.

In Rosenberg's piece and others in part II of this volume, one will find reflections on remembrance and history. "From the 1950s to the 1970s," she writes, "space held many meanings: it was a symbol-laden arena in which people and nations staged Cold War competitions, a 'star' in the media firmament, an ultimate challenge for scientists and engineers, and an inspiration for artists and designers." In this vein, both Roger Launius and Slava Gerovitch contemplate the "master narratives" of the Soviet Union, Russia, and the United States with regard to the Space Age, finding them inevitably grounded in culture. As Gerovitch concludes, "There can be no 'true' memory, as any act of recollection reconstitutes our memories. As different cultures remember the Space Age, it keeps changing, revealing new symbolic meanings and providing an inexhaustible source of study for historians. By shifting the focus from debunking myths to examining their origins and their constructive role in culture, we can understand memory as a dynamic cultural force, not a static snapshot of the past." Reflecting on the relation between reality, memory, and meaning, Launius offers a provocative premise: "Perhaps the reality of what happened does not matter all that much; the only thing that is truly important is the decision about its meaning. That may well be an intensely personal decision predicated on many idiosyncrasies and perspectives."

Remembrance is not simply a pleasant pastime, something to be contemplated at life's end. Nor is history a luxury, an irrelevance, or a straightforward activity as some people might think. Not without reason does there exist a National Archives in the United States with the words "What is Past is Prologue" scrolled along the top of its impressive façade, this building duplicated in function, if not in detail, in most countries of the world. Not without reason does the Smithsonian Institution strive to display thoughtful commentary in its exhibits, despite criticism from its wide variety of audiences, each with their own interpretations of history. And not without reason does every high school, college, and university teach history. As Wouk said in the context of his novel, "the beginning of the end of War lies in Remembrance." Whether we learn the lessons of history is another matter.

Part III of this volume consists of reflections and commentary, where some of the major themes are once again engaged. Walter McDougall views the 50th anniversary of the Space Age as a melancholic affair, filled with disappointment and unfulfilled hopes, a secondary activity compared to the dominant trends of contemporary history, and in any case too embryonic to judge its significance. John Logsdon disagrees in part, arguing that both the modern nation-state and the global economy depend on space-based systems. The ability to operate in outer space, he contends, is an integral part of modern history. He agrees that the progress of the Space Age has been frustrating in many ways to those who lived through the Apollo era, a level of activity that was not sustainable. Sylvia Kraemer argues there are many competing events that may define the last 50 years more than space exploration, including the Cold War and digital and information technologies. She also argues that the contribution of space activity to globalization has been far greater than its contribution to nationalism. Linda Billings reflects on space exploration in the context of culture, concluding that it means many things to many people, quite aside from dominant official narratives. Nor is this an academic exercise, for she suggests that if space programs are to survive and thrive in the 21st century they need to involve citizens and be aware of the visions they have for a human future in space. On the global level, this resonates with John Krige's statement that "when 'Remembering the Space Age,' we should not shy away from admitting the complexity and diversity of the space effort, nor pretend that the view of the world from Washington is the only view worth recording."

Some in the audience at this 50th anniversary conference thought it should have been more celebratory and described the meeting itself as depressing. Others felt it reflected both the frustrations and the realities of the Space Age. In the end, there seemed to be consensus that human spaceflight has been a disappointment in the aftermath of Apollo, and in that sense the Space Age, if indeed it ever existed, has been a disappointment as well. Such disappointment is no artificial construct of historians; the legendary Wernher von Braun, who thought humans would land on Mars by 1984, would undoubtedly have agreed. Nor is disappointment necessarily a bad attitude; it means vision has outstripped practical realities and that vision may yet drive individuals and nation-states toward new realities.

In common parlance, the title "Remembering the Space Age" carries with it a connotation that we are looking back on something that may have ended. Or maybe it never began; certainly launching Sputnik in and of itself did not constitute a Space Age, and the resulting reaction culminating in the manned lunar landings had ended within 15 years. Communications, navigation, weather, reconnaissance, and remote sensing satellites have been more sustained. But is such space activity, bounded by commercial and practical applications, enough to constitute a Space Age? Or, as several speakers opined, is space science the real core of the Space Age? As John McNeill concluded in his opening paper, it may well be too early to tell whether space activities over the last half century constitute a genuine "Age." We may need more time for better perspective. One thing is certain: if indeed the Space Age exists and if it is to continue, it must be a conscious decision requiring public and political will. Like exploration, each culture must set its priorities, and there are no guarantees for the Space Age.

The reader will find this volume filled with many more provocative thoughts and themes, large and small. However one defines or explains away the Space Age, whether in terms of space science, human spaceflight, applications satellites, or a combination of all of them, it is clear that what we usually refer to as the Space Age has been remembered differently by individual scholars depending on their perspective, by scientists and engineers depending on their specific roles, and by the public depending on their priorities. Moreover, it has been remembered differently depending on when one contemplates these questions. Quite aside from references to "semiotics," "tropes," and other postmodern terms common in the first years of the 21st century, the record presented in this volume is quite different from the perspective 25 years ago, or even 10 years ago.[5] And it will be different 25 years from now. Such is the nature of memory; such is the nature of history.

Steven J. Dick
NASA Chief Historian
Washington, DC
May 2008

---

5. See, for example, *Reconsidering Sputnik: Forty Years Since the Soviet Satellite,* ed. Roger D. Launius, John M. Logsdon, and Robert W. Smith (Harwood Academic Publishers, 2000).

PART I.

NATIONAL AND GLOBAL DIMENSIONS OF THE SPACE AGE

CHAPTER 1

GIGANTIC FOLLIES?
HUMAN EXPLORATION AND THE SPACE AGE IN
LONG-TERM HISTORICAL PERSPECTIVE

J. R. McNeill

In 1667, the poet John Milton, in the final quatrain of *Paradise Lost* reflected upon the exodus of Adam and Eve from the garden of eden. As a believing Christian, Milton understood the biblical story as truth, and thus as the original human voyage of exploration:

> *The world was all before them, where to choose*
> *Their place of rest, and Providence their guide:*
> *They hand in hand, with wand'ring steps and slow,*
> *Through Eden took their solitary way.*

Since the first humans trod this Earth, perhaps 10,000 generations ago, slow wandering steps have formed a characteristic part of the experience of most peoples at one time or another, and for some, migration and exploration has stood at the center of their experience of life. In recent years, wandering and exploration rarely involved literal human steps, but rather technologically sophisticated and organizationally complex efforts to take giant leaps.

My aim in this chapter is to place the whole endeavor of the Space Age into a global historical context. My friend and fellow historian, Felipe Fernandez-Armesto, in his recent book entitled *Pathfinders: A Global History of Exploration*, refers to space exploration as a "gigantic folly."[1] He could be right, but it is too soon to be sure. Folly or not, we can be sure space exploration is consonant with the deepest traditions of our species.

In the pages that follow, I will try to show just how deeply rooted exploration is in human society and will speculate on why that should be so. I will also reflect on some of the global-scale changes since the dawn of the Space Age in 1957, and where space exploration fits in this contemporary history.

---

1. Felipe Fernandez-Armesto, *Pathfinders: A Global History of Exploration* (New York, NY: Norton, 2006), p. 399.

THE REAL GREAT AGE OF EXPLORATION

When I was a schoolboy in Chicago—early in the Space Age—I defied the odds by studying "The Explorers" five different times between third and tenth grade. Maybe it was the buzz surrounding the Apollo program that inspired my teachers year in and year out to include a unit on Marco Polo, Vasco da Gama, Columbus, Magellan and all the rest, usually ending with Lewis & Clark. I memorized the dates of voyages the way my grandfather as a school boy had memorized scripture. While it didn't do me any harm, later in life I felt misled upon learning that da Gama hired a local pilot in Mombasa (today's Kenya) to take him and his ship to India. He didn't really explore anything: there were already people everywhere he went who gave him directions, as well as provided him and his men with supplies. The same was true of Marco Polo, Columbus, Magellan, and the rest. They were visiting lands unfamiliar to them, and certainly took great risks on the sea, but "explored" only in a generous sense of the word.

The real explorers in human history are almost all unknown, anonymous figures, people who explored unpeopled realms. And the real great age of exploration was long ago and lasted for tens of thousands of years.

Our species, *homo sapiens sapiens*, evolved from various hominid predecessors some time around 250,000 to 150,000 years ago, or so the scanty evidence suggests. This happened in Africa, somewhere between the Ethiopian highlands and the South African high veld. We evolved in ways that made us well-suited to the grassland and parkland ecosystems of East Africa. Among other things, we became excellent long-distance walkers.

After the burst of climate change that occurred at the onset of a glacial period and a period of technological advance in toolmaking, a few bands of humans walked out of Africa, probably around 100,000 years ago. They and their descendants probably skirted the shorelines of the Indian Ocean—now underwater because of deglaciation and sea level rise in the past 15,000 years— where food could be scooped up fairly easily in the intertidal zones. They arrived in India perhaps 70,000 years ago, and in China about 67,000 years ago. People first made it to Australia, which required a maritime voyage even in those days of lower sea level, perhaps 60,000 years ago. Others veered off into Europe about 40,000 years ago and into Siberia some 30,000 years ago. The final frontiers in this long saga of exploration were the Americas (15,000 years ago) and Polynesia (4,000 to 1,000 years ago). New Zealand was the last sizeable piece of habitable land to be discovered, around AD 1000 or maybe 1200.[2]

These peregrinations were real exploration. In some parts of the world, these footloose *Homo sapiens* encountered a few *Homo erectus*, whose ancestors had also walked out of Africa perhaps half a million years back. But those

---

2. These dates are rough estimates—the older the date the rougher the estimate—subject to revision by a single new archeological find.

Homo erectus did not have language, or at least not much of it, and could not tell *Homo sapiens* much of anything about the lands they were exploring. In any case, *Homo erectus* soon went extinct wherever our ancestors showed up, a disconcerting fact about our family tree. In Europe, the new arrivals encountered Neanderthals and swiftly swept them into the dustbin of prehistory. So for all intents and purposes, these wandering *Homo sapiens* were exploring unpeopled lands, unfamiliar not only to them, but to everyone alive. The trip to Australia must have been especially challenging—across open water and into a new and exotic biological kingdom with almost no familiar plants or animals. Similarly, exploring north into Siberia took much courage: few edible plants, trackless tundra, and bitter cold (the first humans arrived in the middle of the last Ice Age). They needed warm clothes and skill in very-big-game hunting, as well as a full supply of either optimism or desperation.

These Paleolithic pathfinders knew nothing at all about what lay over the horizon, and no one could tell them, yet they went. Had these individuals had the foresight to leave to posterity letters, diaries, journals, and handsome engravings instead of merely the odd flint or chunk of charcoal, their stories would be well-known and their status as historical icons assured.[3]

MOTIVATION AND EXPLORATION

Why did they do it? Why leave home at all, why walk out of Africa, why sail to Australia? We can't know, but we can make informed guesses. The last chapter of these great explorations was the Polynesian one, and we know more about that than any of the earlier ones. There are oral traditions, such as those maintained by New Zealand's Maori, as well as much more plentiful archeological remains. Linguistic and genetic evidence adds details to the general picture. The Polynesians clearly organized deliberate voyages of exploration, discovery, and colonization. Presumably, despite their legendary maritime skills, many Polynesian voyages ended badly because the Pacific is a big ocean with only a few specks of habitable land. These voyages were very risky undertakings. People accepted the risk presumably because staying at home seemed worse. In some cases, perhaps, island populations grew too large and starvation loomed, inspiring some to take to the sea in search of fertile land or fish-filled lagoons. Oral traditions suggest that, in other cases, conflicts arose, such as between two claimants to a chieftaincy, and one had to go: their island wasn't big enough for the two of them. If the loser was lucky, his followers

---

3. Clive Gamble, *Timewalkers* (Cambridge, MA: Harvard University Press, 1994) remains useful on prehistoric migration. See also Steven Mithen, *After the Ice* (Cambridge, MA: Harvard University Press, 2004).

would accompany him over the horizon, perhaps to find a good uninhabited island somewhere, perhaps to find only endless ocean and early death.[4]

The epic Paleolithic peregrinations probably arose more often from conflict than overpopulation. It is hard to imagine the exodus from Africa owing anything to overpopulation: there were probably well under a million people at the time, a number easily sustained on a continent as big as Africa. So why did they move on? Perhaps they were following game animals. Perhaps a wet phase in the Sahara was ending and they had to find a new home. Perhaps they were consistently curious. Perhaps each incremental movement of people arose from different motives. But quite likely conflict was often involved, as among the Polynesians, and the easiest resolution required some group to move away. In some cases, perhaps, cultural conservatives objected to some changes and, like the Puritans who settled in New England, hived off in order to be able to practice their old ways without harassment. Once they got there, they probably often quarreled, split, and the stronger or luckier drove off the unlucky schismatics, as among the Massachusetts Puritans. In other cases, perhaps, cultural radicals pursued new ways that others found distasteful, so, like the Mormons in the mid-19th century, they were driven over the horizon where they, too probably quarreled and split.

Space exploration also arose from conflict.[5] Although the dynamics were very different from what I have claimed about the Puritans, Polynesians, and Paleolithic peoples, it is clear that the funds provided by the Soviet Union and the United States, beginning just over 50 years ago, would not have been allocated without the Cold War context (or some equivalent unprecedented peacetime mobilization of money and resources). The dog Laika, the first living thing to experience Earth orbit, blasted into space aboard Sputnik II just a year after the October 1956 Suez Crisis and the Soviet invasion of Hungary. Gagarin and Glenn were propelled into space by a climate of anxiety fed by the Berlin crisis and the Cuban missile crisis. Americans and Russians did not explore space because they lost a quarrel or feared hunger, but because they feared they might suffer in the Cold War if they allowed space to be dominated by their rival. President Lyndon Johnson feared the communists would drop bombs on America "like kids dropping rocks onto cars from freeway overpasses."[6] In a less colloquial moment, Johnson claimed that, "Failure to master space means

---

4. Within the sizeable literature on Polynesian history, a good starting point is Patrick Kirch, *On the Road of the Winds: An Archeological History of the Pacific Islands before European Contact* (Berkeley, CA: University of California Press, 2000).

5. Roger D. Launius, "Compelling Rationales for Spaceflight? History and the Search for Relevance." In: Steven J. Dick and Roger D. Launius, eds., *Critical Issues in the History of Spaceflight* (Washington DC: NASA, 2006), pp. 37-70, reviews the American motives for space exploration, emphasizing the political ones.

6. *The Washington Post*, October 2, 2007: A1.

being second best in every aspect, in the crucial area of our Cold War world. In the eyes of the world first in space means first, period; second in space is second in everything."[7] Americans and Russians also found, as many had before them, that virtuoso displays of technological prowess and national resolve served useful propaganda purposes, which seemed especially important during the era of decolonization when the allegiance of billions of people around the world (and their geostrategic resources) was up for grabs. Laika, by the way, lasted four days in space before expiring from heat exhaustion on the 40th anniversary of the Bolshevik Revolution, November 7, 1957.[8]

Space exploration has proved to be a risky venture, although in statistical terms it is probably less so than Polynesian voyaging. Polynesians accepted high levels of risk because they felt they had to: at times, staying put carried unacceptable costs or perhaps even greater risks. Presumably, Paleolithic explorers sometimes arrived at the same calculus. Today, as for the last half century, one of the central questions surrounding space programs is that of risk. How much is prudent to accept when the returns are unmeasurable? Should human lives be risked for uncertain rewards? Although insurance companies and their customers explicitly put monetary values on human lives, and those responsible for air traffic and highway safety do so implicitly, when it comes to space exploration, this calculation is not mainly a matter of money and numbers, but of moral and political positions. For some, the ratio of risk to reward spells gigantic folly; for others, an irresistibly noble calling. Different people, different governments, different eras will hold sharply divergent views on this, and they are probably not easily reconciled or persuaded by mere reason.

In point of historical fact, no human being has ventured beyond low-Earth orbit since 1972.[9] To those in control, the further rewards to distant space travel since then apparently did not justify the risks. Détente, perhaps, diminished the determination behind human space exploration, as did financial difficulties. By the late 1960s, the Soviet economy had begun to flag, and although high oil prices helped prop it up for more than a decade after 1973, the malaise associated with the Brezhnev years did not augur well for renewed commitment to lofty ambitions in space. The American economy (and government revenues) suffered a downturn in 1973 due in large part to high oil prices, raised higher in 1979. So with détente dampening the motives, and economic difficulty undermining the means, ambitions for projects in space waned since the heady early days (c. 1957-1969) when all seemed possible. But, as with Paleolithic and Polynesian

---

7. Walter A. McDougall, "Technocracy and Statecraft in the Space Age: Toward the History of A Saltation," *American Historical Review* 87 (1982): 1025.

8. This is not the only ironic calendrical coincidence of Cold War space history: the day Sputnik I was launched, October 4, 1957, was the day American TV stations launched that paean to normalcy, "Leave It to Beaver."

9. Launius, "Compelling Rationales," p. 69.

explorers, such reluctance is always provisional. Conditions will change, political resources will ebb and flow, the premium on human safety will evolve. Sooner or later, someone in power somewhere will consider, once again, that the game is worth the candle. Perception of acceptable risk is not merely a calculation of probabilities costs, and benefits; it is also a cultural choice and always subject to reconsideration.

CHALLENGE AND LIBERATION IN EXPLORATION

When humans first left East Africa for the wider world, they experienced both a challenge and a liberation. Their new environments were unlike the ones to which they had been slowly attuned by biological evolution. A lot of their accumulated wisdom presumably applied less well to the shores of the Indian Ocean and its hinterlands than it had to the savannas of East Africa or the Nile valley. However, over time they developed new wisdom appropriate to their new surroundings. They adapted biologically in small ways in accordance with the novel pressures of their new environments, such as gradual variations in skin color to harvest more vitamin D in higher, sun-starved latitudes. In cultural and in biological ways, they met the challenges of migrating into unpeopled realms.

The liberation consisted, in the first part, of escaping the pathogenic load that had evolved among their ancestors. Countless pathogens had had plenty of time to adapt to life within and among hominids in the long haul of evolution in East Africa. Not all of these pathogens, however, made the trip out of Africa. Some could not handle the cooler temperatures of Eurasia. Others, by sheer chance, had not been along for the ride when the migrants left and could not catch up. Thus humans, upon arrival in Asia, entered into a golden age of health that would last some 90,000 years until the transition to agriculture— farming was a great leap backward as far as health was concerned. To judge from skeletal evidence, the first Eurasians suffered much less from infectious diseases than either their ancestors in Africa or their farming descendants. They did not, it seems, live much longer lives: accidents, violence, and abandonment of infants and toddlers kept life expectancy at birth around 30 years. One had to be healthy enough to walk in those days.

The liberation had a second aspect to it. In East Africa, while the foraging and gathering was probably good, the hunting was probably bad. All the big game there had had plenty of time to develop appropriate suspicion of upright, fire-wielding, projectile-throwing creatures, thus limiting the success of hunters. But in Eurasia, and later in Australia, the Americas, and New Zealand, people arrived amid populations of naïve wildlife. Hunting was comparatively easy when the prey took no notice of hunters until they were well within spear-throwing range. The world outside of Africa was a happy hunting ground until selection weeded out the unsuspicious or until the choicest prey grew scarce.

If humans are to leave Earth behind and settle elsewhere in the universe, they will experience something of the same challenges and perhaps the same liberation. They will need to adjust their culture to their new environments, jettisoning all that which was applicable only on Earth and devising new formats appropriate to the far corners of the universe. They will evolve biologically according to the pressures of their new surroundings, whatever those might be. For example, gravity of different strengths from what we have known on Earth would presumably encourage different sorts of bodies. Since migration around the globe led to numerous small biological adaptations in humans over the past 100,000 years (and among other animals as well), it stands to reason that space colonization would transform our bodies, too. Indeed, in short order humans elsewhere in space might cease to be humans. Given the vast distances involved, space colonists would cease to interbreed with Earthbound populations. Only if space colonies were to consist of glorified versions of Biosphere II, hovering in near space, could the biological oneness of humankind be preserved for long.

The biological evolution of space colonists, as with those of us here on Earth, might in time become a matter of conscious design through genetic manipulation more than of natural selection of the sort characteristic of us since time immemorial. The genetic and biological diversification of the creatures formerly known as humans would, it seems likely, grow rapidly in the event of exploration outside our solar system. Should that happen, then even after people colonize space, further space exploration would still be into unpeopled realms, strictly speaking, once people ceased to be people.

Their social and cultural evolution would, of course, also be affected by their distant new environments. The migrants out of Africa kept their basic social organization, the small band of 30 to 80 people who were mostly kinfolk, wherever they went. It seems to have adequately served the purposes of nomadic foragers and hunters, whether in Africa, Australia, or Siberia. Only when people settled down and domesticated plants and animals did they find new social formats (villages, chiefdoms, states) more appropriate. Space settlers, once free of the umbilical cord of Earth, would likewise presumably experiment with new social formats, finding alternatives to those we have known here on Earth.

Thus, in social and cultural terms, one could anticipate a liberation from earthly patterns in the event of space colonization. Whether this would also include counterparts to the epidemiological liberation and the happy hunting ground effect seems much less likely, depending a great deal on what exists out there in the colonized environments. Since hunting provides only the tiniest proportion of the food supply among peoples technologically capable of pursuing space exploration, and because our digestive capabilities are calibrated to the things we eat here on Earth, it seems most unlikely that space colonization would involve an analogue to the happy hunting the first Eurasians, Australians, and Americans enjoyed—even if there is something out there to hunt. Epidemiological liberation is another matter.

At first glance it seems reasonable to suppose that leaving earthly ecosystems behind might allow space travelers and settlers to shed much, if not all, of their pathogenic load. The early emigrants from Africa apparently did so, as did the wandering bands that left Siberia for the Americas around 14,000 years ago, founding the indigenous American populations that were for millennia unusually free from infectious disease.[10] However, it is implausible to suppose that all microbes can be left behind, and once in space microbes may behave differently. A new National Academy of Sciences study claims that certain pathogens, salmonella in particular, prosper better in space than on Earth.[11] Moreover, in all likelihood, the human immune system, like our digestive system calibrated for conditions here on Earth, would prove far less useful elsewhere in the universe. For this reason, if for no other, the health liberation that eased emigration from Africa and assisted settlement of the Americas, would probably not help us make our way in space.

What Mattered in History During the Space Age

This speculation about space, evolution, and who will remain really human threatens to get out of control and become its own gigantic folly. Let me return to a historian's *terra firma* and reflect upon the changes here on Earth since Laika's orbital flight.

In terms of health and demography, this last half century has been the most revolutionary in the human career. In 1957, the average life expectancy was about 47. Today it is close to 67.[12] While we have not been able to "close the book on infectious diseases," as the Surgeon General forecasted in the 1960s,[13] we have intervened dramatically in the relations between pathogens and our bodies. Sanitation, vaccines, antibiotics, and other measures have made a huge difference in human health. Some pathogens, such as the smallpox virus, have been ushered into extinction or near extinction. Many of the crucial developments in this story—the germ theory of disease, sewage treatment, penicillin—date to well before 1957. But their application, their spread around the world, their full effect came mainly after that date. Even though this health revolution remains unevenly distributed around the world, indeed unevenly distributed within many of the world's cities, it probably amounts to the single

---

10. See Alfred Crosby, *Ecological Imperialism: The Biological Expansion of Europe, 900-1900* (New York, NY: Cambridge University Press, 1987), pp. 197-198.

11. See for example Gillian Young, "Bacterial Virulence: Return of the Spacebugs," *Nature Reviews Microbiology*, 5, no. 11 (November 2007): 833-834.

12. United Nations data appearing in James C. Riley, *Rising Life Expectancy: A Global History* (New York, NY: Cambridge University Press), pp. 37-38.

13. This quotation is variously dated as 1967 or 1969. See J. R. McNeill, *Something New Under the Sun* (New York, Norton, 2000), p. 201.

greatest social change of the last half century. Whether it can be maintained indefinitely is an interesting question that depends chiefly on the ongoing arms race between pathogenic evolution and human efforts at disease control.

One result of the health revolution since 1957 is the global population explosion. The world had about 3 billion people in 1957; today it has more than twice that number. Put another way, it took hundreds of thousands of years for human population to add its first 3 billion, but only 50 years (more like 47 actually) to add the second 3 billion. Whereas for most of human history the annual population growth rate remained well below 0.01 percent, in the 1960s and early 1970s, it briefly attained 2.1 percent per annum. Now it is close to 1.3 percent annually. The last 50 years has been one great spike in population growth rates, unprecedented in our history and destined to end soon.[14] No other primate, perhaps no other mammal, has ever done anything like this in the history of life on Earth. Consider this: roughly 10-15 percent of the years lived by people and their hominid ancestors going back four million years have been lived after 1957.[15] A memorable way to visualize it comes from the Italian historian Carlo Cipolla: if post-1957 population growth rates had obtained from the dawn of agriculture 10,000 years ago to the present, Earth would now be encased in a ball of squiggling human flesh expanding outwards into space with a radial velocity greater than the speed of light, gobbling up planets and stars in its path.[16] (Just as well that didn't happen I suppose, even if it might have saved us the trouble of space exploration.) What did happen was remarkable enough.

Connected to this stunning growth of population is the sudden urbanization of our species. For our first few hundred thousand years on Earth, our characteristic habitat was savanna grasslands and parklands, riverbanks, and shorelines. For a brief span, maybe 7000 or 5000 B.C. to A.D. 2000, the farming village formed the standard human habitat. But now, for the first time, the typical human animal has become a city dweller. In 1800, about 3 percent of the world's population lived in cities, and only one city, Beijing, topped one million. By 1957, about 30 percent of us lived in cities, close to a billion people in all. And today, more than half of us, over three billion souls, are urbanites, and the world has some 468 cities with more than a million people.[17] In 1957, only one urban area, New York, housed upward of 10 million people. Now there are about 25 such megacities, the largest of which—Tokyo/Yokahama—

---

14. United Nations data appearing in Angus Maddison, *The World Economy: Historical Statistics* (Paris: OECD, 2003), pp. 255-256.

15. This figure is adapted from calculations made by J. N. Biraben, "Essai sur l'évolution de nombre des homes," *Population*, 34 (1979): 13-24; and J. Bourgeois-Pichat in, "Du XXe au XXIe siècle: Europe et sa population après l'an 2000," *Population* 43 (1988): 9-42.

16. Carlo Cipolla, *An Economic History of World Population* (Harmondsworth, UK: Penguin, 1978), p. 89.

17. According to Thomas Brinkmann's Web site at *http://www.citypopulation.de/World.html* (accessed September 4, 2007).

is home to some 33 million people, roughly the population of the entire United States at the time of the Civil War. Cities everywhere used to serve as a check on population growth because their infectious diseases killed people faster than others were born. But in the last few decades this has changed and cities are no longer demographic black holes, but instead hothouses of further growth.

This is, to put it mildly, a bizarre transformation. It is less conspicuous in the United States, where half the population was urban by about 1920, than in Asia and Latin America, where things happened later and faster. In national terms, the fastest large-scale urbanizations in world history were those of the Soviet Union in the 1930s (while building socialism) and China since 1980 (while dismantling it). The urbanization of our species surely carries tremendous significance in ways not yet fully apparent. We have built new environments and new habitats while simultaneously populating them and leaving behind the milieux that formed us and our institutions.

One of the reasons that cities could grow as they did and do is the radical changes in energy use witnessed in our times. Before the era of fossil fuels, cities in temperate latitudes, say North China or Europe, needed to command an area of forest some 50 to 200 times their own spatial size to meet their fuel wood needs.[18] This, together with limits to agricultural efficiency, constrained urban growth. Fossil fuels broke this constraint, and helped break the ones on agricultural efficiency. Since 1957, global energy use has almost tripled, largely as a result of the globalization of oil use. Oil was a small part of the energy mix outside of North America until the 1950s. What China and India are doing now in terms of deepening energy and oil appetites, was done by Western Europe and Japan on a smaller scale from the mid-1950s to the 1970s.[19]

Again, as with urbanization, the significance of fossil fuels since the 1950s is less conspicuous in the American context than elsewhere because they became important in the United States earlier. But in global terms, it is only after the 1950s, with the opening of the so-called elephant fields in Saudi Arabia, and then those in Western Siberia, that cheap energy became routine. With cheap oil, automobiles became the normal accoutrements of middle-class and, in richer countries, working-class life. Furthermore, transportation of goods around the world became far more practical, leading to ever more complex divisions of labor and levels of specialization that enabled larger and larger numbers of people to live lives of ease instead of near-universal grim and grinding toil.[20]

---

18. Vaclav Smil, *Energies*, (Cambridge MA: MIT Press, 1999), p. 118.

19. Useful histories of energy include Vaclav Smil, *Energy in World History* (Boulder, CO: Westview Press, 1994); Alfred Crosby, *Children of the Sun: A History of Humankind's Unappeasable Appetite for Energy* (New York, NY: Cambridge University Press, 2006).

20. This is explained for Europe in Christian Pfister, *Das 1950er Syndrom: Das Weg in die Konsumgesellschaft* (Bern, Switzerland: Paul Haupt Verlag, 1995).

Cheap energy is probably the single most important factor behind the spectacular economic growth of the last half century. For most of human history, the global economy grew at a snail's pace and, indeed, often shrank. The period since the middle 1950s, however, has chalked up by far the fastest growth rates ever posted. In the last five decades, the global economy has grown by about four percent per annum, twice as fast as during the second quickest era of expansion, which was 1870–1913. In per capita terms, the global economy has nearly tripled since 1957.[21] Of course, this is a very uneven achievement as some populations—in Central Africa for example—have scarcely benefited from this trend, while others, such as those in East Asia or southern Europe, have experienced far higher than average per capita income growth. For its overall growth, and for its wild geographic unevenness, the economic history of the last half century is far and away the most eccentric in the human record. This would be obvious to all if we did not naturally assume that what we have known from our own experience and observation is normal.

The extraordinary histories of population, urbanization, energy use, and economic growth over the past half century have combined to produce the most turbulent times yet in the history of human relations with the biosphere. Since at least the harnessing of fire, humans have had an outsized impact on Earth. That impact grew more widespread and profound in the 19th century when population growth and energy use began to climb at hitherto unprecedented rates. But the impact entered a new, tumultuous phase in the 1950s, so distinctive that the Nobel laureate chemist Paul Crutzen labeled it the "Anthropocene,"[22] the geological epoch dominated by human influence.

Since the dawn of the Space Age, the carbon dioxide concentration in the atmosphere has risen by a fifth, and global climate has begun to warm. Global forest area has declined by about 11 percent and grasslands by 19 percent. Freshwater use has tripled, and global irrigated area is up by 240 percent. Sulfur dioxide emissions have at least doubled (and that counts their decline in the United States and Europe since 1980). Same with methane emissions. Livestock numbers—mixing sheep, goats, pigs, and cattle together, no doubt an Old Testament, as well as methodological, abomination—are up about 170 percent. Cement production is up eight-fold.[23] You get the picture. Despite some improved technologies and

---

21. Maddison, *The World Economy*, pp. 260-262. See also Eric Lambin, *The Middle Path: Avoiding Environmental Catastrophe* (Chicago, IL: University of Chicago Press, 2007), pp. 26-28.

22. See Will Steffen, Paul Crutzen, and John R. McNeill, "The Anthropocene: Are Humans Now Overwhelming the Great Forces of Nature?" *Ambio* 36, no. 8 (2007): 614-621.

23. Data from the database maintained by Kees Klein Goldewijk at *www.mnp.nl/hyde/bdf* (consulted on September 4, 2007).

greater efficiencies in resource use, the economic miracle of the last half century has put unprecedented pressures on the biosphere that sustains all life.[24]

In the fullness of time, this environmental turbulence may come to appear the most important thing in the history of our times, more so than the Cold War; decolonization and the end of the British, French, Soviet and other empires; the growing emancipation of women; the rise of terrorism; the rise of China; the resurgence of political religion; the splitting of the atom; the decipherment of the human genome; or globalization. That remains to be seen: what is important about a given era depends entirely on what happens next. Should the stresses and strains upon the biosphere turn out one day to have been a mere tempest in a teacup, then this suggestion will have proved wrong. But if they build over time and prove disruptive in human affairs, then they will seem more meaningful, in time, than what preoccupied those alive in the second half of the 20th century.

THE PLACE OF SPACE EXPLORATION IN THE SPACE AGE

Given all these developments of the last 50 years that I claim are unprecedented, remarkable, revolutionary, and so forth, where do space exploration and space programs fit in? I am tempted to take refuge in the wisdom of Zhou Enlai (1898-1976), Mao Zedong's urbane foreign minister. French journalists in the 1960s asked Zhou what he thought was the significance of the French Revolution of 1789. Zhou paused thoughtfully and said that "it is too soon to tell."[25]

It is in fact too soon to tell what the real significance of the Space Age may be. At the moment, space exploration, space flight, and space research, all seem, at most, secondary next to the dominant trends of contemporary history. Moreover, nothing to do with space seems central in the sense that, had there been no Space Age, no Gagarin or Glenn, no Moonshot, no Hubble Telescope, no Laika, everything else probably would have unfolded much the way that it did. Some things would have been a bit different without spy satellites, communications satellites, weather satellites, Earth-observation satellites, and so forth. Hurricane Katrina (2005) and other weather disasters could have been even worse had we not known in advance what was coming. Figuring out the ozone hole over Antarctica would have taken longer.[26] But I am skeptical of the

---

24. An excellent study that shows the interplay of economic expansion and increased efficiencies in the Spanish national economy is Oscar Carpintero, *El metabolismo de la economía española: Recursos naturales y huella ecológica (1955-2000)* (Madrid: Fundación César Manrique, 2005).

25. This phrase is variously reported, for example, as "too early to tell" in Wikiquote (*en.wikiquote.org/wiki/Zhou_Enlai*). In any case, Zhou spoke with French journalists in French as he had studied in France for three years in his youth.

26. See Ray A. Williamson and Henry R. Hertzfeld, "The Social and Economic Impact of Earth Observing Satellites." In: Steven J. Dick and Roger D. Launius, eds., *Societal Impact of Spaceflight* (Washington DC: NASA, 2007), pp. 237-266.

view that, for example, spy satellites prevented the Cold War from turning into World War III. The big things would *probably* be much the same, for better or for worse. I write "probably" in italics as a way to convey uncertainty because I am conscious that there are many things about space programs that I do not know. Furthermore, questions of causation in counterfactual scenarios are inherently unknowable, even for the best informed.[27] Had hundreds of billions of dollars and trillions of roubles not been spent on space, what might they have been used for? We can't know, but my guess is nothing out of the ordinary, that is, a little more of both guns and butter.

Perhaps space programs indirectly affected the big trends, even if spy satellites cannot be credited with preventing World War III. Could, for example, the current surge of globalization have derived some of its momentum from an enhanced awareness that we are all in the same boat, all stuck on the same small blue dot spinning through the darkness? Or could it owe something to instantaneous communications via satellites?[28] My view is the best answer is: yes, but not much. If no one had ever seen photos of Earth from space, and if information from India and Indonesia still arrived by telegraph and took a day or two to reach other continents instead of a second or two, would globalization be substantially different?

Space programs, of course, had spinoffs that affected contemporary history. The two most consequential so far are communications satellites and (very indirectly) the Internet. Nearly two-thirds of all satellites are used for communications,[29] and they have dramatically lowered the time and cost required for long-distance communications. The Internet arose from the Defense Advanced Research Projects Agency (more familiarly known as DARPA), which itself was created in response to the successful launch of Sputnik. These are both developments of consequence in today's world. But the Internet would likely have evolved, in somewhat different ways no doubt, even without DARPA. And in the absence of communication satellites, what they now transmit would likely go via the Internet (as, increasingly, long-distance phone calls do now). These musings reinforce the conclusion that space programs changed the history of our times, but not (yet) in any fundamental ways. Contemporary history, however, will inevitably look different to those no longer in the middle of it.

Space exploration, as opposed to the totality of space programs, could well be relegated to the status of historical footnote if, in the years ahead, exploratory probes are shut down. Satellites in near orbit are surely here to stay for a while,

---

27. For a more favorable assessment of the significance of space programs, see Erik M. Conway, "Overview: Satellites and Security: Space in Service to Humanity." *Societal Impact of Spaceflight*, pp. 267-288.

28. James A. Vedda, "The Role of Space Development in Globalization." *Societal Impact of Space Flight*, pp. 193-206.

29. *The Washington Post,* October 2, 2007: A1, A6.

as they serve several useful purposes, and some of them at least are profitable. But exploration programs are another matter: they are especially expensive and, since they probably won't cure cancer or defeat terrorism, they are at high risk of being phased out by Congress and its equivalents in other lands when money gets tight. If so, in time space exploration will be forgotten, a dead end, a historical cul-de-sac. On the other hand, it could be that space exploration will thrive, find new budgetary champions in the corridors of power, perhaps in China if not elsewhere. The likely endurance of geopolitical rivalry means space exploration programs will probably have some appeal, partly practical, and partly for propaganda value. It could well be, given the appreciation of the risks involved, that robotic space exploration will have a long future but human space flight will not. This, I imagine, is more likely to be the case if the sponsors are aiming at practical benefits rather than rewards in terms of prestige and propaganda, for which heroic humans still, and perhaps always will, carry outsized value.

One way to look at the experience of space exploration, and one justification for its endless continuation, is to see it as a species of expeditionary science. Past rulers have often sent out scouts, spies, and scientists to take inventory of the resources and peculiarities of other lands. In the 18th century, Britain and France competed for geopolitical dominance in several parts of the world, and in that context sponsored scientists and scientific expeditions to gather useful information, whether about medicinal plants, trees suitable for naval timber, or a thousand other things that might come in useful one day. When Napoleon conquered Egypt in 1798, he loosed a team of scholars and scientists upon the country to ransack it for information (and art) of all sorts. From Russia to Spain, all European states with overseas interests sponsored expeditionary science on some scale, as in time did the United States. When Jefferson purchased half a continent from Napoleon in 1803, he bankrolled Lewis and Clark to take a preliminary inventory of what he had bought. During the Cold War, the United States and the U.S.S.R. sponsored scientific expeditions on a much more lavish scale to the polar regions and deep beneath the seas. Their space programs were, among other things, part of this tradition.

Space exploration may survive on one or another basis, but it still will not loom large in terms of human history unless something really new and interesting happens, the sort of thing people in the space business probably dream about—finding intelligent and agreeable (or at least neutral) life out there or colonizing new corners of the universe—or probably have nightmares about—developing effective space-based weapons suitable for use against earthly enemies or finding intelligent but hostile life out there. If any of these things happen, then the first 50 years of space exploration will look like the beginning of something of epic significance. If they don't, it will look like a small step for mankind that led nowhere, and did not amount to much in the balance before being consigned to the dustbin of history. It is indeed too soon to judge whether the whole enterprise is a gigantic folly diverting money and talent from more urgent applications, a noble calling consonant with our deepest nature, or something else altogether.

CHAPTER 2

SPACEFLIGHT IN THE NATIONAL IMAGINATION

Asif A. Siddiqi

INTRODUCTION

Few would recount the history of spaceflight without alluding to national aspirations. This connection between space exploration and the nation has endured both in reality and in perception. With few exceptions, only nations (or groups of nations) have had the resources to develop reliable and effective space transportation systems; nations, not individuals, corporations, or international agencies, were the first actors to lay claim to the cosmos. The historical record, in turn, feeds and reinforces a broader public (and academic) consensus that privileges the nation as a heuristic unit for discussions about space exploration. Historians, for example, organize and set the parameters of their investigations along national contours—the American space program, the Russian space program, the Chinese space program, and so on. We evaluate space activities through the fundamental markers of national identity—governments, borders, populations, and cultures.

As we pass an important milestone, moving from the first 50 years of spaceflight to the second, nations—and governments—retain a very strong position as the primary enablers of spaceflight. And, in spite of increased international cooperation, as well as the flutter of ambition involving private spaceflight, there is a formidable, and I would argue rising, chorus of voices that privilege the primacy of national and *nationalistic* space exploration. The American and Russian space programs remain, both in rhetoric and practice, highly nationalist projects that reinforce the notion that space exploration is a powerful vehicle for expressing a nation's broader aspirations. Similarly, second tier space powers such as China, Japan, and India, which have long been spacefaring nations, have more recently strengthened the link between nationalism and competence in space activities. The evidence from the past 50 years of spaceflight convincingly counters utopian notions—expressed in television, film, fiction, and journalism—that as spaceflight becomes mature, national space programs will disappear, and all spacefaring countries will come together to work towards a shared set of objectives that have global resonance.

Despite the fundamental and enduring nature of the relationship between space exploration and the nation, we know very little about the manner in which nations articulate their engagement in space activities. My goal in this essay is to offer some preliminary thoughts on the broad patterns that characterize the public rhetoric surrounding national space programs, patterns that are common across different national contexts. Here, I define "public rhetoric" to include the discourse generated by governmental agencies, journalists, historians, and public commentators, i.e., those that elucidate and establish the contours of public debate over space exploration in particular national contexts. I do not claim that this discourse reflects or approximates the "real" relationship of spaceflight to national aspirations, i.e., that space exploration can only be understood in terms of the nation. On the contrary, I strongly believe that the immutable association in the public eye of spaceflight with the nation has helped to *obscure* important non-state processes, an understanding of which might offer valuable insights in analyzing the history of space exploration.[1] I do, however, believe that the language describing space exploration has certain semiotic characteristics that communicate persistent ideas about the history of spaceflight that repeat across entirely different cultures and contexts. These ideas are important to discern since they serve as a filter for the public understanding of spaceflight and consequently contribute to the public enthusiasm (or lack of) for space exploration in general.

The evidence suggests that through the first 50 years of the Space Age, all spacefaring nations have used four different tropes—linguistic constructs dependent on symbols—to articulate their space programs to the broader public. These four tropes, which take the form of particular rhetorical strategies, continue to be fundamental to the way that the project of space exploration has been articulated in both official and unofficial discourses; governmental agencies, journalists, historians, public commentators and the lay public in spacefaring nations have consistently invoked these archetypes to construct a master narrative of the history of space exploration. They are: the myth of the founding father, the claim of indigenous creation, the connection between spaceflight and national identify, and the essential need to justify space activities. In elaborating these tropes, I use as examples the five nations which have achieved the domestic capability to launch objects into Earth orbit and still retain that capability—the Soviet Union (achieved orbit in 1957), the United States (1958), Japan (1970), China (1970), India (1980), and Israel (1988). Two European nations which once had that capability—France (1965) and Great Britain (1971)—have relinquished it. The former folded their efforts into the European Space Agency (ESA) while the latter saw no value in having such a

---

1. I make this point in my "Competing Technologies, National(ist) Narratives, and Universal Claims: Revisiting the Space Race," paper presented at the NSF-sponsored workshop of the Society for the History of Technology, October 18, 2007, Washington, DC. The paper can be accessed at *http://fiftieth.shotnews.net/?page_id=23*. (accessed February 29, 2008).

capability. ESA still remains the only multinational organization to develop its own satellite launch capability, having achieved that ability in 1979.[2]

FOUNDING FATHERS

The first trope of a national space history is that of the "founding father."[3] Each space program arrives in the historical record with a singular figure whose determinations mirror and telescope the spacefaring ambitions of the nation in question. For the Soviet Union, there was Sergei Korolev (1906-1966), for the United States, Wernher von Braun (1911-1977), for Japan, Hideo Itokawa (1912-1999), for China, Qian Xuesen (1911-), for India, Vikram Sarabhai (1919-1971), and for Israel, Yuval Ne'eman (1925-2006).[4] In some cases, their claims as founding fathers are contested—especially in the case of von Braun— but the commonalities between them are striking. Each of these individuals embodies a unique combination of dualities: they are always both capable and visionary, brilliant engineers and unequalled managers, and comfortable with the topmost levels of power and yet accessible to the rank-and-file technician. There are early traumas typically associated with each, ordeals that were physical, moral, or professional. For example, Korolev served a sentence in the Gulag, von Braun never fully escaped the moral quandaries of being associated with the Dora labor camp in Nazi Germany, and Qian's life and career were disrupted by the Red Scare in the 1950s when he was deported to China on charges of being a communist sympathizer. In all cases, these men were seen as overcoming these adversities to achieve prominence later in their lives. For those reconstructing narratives of national space programs, these traumas become metaphors for the uphill battles faced by the space programs themselves.

---

2. Although I do not focus on them, the same patterns also apply to those countries that are close to achieving a domestic capability to launch satellites into orbit but have not yet done so: Brazil, North Korea, Iran, and South Korea. In addition, I do not explore the strategies of those dozens of nations that have developed or purchased satellites but lack the expertise or resources to launch them themselves and, therefore, pay other nations or agencies to do so.

3. It goes without saying that there were no women founders of space programs; the history of space exploration has been dominated by men in all nations, partly because of the substantive obstacles faced by women in pursuing higher education in the applied sciences or engineering. On the other hand, women in large numbers did contribute to space programs globally at mid- and lower-levels of management (e.g., as computer operators, medical personnel, draftspersons, administrative staff, custodial laborers, and daycare workers). Because social history has not been a concern for space historians, these women and their contributions remain largely invisible in most space history narratives.

4. For useful biographies of some of these individuals, see Iaroslav Golovanov, *Korolev: fakty i mify* (Moscow: Nauka, 1994); Michael J. Neufeld, *Von Braun: Dreamer of Space, Engineer of War* (New York: Alfred A. Knopf, 2007); Iris Chang, *The Thread of the Silkworm* (New York: Basic Books, 1995); Amrita Shah, *Vikram Sarabhai: A Life* (New Delhi: Penguin Viking, 2007).

What purpose does the founding father trope serve? There is hardly a historian who would agree that Korolev single-handedly founded the Soviet space program, yet his epic biography completely overshadows the mention of many other individuals who made critical contributions to the emergence of the Soviet space program. Here it is important to distinguish between formal academic history and the popular notion of history that becomes part of the collective memory of a nation. With the former, historians are drawn to complexities and the messiness of yesterday; with the latter, our predilection is to distill complexities down to broad themes, personalities, and events that are often deterministic and teleological in nature. Thus, one purpose of the founding father archetype is to reinforce deterministic explanations for space history (e.g., "Korolev did X, therefore the Russian space program is like Y").

The founding father archetypes did not arrive out of a vacuum but rather drew upon a longer tradition of similar archetypes. Most European nations, for example, reinforce narratives that they have founding fathers for particular scientific and applied scientific fields such as physics, chemistry, biology, mathematics, computer science, etc. These narratives center around an individual who is not only a deep thinker but also a builder of institutions, as well as an individual who bequeathed a substantial system (of research, education, etc.) for the good of the nation. In that sense, the founding father narratives of space exploration also parallel and mirror narratives about the founding of the nation itself, which are often tied to singular individuals who embodied some of the same kinds of qualities. Thus, these founding fathers represent not only the space program but also become key figures in nation building. By association, our conceptions of the founding father archetypes attach national space programs to imperatives, challenges, and triumphs associated with the founding of the nation. As a result, to many, the space program acquires a level of gravitas typically associated with concerns about the future of the nation.

INDIGENITY

All national space program narratives depend on the claim that its achievements were native in origin. In other words, the space history of each country assumes that nations are airtight constructs where immutable borders overshadow transnational flows and fixed delineations trump the fluid nature of both identities and knowledge. There are obvious reasons why the appeal of a particular space program depends on the notion of home-grown expertise: such accounts bolster national claims of competence, both to domestic and international audiences. Indigenous technologies—or at least those that are represented as indigenous—serve as surrogates for the projection of national prowess, a phenomenon that dates back at least to the late 19th century when both Great Britain and Germany began to assert their standing on a global stage through accomplishments in science and technology.[5]

In the case of the space powers, each of their achievements served to place them on a global stage. Much like the acquisition of nuclear capability—more prosaically termed "going nuclear"—the domestic capability to deliver objects into Earth orbit secures a powerful and symbolic status that is also discrete since it divides "before" and "after" as being completely different. The symbolic power of such moments derives from the way a single launch can represent a convergence of many national aspirations—pride in history, a consensus that the present is a moment to be celebrated, and a confidence in a bright tomorrow. In 1980, when India launched its first satellite into orbit, Prime Minister Indira Ghandhi noted in a speech to the Indian parliament that "This is a great day for India and for Indian science." Mass media response in the West was predictably reductive but couched the event as a landmark: the *Washington Post* reported, for example, that it was "a remarkable achievement for a country that still uses bullock carts as a prime mode of transportation."[6]

From the Indian perspective, it was important to emphatically underscore the value of indigenity and the issue of ownership: the Indian space program was, above all else, Indian. Participants of the Indian space program continue to emphasize this aspect of the development of their first satellite launch vehicle, the SLV-3, attributing the mastery of this capability both to the high level of existing Indian expertise and the circumstances generated by draconian technology proliferation controls which forced Indian engineers to "go it alone."[7] Even though the development of the SLV-3 actually predated the enforcement of the Missile Technology Control Regime (MTCR) that limited international flows of "sensitive" missile technology to selected countries, the current existence of such controls serves to embolden ahistorical and disingenuous lines of argument and, in fact, obscures the significant international collaboration that led to the SLV-3 rocket.[8]

Claims of indigenity are not monolithic across nations. In the more mature space powers, the tone of these assertions communicate unquestioned celebrations of national character, while in the "newer" space powers, they come across as preemptive responses to accusations of clandestine (or otherwise) appropriation

---

5. Bernhard Rieger, *Technology and the Culture of Modernity in Britain and Germany, 1890–1945* (Cambridge, UK: Cambridge University Press, 2005).

6. "India Becomes 6th [*sic*] Country to Put Satellite into Orbit," *Washington Post*, July 19, 1980. India was actually the seventh nation to put a satellite into orbit using its own rocket, and the eighth if one includes the European Space Agency.

7. See for example, B. N. Suresh, "History of Indian Launchers," IAC-07-D2.2.01, paper presented at the 58th International Astronautical Congress, Hyderabad, India, September 24-28, 2007.

8. The history of international contribution to the SLV-3 has been all but forgotten from the "official" record of its development. For a still-valuable historical work that explores the development of Indian launch vehicles, see Gopal Raj, *Reach for the Stars: The Evolution of India's Rocket Programme* (New Delhi: Viking, 2000).

of technology from other nations. An example of the former is the United States, where the achievements of the American space program—particularly the Apollo lunar landings—represent the achievements of Americans, and not, for example, Germans or Canadians.[9] As the author of a very popular book on Apollo recently noted:

> Free competition motivated American workers whose live-lihoods were related to the quality and brilliance of their work, and we saw extraordinary, impossible things accom-plished by ordinary Americans. The American flag on the Moon is such a powerful symbol because it is not a vain one. America, like no other nation, *was* capable of the Moon.[10]

Soviet and Russian commentators, including veterans, have long made similar pronouncements in relation to the achievements of Sputnik and Gagarin, albeit, in the backdrop of latent suspicions (especially in Europe and the United States) that the help of German engineers kidnapped after World War II was critical to the spectacular early successes of the Soviet space program.[11]

Claims of fully indigenous space technology are often motivated by accusations from abroad that this technology was "borrowed"; such allegations themselves focus mostly on non-Western nations. In other words, while the mature Western programs are largely insulated from charges of benefiting from foreign technological expertise, both new and mature non-Western programs are continually dogged by such accusations—usually emanating from the West—prompting a generally defensive posture that requires repeated assertions about domestic expertise. Through the entire period of the Cold War, for example, Soviet space achievements were continually marred by Western claims that the Soviets benefited from the "other Germans" or that they used

---

9. Both Germans and Canadians, naturalized as U.S. citizens by the early 1960s, made significant contributions to the Apollo program. For the Canadian contribution, see Chris Gainor, *Arrows to the Moon: Avro's Engineers and the Space Race* (Burlington, Ontario: Apogee Books, 2001). There are a vast number of books on the German contribution. For a representative example, see Frederick I. Ordway, III and Mitchell R. Sharpe, *The Rocket Team* (New York, NY: Cromwell, 1979).

10. David West Reynolds, *Apollo: The Epic Journey to the Moon* (New York, NY: Tehabi, 2002), p. 257.

11. Soviet rocketry veteran Boris Chertok, who represents a "mainstream" voice within the Soviet space history community, concedes that German help was important in the immediate postwar years but dismisses any notion that this help was essential to the early successes of the Soviet space program. See Boris Chertok, *Rockets and People*, ed. Asif A. Siddiqi (Washington, DC: NASA, 2004). On the other hand, a number of German writers, without much convincing evidence, have recently attributed most of the early Soviet successes in rocket design to Germans. See for example, the three-part article by Olaf Przybilski, "Die Deutschen und die Raketentriebwerksentwicklung in der UdSSSR," *Luft- und Raumfahrt* no. 2 (1999): 30-33; no. 3 (1999): 28-33; and no. 4 (1999): 33-40.

technology stolen from the U.S. space program through skillful spying.[12] Similarly, Western commentators, both official and independent, continue to express concern about possible Chinese use of sensitive American technology for use in the development of their ballistic missiles and launch vehicles.[13] While such expressions are linked to concerns about the global proliferation of potentially harmful technology, they also communicate an implicit message about the inability of certain nations to innovate without outside help. Not surprisingly, such a stance tends to embolden and fortify the opinions of the scientific elite in non-Western nations who reject the notion that they are not capable enough to master the technology of space exploration. Affirmations of domestic competence emanating from Chinese or Indian scientists and engineers challenge the unquestioned assumption that there is an arbitrary line in history that divides those who are innovators (i.e., Western nations) and those who are proliferators (i.e., non-Western nations).[14] As such, in the non-Western world, claims of indigenity serve not only to boost national pride but are also vehicles for affirming a kind of revisionist and non-Orientalist historical thinking that decenters and deprivileges the West as the de facto basis for all discussions of spaceflight.

SPACE AS AN EXPRESSION OF NATIONAL IDENTITY

Each national space program is also articulated both in contemporaneous times and in retrospect as an expression of a nation's identity. In other words, discussions about space exploration across extremely different national

---

12. The most famous example of Soviet "copying" was the case of the Buran space shuttle. See John Noble Wilford, "Soviet Design Appears in Debt to U.S. Shuttle," *New York Times*, November 16, 1988. For a careful and recent analysis of the possibility of Soviet appropriation of U.S. technology in relation to the Buran, see Bart Hendrickx and Bert Vis, *Energiya-Buran: The Soviet Space Shuttle* (Springer: Chichester, UK, 2007), pp. 82-85.

13. For the controversial and error-ridden report issued by the U.S. House of Representatives on China's alleged efforts to obtain technological information covertly from the United States (including those related to space technology), see the *Report of the Select Committee on U.S. National Security and Military/Commercial Concerns with the People's Republic of China* (more commonly known as the "Cox Report") at *http://www.house.gov/coxreport/* (accessed February 29, 2008).

14. Itty Abraham makes this argument about the arbitrary nature of the definition of nuclear proliferation in "The Ambivalence of Nuclear Histories," *Osiris* 21 (2006): 49-65. Hugh Gusterson similarly describes a moral distinction made by Westerners in terms of the acquisition of nuclear weapons. He writes: "There has long been a widespread perception among U.S. defense intellectuals, politicians, and pundits—leaders of opinion on nuclear weapons—that, while we can live with nuclear weapons of the five official nuclear nations for the indefinite future, the proliferation of nuclear weapons to nuclear-threshold states in the Third World, especially the Islamic world, would be enormously dangerous. This orthodoxy is so much a part of our collective common sense that, like all common sense, it can be usually stated as simple fact without fear of contradiction." See Hugh Gusterson, "Nuclear Weapons and the Other in the Western Imagination," *Cultural Anthropology* 14 (1999): 111-143.

contexts almost always include the notion, implicitly or explicitly, that there is something fundamental in the national character that gives force to the urge to explore space. Such expressions use three different rhetorical strategies that are not necessarily mutually exclusive: first, they involve a suggestion that space exploration represents a logical and further expression of deep-rooted cultural traits; second, they underscore national space achievements as a natural outcome of historical events; and third, they couch the space program as a vehicle for communicating a nation's prowess in science and technology.

Both the United States and the Soviet Union had deep-rooted traditions that suggest antecedents for their respective 20th century space programs. In the former case, there are any number of archetypes that justify and underlie the spacefaring activities of the United States. These are dominated by the notion of exploring the Western frontier and its attendant links to the idea of freedom: the freedom to explore, the freedom to settle, and the freedom to move again into the unknown. The "frontier thesis," as first cogently articulated by historian Frederick Jackson Turner in the late 19th century was a powerful statement of American exceptionalism, and as an analogy, it has proved remarkably resilient for many different American endeavors, including, of course, the space program.[15] In American space exploration, many commentators saw not only how engagement with the frontier shaped American society and culture but also how American society and culture shaped the frontier itself. American exploration—from Lewis and Clark to the Apollo program—was acting both on a generic human impulse to seek knowledge and a deep-rooted American urge for inquiry, exploration, and the freedom of wide open spaces.[16] Commentators as varied as rocket engineer Wernher von Braun, space visionary Gerard K. O'Neill, and space advocate Robert Zubrin all have couched their arguments with a distinctly American spin—ingenuity, frontier, freedom—in their search to advance the cause of human survival in the form of human colonization of the cosmos.[17]

As with Americans, many Russians also argue for deep-seated autochthonous urges for space exploration. In a recent article, a prominent Russian philosopher argued that the ideas of Konstantin Tsiolkovskii—the founding theorist of Soviet space exploration—provides the basis for a "Russian

---

15. For Turner's original works, see John Mack Faragher, ed., *Rereading Frederick Jackson Turner: The Significance of the Frontier in American History and Other Essays* (New Haven, CT: Yale University Press, 1994); George Rogers Taylor, *The Turner Thesis: Concerning the Role of the Frontier in American History*, 3rd ed. (Lexington, MA: Heath, 1972). For a more contemporary critique, see Richard Slotkin, *Gunfighter Nation: The Myth of the Frontier in Twentieth Century America* (New York, NY: Atheneum, 1992).

16. For an excellent summary of these themes as they relate to American space exploration, see Roger D. Launius, "Perfect Worlds, Perfect Societies: The Persistent Goal of Utopia in Human Spaceflight," *Journal of the British Interplanetary Society* 56 (2003): 338–349.

17. Howard E. McCurdy, *Space and the American Imagination* (Washington, DC: Smithsonian Institution Press, 1997).

national idea," an alternative to a "Europeanized" Russia that is part of the global system of capitalism and dependency. Tsiolkovskii, the author argued, had shown that the true destiny of Russians, like no other nationals on this Earth, resided in space, a place that transcends borders and nations.[18] While some would argue that this line of thinking is rooted in the Marxist-Leninist utopian thinking unleashed by the Russian Revolution of 1917, such ideas of technological utopianism can actually be traced further back to the mystical and occult pre-Revolutionary philosophy known as Cosmism, a tradition that was made up of a hodgepodge of Eastern and Western philosophical traditions, theosophy, panslavism, and Russian Orthodox thinking. The outcome was a nationalist and often reactionary philosophy that, in spite of its reactionary tenets (or perhaps because of it), continues to attract the attention of many Russian nationalist intellectuals in the post-Communist era.[19] The cause of Cosmism was "liberation from death," a goal that would be achieved by human migration into space that would allow humans to reanimate the atom-like particles of all those who had already "died" in the previous hundreds of thousands of years. The eccentric late 19th century Russian philosopher Nikolai Fedorov, who articulated much of this philosophy before anyone, wrote in 1905 that "[the] conquest of the Path to Space is an absolute imperative, imposed on us as a duty in preparation for the Resurrection. We must take possession of new regions of Space because there is not enough space on Earth to allow the coexistence of all the resurrected generations. . . ."[20] In present-day Russia, the philosophy of Cosmism holds a deep sway among many commentators, especially those who meditate on the meaning of Russian space exploration.[21]

Spaceflight is also linked to national identity through history. Most spacefaring countries, for example, claim pre-modern historical events as part of their narrative of space exploration. Such arguments rooted in history lay claim to the idea that the nation's path to space was preordained and inevitable, and that the modern space program is but a continuation of activities stretching

---

18. L.V. Leskov, "K. E. Tsiolkovskii i rossiiskaia natsional'naia ideia," *Zemlia i vselennaia* no. 4 (1998).

19. For links between modern Russian Cosmism and post-Soviet Russian nationalism, see James P. Scanlan, ed., *Russian Thought After Communism: The Recovery of A Philosophical Heritage* (Armonk, NY: M. E. Sharpe, 1994), pp. 26–28. See also Michael Hagemeister, "Russian Cosmism in the 1920s and Today" *The Occult in Russian and Soviet Culture*, ed. Bernice Rosenthal (Ithaca, NY: Cornell University Press, 1997), pp. 185–202.

20. S. G. Semenova and A. G. Gacheva, eds., *N. F. Fedorov: Sobranie sochinenii v chetyrekh tomakh*, 4 vols. (Moscow: Progress, 1995–2000). For a detailed exposition on the role of Cosmism in the origins of Soviet space exploration, see Asif A. Siddiqi, *The Red Rockets' Glare: Soviet Imaginations and the Birth of Sputnik* (Cambridge, UK: Cambridge University Press, forthcoming).

21. For a small sampling of works on Russian Cosmism since the early 1990s, see L. V. Fesenkova, ed., *Russkii kosmizm i sovremennost'* (Moscow: IFAN, 1990); S. G. Semenova and A. G. Gacheva, eds., *Russkii kosmizm: antologiia filosofskoi mysli* (Moscow: Pedagogika-Press, 1993); O. D. Kurakina, *Russkii kosmizm kak sotsiokul'turnyi fenomen* (Moscow: MFTI, 1993); O. Ia. Gelikh, ed., *Kosmizm i novoe myshlenie na Zapade i Vostoke* (St. Petersburg: Nestor, 1999).

back centuries that embody similar sensibilities. In non-Western nations, there is also a specific pattern of linking contemporary space programs with events that predate Western modernity. Chinese writers, for example, are eager to emphasize the importance of China as the birthplace of rocketry in the 13th century, while Indian writers similarly stress the importance of Tippu Sultan's rockets from the late 18th century as a harbinger of the future.[22] In these narratives, Tsiolkovskii, Goddard, and Oberth are all peripheral.

Finally, national identity is linked to spaceflight as an expression of national technological competence. Since the very first satellites, space exploration has served as a reminder to both domestic and international audiences of a nation's mastery of science and technology, not too dissimilar from other technological metrics of late 20th century modernity such as nuclear power, computing, and biotechnology. Already by the late 19th century, and especially in the light of experiences during the Great War, technology had assumed a fundamental role in the projection of national prowess, contributing to and joining the other measures of global dominance such as imperial adventurism, military assets, and industrial growth. In his study of the role of technology in the creation of modernity in early 20th century Britain and Germany, Bernhard Rieger notes that:

> [t]echnological innovations not only underpinned the competitiveness of national economies as well as both countries military might; a large range of artifacts also became national symbols and prestige objects that signaled international leadership in a variety of engineering disciplines.[23]

A half a century later, especially after the launch of Sputnik in 1957, the connections between technology and national prowess became fully established. And just as Sputnik marked a particular historical moment that attached the notion of technological competence to the Soviet Union, Apollo did the same for the United States. I would argue that the most enduring aspect of the iconography of Apollo has been to set a benchmark for technological competence in American

---

22. For the Chinese references, see Brian Harvey, *China's Space Program: From Conception to Manned Spaceflight* (Berlin: Springer, 2004). For India, see A. P. J. Abdul Kalam, *Wings of Fire: An Autobiography* (Hyderabad: Univ. Press, 1999); S. Krishnamurthy and B. R. Guruprasad, "On the Nature and Significance of Tipu Sultan's Rockets from a Historical Perspective," IAC-07-E4.4, paper presented at the 58th International Astronautical Congress, Hyderabad, India, September 24-28, 2007.

23. Rieger, *Technology and the Culture of Modernity in Britain and Germany, 1890–1945*, p. 224. In a similar vein, see Guillaume de Syon, *Zeppelin!: Germany and the Airship, 1900–1939* (Baltimore, MD: Johns Hopkins University Press, 2002); Peter Fritzsche, *A Nation of Fliers: German Aviation and the Popular Imagination* (Cambridge, MA: Harvard University Press, 1994); Gabrielle Hecht, *The Radiance of France: Nuclear Power and National Identity after World War II* (Cambridge, MA: MIT Press, 1998).

culture, as underscored in the oft-repeated lament that begins, "If we could send a man to the Moon, there's no reason we can't. . . ." The later second-tier space powers have deployed this fundamental link between national prowess and space technology in similar ways. For emerging global players such as China and India, space exploration represents one of a constellation of important ways with which to announce their "arrival" as global powers: upon the launch of their first lunar probe, for example, Chinese space scientist Ouyang Ziyang noted that, "[a]s lunar exploration embodies our overall national strength, it is very significant for raising our international prestige and our national unity."[24] The media hype over a possible Asian space race among China, Japan, and India in recent times is one symptom of this belief in "raising" prestige on a global level; the overtly nationalist rhetoric about the meaning of space exploration for the youth of the nation—as was seen with the domestic coverage of cosmonaut missions from Malaysia and South Korea—was another.[25]

JUSTIFICATIONS

The fourth dimension of the public articulation of national space programs are best described as justifications. Space exploration—especially the kind that involves developing a domestic space transportation system—requires enormous investments in resources. As such, articulation of any particular space event, whether in real time or in retrospect, demands a variety of rationalizations not only to justify but also to explain the event. Historically, most other major and mature technological systems of the 19th and 20th centuries, especially ones that have developed over a period of a half a century (such as urban electrical systems, air travel, high speed rail, telephone networks, and television systems) have not required the kind of concomitant justifications that are *de rigueur* in discussions about space travel. While the benefits of these other systems—in the form of social welfare or profit or both—have been seen self-evident, in the case of space travel, social benefits and material gain continue to be issues of debate

---

24. Jim Yardley, "China Sends Its First Probe for the Moon Into Space," *New York Times*, October 25, 2007.

25. Both Malaysia and South Korea paid the Russian Space Agency to launch individuals from their respective nations into orbit on board a Soyuz spacecraft for short visits to the International Space Station. See Azura Abas and Nisha Sabanayagam, "First Malaysian in Space: Angkasawan to Inspire Schoolkids," *New Straits Times Online*, October 11, 2007, *http://www.nst.com.my/ Current_News/NST/Thursday/Frontpage/2057731/Article/index_html* (accessed February 29, 2008); "Malaysians over the Moon as Their Astronaut Blasts into Space," *Space Travel: Exploration and Tourism*, October 10, 2007, *http://www.space-travel.com/reports/Malaysians_over_the_moon_as_ their_astronaut_blasts_into_space_999.html* (accessed February 29, 2008); Cho Jin-Seo, "Sputnik and Arirang: 50 Years of Space Exploration and Korea," *Korea Times*, October 8, 2007, *http://www. koreatimes.co.kr/www/news/tech/2007/10/129_11545.html* (accessed February 29, 2008).

rather than unquestioned axioms. As a result, discussions surrounding national space programs have remained inseparable from invocations of justifications.

Historian Roger Launius has described the various rationales put forth justifying the cause of space exploration: survival of the species, national pride, national security, economic competitiveness, and scientific discovery.[26] To these five, I would add "benefits to the populace" as a sixth set of justifications. These justifications are central to space narratives because they preemptively try to insulate discussions about space travel from critiques both internal (i.e., domestic and institutional) and external (i.e., international and public). Without dispensing judgment on the validity of these justifications, it is clear that they play a critical role in the discourse about space exploration, one that is so deeply ingrained that we hardly even think it odd that there should be any suggestion that we not have to justify spaceflight.

Justifications for spaceflight have been historically contingent; different historical periods required different justifications to be accentuated. Moments of perceived crisis, for example tend to privilege some justifications over others. In the initial collective national anxiety following Sputnik, the *raison d'être* of the American space program was framed in discourses of national pride and national security. These justifications were particularly effective in the 1960s, the former for Apollo and the latter for various military and intelligence space projects. The other three justifications—economic competitiveness, survival of the species, and scientific discovery—were at the forefront in the post-Apollo years when the American space program was more mature but also more directionless in the inevitable letdown after the Moon landings.

The crisis of the post-Apollo years—in the aftermath of a costly foreign war, an energy crisis, and a space program without a vision matching Apollo—generated enormous discussion about the practical costs and benefits of the space program.[27] As indifference to the space program mounted in the 1970s, NASA sought to attract positive attention to its cause by emphasizing the rewards of space exploration, benefits beyond the clichés of Tang, Teflon, and Velcro—none of which were developed by NASA but which had become comedic counterpoints to the perceived majesty of Apollo. The Agency also devoted significant resources to advertising its efforts to transfer the benefits of space travel to taxpayers; in 1962, it created the Technology Utilization Program, and, since 1976, it has published the annual *Spinoff* volume. What is the purpose of preparing this publication? According to NASA:

26. Roger D. Launius, "Compelling Rationales for Spaceflight: History and the Search for Relevance" in *Critical Issues in the History of Spaceflight*, eds., Steven J. Dick and Roger D. Launius (Washington, DC: NASA, 2006), pp. 37-70.

27. For a lengthy discussion of how the writing of American space history was also affected by the rise and fall of Apollo, see Siddiqi, "American Space History: Legacies, Questions, and Opportunities for Further Research" in *Critical Issues in Space History*, pp. 433-480.

it is a convincing justification for the continued expenditure
of NASA funds. It serves as a tool to educate the media and
the general public by informing them about the benefits and
dispelling the myth of wasted taxpayer dollars. It reinforces
interest in space exploration. It demonstrates the possibility
to apply aerospace technology in different environments. It
highlights the ingenuity of American inventors, entrepreneurs, and application engineers, and the willingness of a
government agency to assist them. And finally, it continues
to ensure global competitiveness and technological leadership by the United States.[28]

One striking aspect of these justification narratives is that they have been
deployed in support of space programs regardless of the nature of the political
system in question: nations that are vibrant democracies use the same kind of
justifications as those nations where large portions of the popular are politically
disenfranchised. For example, while the Chinese space program has no
immediate counterpart to NASA's Commercial Technology Program, it does
frequently articulate very similar justifications about its own growing space
program. In a white paper on the Chinese space program prepared in 2000, the
foremost rationale of the Chinese space program was laid out as such:

> The Chinese government attaches great importance to the sig-
> nificant role of space activities in implementing the strategy of
> revitalizing the country with science and education and that
> of sustainable development, as well as in economic construc-
> tion, national security, science and technology development
> and social progress. The development of space activities is
> encouraged and supported by the government as an integral
> part of the state's comprehensive development strategy.[29]

China's democratic neighbor, Japan, has communicated similar rationales,
albeit ones that have changed over the decades with the evolution of the
Japanese economy and industry. If in the 1970s and 1980s the space program
was rationalized by the need to keep the Japanese economy competitive and
its industry robust, by the early 2000s the justifications for space exploration
incorporated a new motive: the security of the Japanese people from natural
disasters and global environmental degradations. Perhaps responding to the
perception that the Japanese public "is becoming increasingly skeptical of

---

28. *"History of Spinoff,"* http://www.sti.nasa.gov/tto/spinhist.html (accessed February 29, 2008).

29. Information Office of the State Council, "White Paper on China's Space Activities," http://
english.peopledaily.com.cn/features/spacepaper/spacepaper5.html (accessed February 29, 2008).

claims that the space program will produce major economic benefits," the Japan Aerospace Exploration Agency (JAXA) issued a 20-year vision statement in 2005.[30] In it, the Agency emphasized goals that were reiterated by JAXA President Keiji Tachikawa in an annual message:

> I feel that Japan's space program can contribute more to the safety and security of the Japanese people. I hope that JAXA will actively bear responsibility to follow this lofty goal and space development leads to greater safety and security for all mankind, from our daily lives to emergency situations.[31]

Tachikawa's message is emblematic of a general shift in justifications characteristic of all the major global space programs, one that equates a concern for the welfare of the environment with important social benefits. All national space programs—both major and minor—now pay lip service to critical environmental issues such as global warming, deforestation, land erosion, earthquake prediction, and disaster warning. Such rationales have begun to augment and replace Cold War-centered justifications that centered largely around prestige and national security.

The justification tropes, then—whether arguing for survival of the species, national pride, national security, economic competitiveness, scientific discovery, or benefits to the populace—serve to provide a foundation for which to discuss the very possibility of space exploration. Because of its extremely high costs and attendant high risks, nations have had to frequently and insistently justify the existence of space programs; thus, justifications are not simply extraneous rhetoric but have become intrinsic to our future visions of space exploration.

CONTESTED VISIONS

Each of these four elements that form the core of space exploration narratives—the founding fathers, the notion of indigenity, connecting spaceflight with national identity, and the need for justifications—are contested and mutable. In each case, there are actors who seek to displace or destabilize the master narratives.

Perhaps the most rancorous disagreements have been over the founding father archetypes and the claims of indigenity. In the former case, the U.S. space program is somewhat of an anomaly. A plausible candidate for a founding father is the rocketry pioneer Robert Goddard who designed, built, and

---

30. The quote is from Steven Berner, *Japan's Space Program: A Fork in the Road?* (Santa Monica, CA: RAND Technical Report TR-184, 2005), p. 30.

31. "Message from President of JAXA," *http://www.jaxa.jp/about/president/index_e.html* (accessed February 29, 2008).

launched America's first liquid propellant rocket in 1926.[32] Despite Goddard's quite significant technical achievements in rocket development in the interwar years, however, he had little or no influence on the birth of the American space program, having passed away in 1945. And although his place in the pantheon of original space visionaries is secure, his contributions to spaceflight in the American context have been overshadowed by those of Wernher von Braun.

For many reasons, von Braun does not fit the typical mold of the founding father: he was originally German, he did not "found" the American space program, and he had little or no influence on the development of U.S. spacecraft. Yet he and his biographers, based upon his undeniably significant achievements, have positioned him—some would say very successfully—as one of the most iconic, if not *the* most iconic non-astronaut figure in the history of the American space program.[33] The fact that rockets designed under von Braun's direction launched the *first* U.S. satellite, the *first* American into space, and the *first* American to the Moon are important touchstones in his legacy; arguably, all of these achievements are overshadowed by von Braun's charisma and larger-than-life charms as a public figure in the 1950s and 1960s. Besides the astronauts, no individual in the public eye during that time personified the ingenuity, daring, and resourcefulness required to send humans to the Moon than Wernher von Braun.

Von Braun's legacy has been a contested one. Within the historical community, disagreements have raged over his alleged complicity with the forced labor at Dora during World War II.[34] Another debate has centered on his proper place in the history of the U.S. space program: for many years, von Braun's "rocket team" was square and center in the American space narrative that began with the capture of V-2 rockets at the end of World War II and ended with Apollo 11. A group of influential historians invested in maintaining von Braun's legacy have ensured the continuing prominence of this narrative (often called the "Huntsville School" of historiography), one that traces the roots of the American space program, particularly the Apollo project, to the V-2 rocket and its brilliant designers in Germany during the interwar years. In this narrative, which has had a near-impervious hold on the public perception of the American space program, the so-called German rocket team who were

32. David A. Clary, *Rocket Man: Robert H. Goddard and the Birth of the Space Age* (New York, NY: Hyperion, 2003).

33. For the many sympathetic and often hagiographic biographies of von Braun, see Erik Bergaust, *Wernher von Braun: The Authoritative and Definitive Biographical Profile of the Father of the Modern Space Age* (Washington, DC: National Space Institute, 1976); Ernst Stuhlinger and Frederick I. Ordway, III, *Wernher von Braun, Crusader for Space* (Malabar, FL: Krieger, 1994); Bob Ward, *Dr. Space: The Life of Wernher von Braun* (Annapolis, MD: Naval Institute Press, 2005).

34. Michael J. Neufeld, "Wernher von Braun, the SS and Concentration Camp Labor: Questions of Moral, Political, and Criminal Responsibility," *German Studies Review* 25, no. 1 (February 2002): 57-78.

brought to the United States in the aftermath of World War II played a singular and critical role in taking America to space and eventually to the Moon.[35] Although there has been a stream of recent scholarship highlighting more indigenous sources of innovation in the American context—such as the Jet Propulsion Laboratory and Reaction Motors—there continues to be a large divide between historians' understanding of the role of von Braun in the early U.S. space program and laypeople's perception of the same topic.[36]

Perhaps the most contested aspect of national space history narratives is the issue of indigenity. Every single space power has made a claim for indigenous origins of expertise, technology, and competence, and for every one of these claims, there exist counter-claims. In the American case, there are competing schools centered on German and more homegrown contributions. Similar arguments over German help have raged over the birth of the Soviet space program. The "second-rank" space powers all have comparable disputes over their stories of origin. We find obvious parallels in claims made for the development of atomic energy by various nations. At least one recent scholar of the history of atomic energy has begun to question the hermetically sealed nature of these nation-centered narratives. Writing on the history of nuclear power, historian Itty Abraham has noted that "practically no state travelled alone."[37] He adds:

> One of the most enduring tropes of nuclear histories is the idea that atomic energy programs are always national programs. The close relation between nuclear power and national power has led to the assumption that, for reasons of security especially, nuclear programs must be uniquely identified with particular countries. Official histories and scientists encourage this belief, for obvious parochial reasons, but it is rarely true. No atomic program anywhere in the world has ever been purely indigenous . . .[38]

---

35. For an erudite analysis of the Huntsville School, see Roger D. Launius, "The historical dimension of space exploration: reflections and possibilities," *Space Policy* 16 (2000): 23-38.

36. For von Braun-centered works embodying the Huntsville School, see, for example, Willy Ley, *Rockets, Missiles, and Men in Space* (New York: Viking Press, 1968); Ordway, III and Sharpe, *The Rocket Team*; Wernher von Braun, Frederick I. Ordway, III, and Dave Dooling, *History of Rocketry and Space Travel* (New York: Thomas Y. Cromwell, 1986); Ernst Stuhlinger, Frederick I. Ordway, III, and Wernher von Braun, *Crusader for Space*, 2 vols. (Malabar, FL: Robert E. Krieger, 1994). For syntheses that take a more balanced approach to U.S. space history, see T. A. Heppenheimer, *Countdown: A History of Space Flight* (New York: John Wiley & Sons, 1997); William E. Burrows, *This New Ocean: The Story of the First Space Age* (New York: Random House, 1998).

37. Itty Abraham, *Making of the Indian Atomic Bomb: Science, Secrecy, and the Postcolonial State* (London: Zed Books, 1998), p. 9.

38. Abraham, "The Ambivalence of Nuclear Histories." See also his "Notes Toward a Global Nuclear History," *Economic and Political Weekly* 39 nos. 46-7 (November 20, 2004): 4,997-5,005.

The available evidence points strongly to similar processes of knowledge flows in the evolution of ballistic missiles and space technology.[39] Every nation engaged in this technology has been a proliferator and has benefited from proliferation; this process of proliferation already began in the 1920s when an informal and international network of spaceflight enthusiasts in Europe—particularly in Germany, Austria, France, Poland, Great Britain, and the Soviet Union—and the United States, generated the first substantive exchange on topics related to rocketry and space exploration.[40] The development of sophisticated German ballistic missiles in the 1930s benefited from this discourse as did parallel but less ambitious Soviet efforts to build rockets. In the aftermath of World War II, the remainder of the German missile program, the most developed effort at that point, then fed into several different postwar missile programs, including those of the United States, the Soviet Union, France, and Great Britain. The Soviet Union in turn passed both German and "indigenous" technology to the Chinese while the Americans did the same to the Japanese. By the mid-1970s, the "space club" included all of the countries, joined in the 1980s by India and Israel who depended on flows from the United States and France respectively. Europe itself—in the form of international agreements—had many cooperative efforts that blurred distinctions of ownership, even as it gained the "indigenous" capacity for space activity in 1979.

CONCLUSIONS

The public awareness of spaceflight as an endeavor fundamentally associated with nations will remain unchanged for the foreseeable future. This relationship depends on a number of factors that are unlikely to alter soon; these include the perception of a powerful relationship between science and technology and nationalism; and an understanding of the high costs of space exploration that have impeded non-state actors in investing in such activities. In the latter case, the promise of private spaceflight remains only a promise; even if the sector develops into a vibrant industry in the next decade or so, private spaceflight will represent a very small portion of the overall space projects of any given nation. In perception at least, the major space projects such as human spaceflight and deep space exploration—executed by federal agencies such as NASA—will dominate. And while the creation, maintenance, and expansion of the ISS represents a striking case of international cooperation on a global scale, it is too early to say whether the ISS will serve as a harbinger of future international cooperation; it might well be remembered as a historical anomaly

---

39. For an ahistorical but useful and recent take on space technology transfers, see Mike H. Ryan, "The Role of National Culture in the Space-Based Technology Transfer Process," *Comparative Technology Transfer* 2 no. 1 (2003): 31-66.

40. Siddiqi, *Red Rockets' Glare*.

rather than as a precedent for future international cooperation. President George W. Bush's announcement of a new Vision for Space Exploration (VSE) that mandates a termination of American activities involving the ISS sometime around 2016 suggests that, on a tangible level, the most powerful and capable spacefaring nation on the globe is rejecting a global cooperative vision of human spaceflight in favor of a unitary national imperative.[41] There are many complex geopolitical, technological, and cultural reasons for taking this path, but from the perspective of public rhetoric and public understanding of the future of spaceflight, the VSE has unambiguously reinforced the link between the nation and spaceflight.

I have argued that there are four elements ubiquitous in the public conception of any national space program: the iconography of a founding father, the claim of indigenity, the link with national identity, and the necessity of justifications. It is doubtful that any of these four rhetorical archetypes will recede in importance in the near future. Barring a fundamental change in the link between the projection of national prowess and science and technology, there is little chance that we will see the founding father trope disappear or claims of indigenity recede. And unless space exploration becomes cheap or immensely profitable—a distant possibility—we may not soon see any need to reduce or eliminate the need for justifications in considering the topic of national space travel. On the other hand, there is a probability that public discussions about national space programs will accrue other characteristics, including, paradoxically, an appeal to a global imagination. There are already a few singular achievements in the history of spaceflight that could be described in terms of universal import, i.e., achievements of a national space program that have relevance to the people of the Earth itself. These undertakings would include the launch of Sputnik (the first human-made object in orbit), the mission of Yuri Gagarin (the first human in space), and the landing of men on the Moon (the first humans on another planetary body). One might also include the flotilla of robotic spacecraft sent out to deep space, to the inner and outer planets, and ultimately out of the solar system. On some level, these spacecraft represent artifacts that transcend national ownership.

I believe that significant global firsts and the capability to exit near-Earth space can be construed as benchmarks for a national space program to rise to a new level and claim global significance. Until now, only two nations have achieved that capacity: the former Soviet Union and the United States. The

---

41. "President Bush Announces New Vision for Space Exploration Program," *http://www.whitehouse. gov/news/releases/2004/01/20040114-3.html* (accessed February 29, 2008); Marcia S. Smith, *Space Exploration: Issues Concerning the "Vision for Space Exploration,"* CRS Report for Congress RS 21720, revised June 9, 2005, *http://opencrs.com/getfile.php?rid=51025* (accessed February 29, 2008); Carl E. Behrens, *The International Space Station and the Space Shuttle*, CRS Report for Congress RL33568, revised November 9, 2007, *http://opencrs.com/getfile.php?rid=59204.*

language of global significance has been deployed frequently by commentators to characterize a few singular achievements—Sputnik, Gagarin, and Apollo being the most obvious ones—since the beginning of the space era in 1957. Arguably, some other nations or international agencies, including the European Space Agency (ESA) and Japan, can make a claim to have performed acts with comparable significance, particularly in the area of planetary exploration.[42] And although China has a vibrant and diversified space program, until now it has only repeated actions done by others. But as more nations begin to become vibrant space powers capable of achieving critical "firsts" in the history of space exploration and equally capable of sending their handiwork out into deep space, we will probably see a rise in the kind of rhetoric we saw during the times of Apollo. In that sense, we may be soon witness to an interesting rhetorical clash between the national and the global—and at this point, it remains to be seen how that tension will play out.

---

42. ESA has directed and participated in a number of ambitious and path-breaking deep space exploration projects, including missions to Halley's Comet (Giotto, launched in 1985), Mars (Mars Express, 2003), the Moon (SMART 1, 2003), minor planets (Rosetta, 2004) and to Saturn's moon Titan (Huygens, 1997). Similarly, Japan has implemented a modest series of deep space missions since the 1980s including missions to Halley's Comet (Sakigake and Suisei, both 1985), the Moon (Hiten in 1990, Kaguya in 2007), the minor planets (Hayabusa, 2003), and Mars (Nozomi, 1998). See Asif A. Siddiqi, *Deep Space Chronicle: A Chronology of Deep Space and Planetary Probes, 1958-2000* (Washington, DC: NASA, 2002).

CHAPTER 3

BUILDING SPACE CAPABILITY THROUGH EUROPEAN REGIONAL COLLABORATION

John Krige

On September 26, 2007, the widely distributed daily *USA Today* published a special feature on the dawn of the Space Age. It devoted more than a full page to the launch of Sputnik and to the conquest of space, more than a week before the 50th anniversary of the Soviet achievement. *USA Today*, obviously trying to steal a march on its competitors, also wanted to intervene in current debates on American space policy, as the title of the feature, "Lost in Space," made clear. *USA Today*'s approach was dominated by two themes: the U.S. vs. Soviet competition in the space race, of which the newspaper gave a blow-by-blow chronological summary, and which ended victoriously when Neil Armstrong stepped onto the Moon; and the frustration of "those who were involved at the beginning and others who are key to future explorations"—pioneers and visionaries who were concerned that the United States had no long-term and sustainable space policy. The feature in *USA Today* thus provided both an historical and a policy-oriented intervention, an attempt to define a past and to use that representation of the past to shape the future.

I do not draw attention to this article because I deem it to be representative; indeed, a thorough, comparative analysis of how the launch of Sputnik and its aftermath were depicted in the world's press 50 years later awaits scholarly attention and will, I am sure, be most illuminating. It interests me because it embodies some of the typical traps that lie in wait for those of us who set out to "Remember the Space Age." Three of these are particularly striking.

Firstly, *USA Today* shrunk the content: the Space Age is reduced to human space flight and the competition for space firsts between two superpowers. While this focus is understandable in a popular daily newspaper, it is also regrettable It is understandable since human spaceflight is a feature of the conquest of space that continues to inspire the public's imagination. It is regrettable because people are not becoming educated about the other dimensions of space (i.e., space for science, space for applications both civil and military, and space as a means for building high-tech industry and national competitiveness in the aerospace sector). Public support for human space exploration may be, according to Roger Launius in the same edition of *USA Today*, "a mile wide and an inch deep." But it is surely our task as scholars to criticize this obsession with human spaceflight—however important it may be to maintaining NASA's momentum—and draw the public's attention to the many other reasons for a major technological nation to have a space program. In remembering the Space Age, we must uncouple the conquest of space from the always-contested domain of human space exploration in order to recognize that that conquest has multiple dimensions that range from stimulating basic science and engineering to national security applications.

Secondly, in the article in *USA Today*, the history of the Space Age shrinks geographically: the commemorative article is entirely Americo-centric. Even though Soviet feats are mentioned, their context is how they impacted the United States and provided the challenge that stimulated the U.S. response. Such an approach is misleading in many ways. For one thing, it completely overlooks the fact that human spaceflight is no longer at the core of superpower rivalry and the associated ideologies of leadership and "domination" that went along with this competition. On the contrary, human spaceflight is increasingly seen as an international, collaborative venture in which America's partners—including its previous Cold war rival—play a critical role. This narrow Americo-centrism also ignores the fact that some major space efforts, such as that typically occurring in Western Europe—my concern here—have never included their own transport system for human spaceflight, nor attempts to compete with the two superpowers in this domain. (The project to develop the space plane Hermes was a brief but quickly abandoned effort to do just this: its rejection reinforces my point.) If we remember the Space Age through the prism of countries other than the United States, human spaceflight assumes an entirely different and far less central significance. The conquest of space is also seen to be driven by concerns other than the competition for "leadership" between two Cold War rivals. It is time that the American public understand that America's ongoing activity in human spaceflight requires genuine partnership in ways that were inconceivable 20 years ago. The article in *USA Today* gives no indication of this context.

Thirdly, in the feature in *USA Today*, the history of the Space Age shrinks in time: it is confined to the first decade or so from the launch of Sputnik in 1957 to the first steps on the Moon in 1969. This narrowing of temporal context is obviously related to the two previous points. Such an approach is acceptable as long as one realizes that the events in that period were driven by an historically specific agenda that was not respected in other domains of space or, indeed, even in the domain of human spaceflight in the U.S. beginning in the 1970s. While this may seem trivial on first blush, it is not so when we consider that serious policy prescriptions may be based on the assumption that the way to redynamize the space program is to reconstruct in the present day the situation that prevailed in the late 1950s and 1960s. These arguments conclude, in effect, that only competition with a rival superpower (and China is the prime candidate looming over the horizon) can imbue the American space program with new vitality. The point I want to stress is simply a variant of the one I made earlier: the visionaries and pundits who complain that the United States is "lost in space" need to start thinking about how to find useful ways of collaborating with other nations who have developed important space programs instead of remaining frozen in an obsolete mental framework dominated by the paradigm of Cold War rivalry.

Perhaps I have devoted too much time to one newspaper article that was never intended to be comprehensive or analytically rich. However, it is a useful and accessible source for making a more general point: all historical analysis is necessarily partial and selective. Efforts to imbue historical accounts with universality are not simply methodologically flawed. They also stifle our critical capacities while dominating our perception of the past and our definition of what the future should look like. When remembering the Space Age, we should not shy away from admitting the complexity and diversity of the space effort nor pretend that the view of the world from Washington is the only view worth recording. Our watchwords should be disaggregation and contextualization. We should emphasize the heterogeneity of space programs and explore the diversity of space policies as they evolved at the national, regional, and global levels. Over the past 50 years the conquest of space has followed different rhythms, been driven by different motives, and had a different physiognomy inside each spacefaring nation and region, as well as between them. That may make for a messy story, from which it is difficult to draw general policy implications. So be it. Surely one of the most important lessons of history is that we must grasp the past in its specificity in order to understand the present and think intelligently about the future.

Three Distinguishing Features of the European Collaborative Space Program

It is time for me to turn away from these warnings and to focus on the European space program, which is the subject of this paper. My central claims are three. Firstly, Western European space projects have contributed to building a regional capability and identity. That identity is embedded in institutions and practices that brought together (mostly) men from separate European nation states and had them work in partnership around scientific and technological projects that their governments were willing to pursue at a collaborative level. The kind of project that was suitable, and the form of collaboration embarked upon, usually did not have immediate strategic significance for these national governments. European identity was not forged around space *tout court,* but around certain space projects that, it was believed, preserved or even advanced the ability both to build Europe and to secure key national interests.

My second main point is that this construction of regional identity was possible because of a combination of shared political, technological, and industrial objectives that were defined in the 1960s and 1970s. At that time, European integration was an important component of foreign policy, the European space science and engineering community was small and inexperienced, and European firms were novices in space technology and, above all, in systems engineering and project management. Space served as a scientific and technological platform around which to build Europe because of the cost, complexity, and industrial challenges it engaged. Integration was a solution to structural weaknesses that many, though not all, European nation states believed they could not overcome by "going it alone."

My third point is that the United States played a key role in constructing and consolidating regional capability and identity. This may seem counterintuitive. It is obvious that many people in Europe, particularly today, regard the United States as an overbearing hegemon that unilaterally tries to impose its political, technological, commercial, and cultural values on friend and foe alike, resorting to force if needed to achieve its objectives. Many in the European space sector share that view, or some variant of it, in regard to the current situation. However those same people are the first to recognize the crucial role played by NASA and the United States in helping Europe get on its feet in the space sector in the first couple of decades following Sputnik. Some put this down to American generosity and to a sense of shared historical and cultural ties. The United States, President Eisenhower once said, "was related by culture and blood to (the) countries (of) Western Europe and in this sense is a product of Western Europe." Similarly, for McGeorge Bundy, a National Security Adviser to President Kennedy, the European peoples were "our cousins by history and culture, by language and religion." If here the personal and the cultural are stressed, on other occasions the political is the focus. For example, when Bundy was asked why he favored the postwar reconstruction of a united and "independent" Europe—since, after all, "great states do not usually rejoice in the emergence of other great powers" —his response was unequivocal: "The immediate answer is in the current contest with the Soviet Union."[1] For Washington, a strong united Europe built on a solid scientific and technological base would bring with it the economic prosperity and political stability essential to maintaining democracy among America's allies. European integration would act as a bulwark against communist expansion on the continent and, through NATO, help take the burden of the defense of the region off the back of the United States. In pursuit of these and related policies, NASA acted as an arm of American diplomacy in the 1960s and 1970s, and, along with the State Department, played a crucial role in fostering a collaborative European space

---

1. Cited in John Krige, *American Hegemony and the Postwar Reconstruction of Science in Europe* (Cambridge, MA: MIT Press, 2006), pp. 254, 255.

program and, in so doing, contributed to the building of a regional European identity.

In the remainder of this paper, I want to flesh out these claims in more detail and lay bare the poles around which the European space program has been built. In addition, I want to emphasize the very different shape the program has assumed as compared to the American program. The Space Age, as I insisted earlier, is not all of a piece, and is certainly not to be collapsed into those highly visible and exotic features that so often dominate the public debate and the public face of the American space program.

THE RELATION OF THE CIVIL AND THE MILITARY

Walter McDougall has claimed that, when NASA was launched, the separation of military and civilian activities was increasingly artificial in the age of scientific warfare and total Cold War. Even scientific programs, under a civilian agency, were tools of competition in so far as an image of technical dynamism was as important as actual weapons. The space program was a paramilitary operation in the Cold War, no matter who ran it. All aspects of national activity were becoming increasingly politicized, if not militarized.[2]

McDougall was of course deeply aware of the technological, political, industrial, and cultural dimensions of superpower rivalry, In addition, he was disturbed by what he saw as the corresponding militarization of every facet of American life in an age of what Eisenhower called "total cold war."[3] But even if we insist on drawing the distinction between the civilian and the military more finely than he did, there is no doubt that space was and is fundamental to national security, notably during the Cold War. As Paul Stares pointed out in 1985, about two-thirds of all satellites launched in the first 25-odd years after Sputnik by the United States and the Soviet Union were for military purposes. The fiscal year (FY) 1984 U.S. military space budget alone was about $10.5 billion in current dollars—about half of the total American space budget.[4] The apparatus of the national security state will ensure the future of spaceflight in the United States for multiple forms of reconnaissance whether or not there is a moonbase or a mission to Mars. In fact, one may go so far as to say that all major space programs are, to some extent or another, parasitic on governments recognizing the military potential of space. Without that military dimension, they would never be willing to invest the billions of dollars of taxpayers' money

---

2. Walter A. McDougall, . . . *the Heavens and the Earth. A Political History of the Space Age* (Baltimore, MD: Johns Hopkins University Press, 1985), p.174.

3. Kenneth Osgood's *Total Cold War. Eisenhower's Secret Propaganda Battle at Home and Abroad* (Lawrence, KS: University Press of Kansas, 2006) described the many dimensions of this phrase.

4. Paul B. Stares, *The Militarization of Space. U.S. Policy, 1945-1984* (Ithaca, NY: Cornell University Press, 1985), p. 14.

needed to establish and to maintain a major presence in space and, above all, to acquire independent access to space.

That said, the situation has to be nuanced. Certainly Britain and France embarked on space programs in the 1950s and 1960s that owed much to their military ambitions (and, in the case of France, to a considerable influx of technical personnel from ex-Nazi rocket programs).[5] Both countries were medium-sized, technologically dynamic powers that sought to maintain their global influence as their empires withered. Both sought independent nuclear deterrents and their appropriate delivery systems, and both were in a position to deploy engineering skills, hardware, and production techniques acquired in laboratories, design shops, testing grounds, and industries for both civilian and military purposes. It is also true that the earliest experiments in the upper atmosphere with sounding rockets were only possible due to the military infrastructure, be it at Woomera in South Australia for the British, at Hammaguir in the Sahara for the French, or in Sardinia for the Italians.[6]

All the same, as the collaborative European space program began to take shape, a distinct effort was made to distance it from the military. One of the reasons for this was the personalities and priorities of the main protagonists of a joint European effort.[7] These were not government officials but cosmic ray physicists turned scientific statesmen, one Italian (Edoardi Amaldi), the other French (Pierre Auger). Amaldi and Auger were among the founders of the European Organization for Nuclear Research (CERN), a particle physics laboratory established in Geneva in 1954. Both firmly believed that the only way that European "big" science and technology could compete with the United States was if governments pooled their resources (financial, industrial, and skilled) in collaborative efforts. Both men had strong support in the highest level of national administrations where senior bureaucrats saw promising careers in joint European scientific and technological activities. Both deplored the militarization of scientific research in Cold War America, and both were extremely concerned by the proposals, emanating from the newly-formed NATO Science Committee directed by Fred Seitz, that NATO should take the initiative and build a European satellite. In short, the first push for a European

---

5. For Britain, see Harrie Massey and M. O. Robins, *History of British Space Science* (Cambridge: Cambridge University Press, 1997). For France, see France Durand-de Jongh, *De la fusée Véronique au lanceur Ariane. Une histoire d'hommes, 1945-1979* (Paris: Stock, 1998).

6. For the military significance of early sounding rocket work, see David H. de Vorkin, *Science with a Vengeance. How the Military Created the US Space Sciences after World War II* (New York, NY: Springer, 1992).

7. The story is told in detail in John Krige and Arturo Russo, *A History of the European Space Agency, 1958–1987*, Vol. 1 *The Story of ELDO and ESRO, 1958–1973* (Noordwijk: European Space Agency Special Publication-1235, 2000). For the later period, John Krige, Arturo Russo, and Lorenza Sebesta, *A History of the European Space Agency, 1958–1987*, Vol. 2 *The Story of ESA, 1973–1987* (Noordwijk: European Space Agency Special Publication-1235, 2000).

space effort was made precisely by people who were determined that it should be free from military control and independent of military funding.

These ambitions dovetailed with those of the potential member states of any future European space organization. Collaborative programs in the military sphere were politically impossible. For the British, to do so would mean jeopardizing their special relationship with the United States and the privileged access to military technology that that gave them. For the French, it would mean diluting sovereignty at the very moment when President de Gaulle was in no mood to be tied by obligations to European partners that might restrict his development of France's independent strike force (or *force de frappe*). For its part, Germany had to tread with extreme caution in all areas of potential military significance. The memories of two world wars and the role of Germany in provoking them were still fresh. Indeed, the "double containment" of Soviet communism and German nationalism and militarism was another important reason why Washington and its continental allies strongly supported the emergence of a supranational, integrated Europe. A European space effort was not and could not be dictated by military considerations. Its rationale would need to lie elsewhere.

The history of European launchers confirms this point. Amaldi and Auger originally envisioned an organization similar to NASA for Western Europe—an organization responsible for developing both launchers and scientific satellites. This plan was rejected by many countries that, while happy to be part of a collaborative scientific research effort, were not willing to work together to build a launcher (baptized Europa and built under the auspices of the European Launcher Development Organization [ELDO]). This was partly due to cost, but it was also because small countries like Switzerland, whose global weight was intimately associated with its posture of neutrality, deemed launchers as being too close to the military end of the civilian-military spectrum for their government to participate in developing such technology.

The military significance of launchers also partly explains the disastrous failure of the Europa program. This is generally, and rightly, put down to the lack of project management in ELDO and the failure to integrate the three stages of the rocket being built separately in Britain, France, and Germany.[8] However, even if Europe had had the requisite project management skills— which it did not—the mutual mistrust between the partners and the reluctance of government and industry to let engineers from other countries have access to domestic missile and rocket technology (believed to be a national strategic asset) sabotaged any serious effort at technological integration. Europe learned the lesson. The successful Ariane launcher was not only built under French industrial and management leadership. It was also a product of the civilian

---

8. Stephen B. Johnson, *The Secret of Apollo. Systems Management in American and European Space Systems* (Baltimore, MD: Johns Hopkins University Press, 2002).

Britain's Blue Streak ballistic missile, designed in 1955 and tested at Woomera test range, Australia. It was used as the first-stage of the European satellite launcher, Europa. (Google Images)

French National Space Agency Centre Nationale des Études Spatiales (CNES) and so financially and formally independent of the French missile program.[9] I could provide multiple examples, but the point is, I hope, clear: space in Europe was predominantly civil, and whenever that boundary risked being blurred, above all in industry, some countries simply did not participate in that particular aspect of space activities.

SCIENCE AND APPLICATIONS

Science has been a preferred site for international collaboration, at least in times of peace and between traditional allies. Yet, contrary to what one might think, science could not directly bear the weight of building a strong European regional capability in the early years of their space program.

One of the main reasons for this was that bilateral programs with the United States were very attractive and even essential alternatives to multilateral programs with European partners. At the Committee on Space Research (COSPAR) meeting in The Hague in March 1959, the American delegate announced that NASA was willing to fly experiments by foreign scientists on U.S. satellites, even going so far as to help scientists in other countries build an entire scientific payload for launching on American Scout rockets. Several factors informed NASA's policy. It was a tangible expression of the requirement specified in the Space Act of 1958 that NASA foster international collaboration. It could trade on the longstanding tradition of international scientific exchange and mobilize networks and institutions already in place that were familiar with working outside national frameworks. It raised no obvious risks to national security, nor of technological exchange—useful work could be done with relatively simple and inexpensive instruments that perfectly embodied the strategy of "clean interfaces" and "no exchange of funds" that quickly became the hallmark of NASA's international programs.[10] Finally, NASA was particularly interested in seeing that a country from the Western bloc be the first to launch a satellite after the superpowers had done so, a position consistent with the all-pervasive logic of Cold War rivalry that marked every aspect of U.S.–Soviet relations in space contained in the earlier McDougall reference.[11] Seen from Europe, where the space science community was small, inexperienced, and

---

9. Emmanuel Chadeau, ed., *L'Ambition technologique: naissance d'Ariane* (Paris: Rive Droite, 1995); Claude Carlier et Marcel Gilli, *Les trente premières années du CNES*. L'Agence Française de lEspace, 1962 – 1992 (Paris: CNES, 1994).

10. Arnold Wolfe Frutkin, *International Cooperation in Space* (Englewood Cliffs, NJ: Prentice-Hall, 1965).

11. John Krige, "Building a Third Space Power: Western European Reactions to Sputnik at the Dawn of the Space Age," in Roger D. Launius, John M. Logsdon, and Robert W. Smith, *Reconsidering Sputnik. Forty Years Since the Soviet Satellite* (Chur: Harwood, 2000), pp. 289-307.

fragmented, NASA's offer was a godsend. Indeed, French space scientists are unstinting in their praise for NASA's generosity and support in these early days and recognize that without it their own program could never have taken root and been as successful as quickly as it was. But it had a downside from a European collaborative perspective. Space science was a fragile platform for integration since so much first had to be done through bilateral arrangements with the United States.

The lukewarm enthusiasm in the 1960s for major collaborative space science projects had tangible effects. The European Space Research Organization (ESRO), established along with ELDO, had a pitifully small budget and had great difficulty in developing any major scientific satellite program of its own. The attempt to build a Large Astronomical Satellite that was sufficiently costly and complex to serve as an integrative glue collapsed ignominiously.[12] The point of having a European-based science program at all was vigorously contested by the French in the early 1970s when the future of ESRO, or at least its mission, hung in the balance. It was only saved by reorienting the organization towards application satellites (to the dismay of the scientific founding fathers) and at the insistence of the British, who demanded that science be made a mandatory component in the new ESA that emerged in 1975. This was a sign of the vulnerability of the collaborative science program, not of its strength. At the time it was feared that, if science was made optional like all the other major programs then being agreed on for ESA, it would simply collapse for lack of political and financial support.

One reason for the assault on science in the early 1970s was a determination in Europe, led by the French, to make launchers and application satellites the backbone of the European space effort. The rationale in Paris combined a Gaullist determination to become independent of the United States with a recognition that space not only had important commercial possibilities, but was a crucial domain in which one could hope to close the technological and managerial gap that has opened up between the two sides of the Atlantic. However, France did not entirely get its way. Germany insisted on building Spacelab in collaboration with the United States, and the science program actually became one of ESA's outstanding domains of activity. The fact remains, though, that European foreign and industrial policy are central drivers of the regional space effort, a point that is so important (and which makes Europe so different from the United States) that it deserves further elaboration.

---

12. John Krige, "The Rise and Fall of ESRO's First Major Scientific Project: The Large Astronomical Satellite (LAS)," in John Krige, ed., *Choosing Big Technologies* (Chur: Harwood, 1993), pp. 1–26.

EUROPEAN SPACE AS AN INSTRUMENT OF FOREIGN AND INDUSTRIAL POLICY

European foreign policy is expressed in its space policy. Space industry, science and technology provide the material infrastructure that lock governments into formal multilateral agreements. The United Kingdom is the exception that proves the rule as far as the integrative urge is concerned. Britain emerged from the war much impoverished with respect to the United States, but it still remained the leading scientific and technological power in Western Europe. It was the first country after the U.S. and the Soviet Union to test both atomic and hydrogen bombs, it had the first commercial nuclear power reactor dedicated to civilian energy production in the free world, and it built the first commercial jet passenger aircraft.[13] By virtue of this leadership and its close alliance with the United States and the Commonwealth, it had little interest in collaborative European space efforts. There was just one brief moment when matters were otherwise. In 1960, after some hesitation, Prime Minister Macmillan decided that the time had come to accept that Britain was no longer a major world power and as a result should draw closer to its neighbors across the English Channel. In June 1961, he deposited Britain's request to join the Common Market. While the six existing members of the club debated the terms and conditions of British entry, two major Franco–British aerospace projects were launched. One was Concorde, the supersonic airliner. The other was the Europa rocket to be built by ELDO. In January 1963, President de Gaulle vetoed British entry, arguing that London was not really committed to European integration and that its inclusion would do little more than serve as a Trojan horse for American interests on the Continent. Concorde survived, but Britain's commitment to ESRO and especially ELDO did not. For the British, space policy became something to be conducted primarily at a national level and through multilateral agreements: they would not be tied into a supranational organization in which their control over programmatic decisions would be diminished. The result was predictable: Britain maintains a strong presence in space science, but it is totally absent from rocketry and has a selective approach to the development of applications.

The place of science and technology in German foreign policy is somewhat unique since here the European option was an essential path back into scientific and technological collaboration with its erstwhile enemies.[14] It also relegitimated technological projects in sensitive areas such as nuclear energy and space. The precedent was set in high-energy physics in 1950 when Isidor I. Rabi from Columbia University proposed at a United Nations Educational, Scientific

---

13. David Edgerton, *Warfare State. Britain 1920-1970* (Cambridge University Press, 2006).

14. Niklas Reinke, *The History of German Space Policy: Ideas, Influences and Interdependence, 1923-2002* (Paris: Beauchesne, 2007). Translated from the German.

and Cultural Organization (UNESCO) meeting in Florence, Italy, that a new European physics laboratory built around big equipment be established and that Germany be included in this venture. This meeting happened just after the West German state had been formed but before it was admitted to UNESCO, when the odor of opprobrium still hung over the German physicists who had stayed behind and worked in Nazi Germany and when research with accelerators was highly restricted by the occupying powers. This change of approach reflected a larger change of tack in U.S. foreign policy. Germany was no longer to be treated as an occupied state and a threat to stability in Western Europe, but as a scientific, technological, industrial, and economic force to be reintegrated back into Europe if its potential for the growth and security of the region and the free world were to be realized. Rabi was party to those debates. For Germany, membership of CERN gave a new legitimacy to its physics community and to research with particle accelerators, in addition to opening the way for its reacceptance into the international scientific community.[15]

Ten years later, Germany's national interests in the space sector were served in precisely the same way when it was admitted to ESRO and ELDO. The allies imposed tight constraints on German rocketry after the war. Many of her rocket scientists and engineers fled the country for fear of reprisals, leaving a demoralized, isolated, and restricted community at home. Some hoped to return one day to build up their national space effort: Von Braun and his team are not to be taken as typical of the German engineering community.[16] The regional European option provided a way back for a nation that had been effectively barred from space pursuits for more than a decade. By allowing the German space program to grow under the auspices of a supranational regime, government, industry, engineers, and scientists could once again embark on building the infrastructure for a major space effort at the national, regional, and international levels.

Space policy is not only a matter of foreign policy in Europe; it is also a matter of industrial policy. European nations are not shy in admitting that they work together in space to build a shared industrial infrastructure and the pool of scientific and engineering skills that will enable them better to position themselves competitively in the global market, notably vis-à-vis the world leader, the United States. It must be said that this consideration also has some weight in the United States, even though it is given far less prominence both in the media and in scholarship. Indeed, as the Cold War moved from confrontation to détente, and the two superpowers sought stability in their separate blocs, arguments other than superpower rivalry had to be found to maintain a major space program that could ensure the future of NASA after the Apollo Moon

15. Krige, *American Hegemony,* chapter 3.

16. Michael J. Neufeld, *Von Braun. Dreamer of Space. Engineer of War* (New York: Alfred Knopf, 2007).

landings.[17] One of those arguments, used by Caspar Weinberger in his famous memo to President Nixon in favor of developing the Shuttle, was that it would save jobs in the aerospace sector.[18] Space policy is tightly linked to industrial strength and competitiveness on both sides of the Atlantic, even though this is more obvious in Europe than in the United States.

It is also more explicit. The Europeans have developed the so-called principle of fair return in which the proportion of money contributed by a government to a collaborative program should be the same as the share of technologically significant contracts that flow back to national industry from that program. This policy provides smaller nations with one of their most important incentives for remaining engaged in space since the industrial leaders are "obliged" to include their firms in European-wide consortia to secure contracts through ESA. It also explains the major contributions to the European space program made by countries that were technologically "lagging" behind the rest of Europe in the 1970s, specifically Spain as it recovered from the drag of the Franco regime.

THE TENSION BETWEEN THE REGIONAL
AND THE NATIONAL

This paper has stressed the importance of space as an instrument for building a regional capability in Western Europe. That process is not "natural" or spontaneous: it requires ongoing work by scientists, engineers, industrialists, and politicians. Regional agreements require that states dilute their sovereignty, industries build transnational consortia, and scientists take deliberate efforts to construct multinational, multi-institutional collaborative payloads and satellites. In short, the European path is not a necessity for many of the major European states: it is an option. That option will be adopted only after careful consideration and sometimes heated debate and power struggles between interest groups both within nations and between them.

The economic historian Alan Milward has argued, somewhat controversially, that the integration of Europe did not occur at major cost to the sovereignty of the nation state.[19] On the contrary, it was compatible with the rescue of the nation state as a major historical actor. For Milward, the European option involved the pursuit of national interest through instruments and institutions in which the benefits of integration were believed to outweigh the costs. This view has many merits. The application of the principle of fair return

---

17. Jeremi Suri, *Power and Protest. Global Revolution and the Rise of Détente* (Cambridge: Harvard University Press, 2003).

18. Reproduced in Roger D. Launius, *NASA. A History of the U.S. Civil Space Program* (Malabar, FA: Krieger, 1994), Reading No. 19.

19. Alan S. Milward, *The European Rescue of the Nation State* (London: Routledge, 1992).

nicely illustrates the point. So does the case of postwar Germany, for whom, as we have seen, the European road was a crucial path back into scientific and technological collaboration in sensitive domains. In the early days when most national programs were weak, the loss of autonomy required by supranational integration was, at least for the larger powers, a useful way to acquire the scientific, technological, and industrial capacity needed for an independent national program (that is, if a government eventually decided that it wanted one). In this sense, then, Milward is right. But his argument must be relativized to take account of the changing circumstances under which national space policies were defined in Western European states with the passage of time and the evolving attraction of "going it alone" or working with select partners as national space programs matured.

The ongoing debates over industrial policy illustrate the changing equilibrium between the national and the regional. When ESRO established the principle of fair return in the 1960s, a coefficient of 0.8 was deemed acceptable. Today, however, member states demand a coefficient very close to unity, itself an expression of the enormous weight that they attach to space as an industrial activity in the high-tech sector. Smaller states such as Belgium are particularly emphatic about this since they cannot dream of having significant national space programs of their own, and they justify their participation in space at the political level by the advantages it brings to domestic industry. Larger states see matters otherwise. Fair return was important to France, Germany, and Italy in the 1960s and Spain in the 1970s. By the 1990s, however, a country like France had developed such a broad-based strength in all dimensions of space that were of interest to it, and its government was so committed to the space effort, that it was in a position to go it alone or to work bilaterally with selected partners. Regional collaboration, with the requirement imposed by the fair return principle—that French firms become integrated into transnational consortia, sometimes with partners far less experienced than themselves—turned from an asset into an albatross, especially if there were firms that could do the same job in France itself. Put differently, the stronger a nation becomes, the less interest it has in collaborating meaningfully with others in supranational projects. The political motivations have to override the centrifugal pull of industrial and commercial benefits and the control over programmatic matters that a purely national or loosely collaborative project allows. Regional and international collaboration in advanced technology is not a taken-for-granted given: it has to be constantly sustained if it is to survive.

THE UNITED STATES AND EUROPEAN INTEGRATION IN SPACE

I have discussed at length the role and the interests of the United States in fostering international collaboration in space with Western Europe at two previous NASA conferences.[20] I do not want to repeat myself here. The central point to bear in mind is that NASA, in consultation with the State Department and other arms of the administration have, loosely speaking, two modes of interaction with their international partners. One involves sharing: the sharing of data, skills, and technology. The other involves denial of these self-same assets. The boundary between the two is fluid, and NASA may often be unwilling to share a particularly advanced version of a technology, but be quite happy with allowing partners access to an earlier, less sophisticated variant (e.g. inertial guidance technology). The boundary also shifts depending on the domestic situation in the United States, the availability of the technology from other nations, and the strength of potential partners. NASA's role, after all, is to promote both international collaboration and American space leadership. On the face of it, these two goals are contradictory unless the partners are relatively weak and pose no threat to American leadership.

Policy fluctuations between sharing and denial have marked NASA–Western European relations over the last 50 years. In the early 1960s, as I have explained, most Western European countries depended on international collaboration with NASA to acquire the basic skills required to kick-start key parts of their space programs. NASA gladly collaborated, and the Europeans enthusiastically appreciated their gesture.[21] This willingness to work with Europe was politically easy in science, which was eminently suited to international collaboration and posed no threat to U.S. leadership; indeed, leadership was made manifest in generosity and openness. But it also extended to more sensitive areas like rocket technology when, in 1966, NASA seriously considered offering a wide-ranging package of technological assistance, including cryogenic technology, to keep ELDO afloat. The proposals defined at this time ingeniously respected national security constraints, furthered U.S. foreign policy

---

20. John Krige, "Technology, Foreign Policy, and International Cooperation in Space," in Steven J. Dick and Roger D. Launius, eds., *Critical Issues in the History of Space Flight* (Washington, DC: National Aeronautics and Space Administration Special Publication-2006-4702), pp. 239–260; John Krige, "NASA as an Instrument of U.S. Foreign Policy," in Steven J. Dick and Roger D. Launius, eds., *Societal Impact of Spaceflight* (National Aeronautics and Space Administration Special Publication-2007-4801).

21. Jacques Blamont, "La creation d'une agence spatiale: les Français à Goddard Spaceflight Center, en 1962-1963," and Jean-Pierre Causse, "Le programme FR1," in Hervé Moulin, *Les relations franco-américaines dans le domaine spatial (1957– 1975),* Quatrième rencontre de l'IFHE sur l'Essor des recherches spatiales en France, 8-9 décembre 2005, Paris, France (Paris: Institut Français de l'Histoire de l'Éspace, in press).

interests in the European theater, and made a substantial contribution to European technological development.

As Europe emerged from adolescence to maturity in the space sector beginning in the 1970s, the relationships with NASA became more strained. The unilateral cancellation by Washington of America's contribution to the International Solar Polar Mission (ISPM) has been recounted too often to bear repeating.[22] Suffice it to say that it left a bitter taste in Europe and damaged U.S.–European collaboration in space science for many years. Far more significant for the present argument, however, was the (alleged) refusal of NASA and the State Department to launch the Franco-German telecommunications satellite, Symphonie. The precise details surrounding the negotiations over the request for this launch in the early 1970s are still the subject of heated controversy and will probably never be resolved.[23] NASA's interlocutors still believe that they imposed no unfair or illegitimate restrictions on providing a launcher for Symphonie. The Europeans, and the French in particular, insist that matters were otherwise and that, as negotiations proceeded, the position in Washington became increasingly untenable. All are agreed on one thing: that the conditions imposed on launching Symphonie, or more precisely perhaps, European willingness to interpret U.S. behavior as a refusal to launch Symphonie, played into the hands of engineers in CNES who insisted that America was not to be trusted and that France (and Europe) had to have their own launcher to guarantee them independent access to space. In other words, politically speaking, Ariane was a child of Washington's perceived denial of launch technology to Europe for its first (experimental?) telecommunications satellite.

The Symphonie affair, combined with others like ISPM, and the recent application of the terms of ITAR (International Traffic in Arms Regulations) to the export of space technology are embittering many people in Europe, and undermining the prospects for constructive U.S.–European space collaboration.[24] Indeed, in several interviews that I had with European space scientists, engineers, policymakers, and senior government officials in summer 2007, there were repeated complaints about the new constraints on international collaboration and technological sharing in the space sector that have been put in place since 9/11 and of the disastrous effects that ITAR is

22. Roger M. Bonnet and Vittorio Manno, *International Cooperation in Space: The Example of the European Space Agency* (Cambridge MA: Harvard University Press, 1994).

23. Richard Barnes, "Symphonie Launch Negotiations" and the comment by Bignier in Moulin, *Les relations franco-américaines.* See also Arnold Frutkin interview by John Krige, Angelina Long, and Ashok Maharaj, Charlottesville, VA, August 19, 2007, (NASA Historical Reference Collection, History Division, NASA Headquarters, Washington DC); André Lebeau interview, Paris, France, by John Krige, June 4, 2007, (NASA Historical Reference Collection, History Division, NASA Headquarters, Washington DC.)

24. David Southwood interview, Paris, France, by John Krige, July 16, 2007, (NASA Historical Reference Collection, History Division, NASA Headquarters, Washington DC.)

having on international space collaboration and the perception of the United States in Western Europe.

From the United States' point of view, dealing with partners involves striking a delicate balance between the outward push for scientific and technological collaboration and the inward pull of national security concerns—the conflict between sharing and denial. Sharing builds alliances and secures American access to foreign skills. Denial alienates allies, and encourages them to develop their own capabilities and to seek partners other than the U.S. These dilemmas often pit NASA and the State Department against the Department of Defense and other bodies concerned with national security. The management of that tension in the next few years will, I am persuaded, have a major impact on the future of the space programs not only in Europe, but throughout the world. In the meantime, it is ensuring that Western Europe's regional capability and identity as an independent player in space is being reinforced. The vulnerable child of the 1960s is the mature adult of the 21st century who now seeks genuine rather than junior partnership with the United States, along with the mutual political respect and technological sharing that that entails.

CHAPTER 4

IMAGINING AN AEROSPACE AGENCY IN THE ATOMIC AGE

Robert R. MacGregor

Much has been written about the 184-pound satellite lofted into the heavens by the Soviet Union on October 4, 1957. The story is an insidiously seductive one; it is the romantic narrative of a small metal ball usurping the assumed technological authority of the United States. The frenzy of the media and the swift political backlash seem almost comical in light of the diminutive physical size of Sputnik.

The launch of Sputnik was one of the most disruptive singular events in the history of the United States.[1] The temptation to label it a discontinuity is strong. The year following the Sputnik launch saw the formation of the Advanced Research Projects Agency (ARPA), the creation of the new post of Special Assistant for Science and Technology to the President and its associated committee (PSAC), the transformation of the National Advisory Committee for Aeronautics (NACA) into NASA, and the National Defense Education Act (NDEA). Walter A. McDougall in . . . the Heavens and the Earth: A Political History of the Space Age traces the roots of technocracy in America to this "spark":

> Western governments came to embrace the model of state-supported, perpetual technological revolution . . . What had intervened to spark this saltation was Sputnik and the space technological revolution . . . For in these years the fundamental relationship between the government and the new technology changed as never before in history. No longer did state and society react to new tools and methods, adjusting, regulating, or encouraging their spontaneous development. Rather, states took upon themselves the primary responsibility for generating new technology.[2]

---

1. For a good overview of the Western reaction to Sputnik see Rip Bulkeley *The Sputniks Crisis and Early United States Space Policy: A Critique of the Historiography of Space* (London: MacMillan Academic and Professional Ltd., 1991).

2. Walter A. McDougall, . . . *the Heavens and the Earth: A Political History of the Space Age* (Baltimore, MD: The Johns Hopkins University Press, 1985), pp. 6–7.

McDougall has since revised his original argument by noting that the space technological revolution was an "ephemeral episode in the larger history of the Cold War, rather than the Cold War having been an episode in the larger story of the march of technocracy."[3] This revisionism addresses the eventual fate of the space technological revolution. It is the purpose of the current essay to revise the story of the birth of that technological revolution. Specifically, it will be argued that the conception of the Sputnik launch as a discontinuity that ushered in a technocratic revolution in modern America does not fit the historical record. The environment in which the Sputnik crisis unfolded in the United States was already saturated with preconceived, technocratic notions of the relation of science to the state. The crystallization of the new Agency that would become NASA was a process that was both simultaneously instigated by a singular event and followed in the footsteps of institutional ancestors. The two are not mutually exclusive; contingency must be embedded in a framework of continuity. The precursor of the space technological revolution was the Atomic Energy Commission (AEC).

"Technocracy" is a contentious term, with definitions running the gamut from a literal etymological interpretation as "the control of society or industry by technical experts"[4] to the idolization of science for propaganda purposes by non-scientific bureaucrats.[5] An attempt at a precise definition is necessarily doomed to failure, but for the purposes of this essay I will adopt McDougall's definition of technocracy as "the institutionalization of technological change for state purposes, that is, the state-funded and -managed R&D explosion of our time."[6] McDougall's definition captures the key features relevant to the current analysis: massive state funding and intentional control of technological development to serve state purposes. There exist a myriad of other possible definitions, which remain outside the scope of the present argument.[7]

---

3. Walter A. McDougall, "Was Sputnik Really a Saltation?" in *Reconsidering Sputnik: Forty Years Since the Soviet Satellite,* ed. Roger D. Launius, John M. Logsdon, and Robert W. Smith (Harwood Academic Publishers, 2000), pp. xviii.

4. *The Oxford English Dictionary* (New York, NY: Oxford University Press, 1989).

5. A famous example in space history is Nikita Khrushchev's shrewd tactical use of spaceflight for internal and external political maneuvering. For an overview of Khrushchev's manipulation of the space program, see Asif Siddiqi *Sputnik and the Soviet Space Challenge* (Gainesville, FL: University Press of Florida, 2003), especially pp. 409–460.

6. McDougall, . . . *the Heavens and the Earth,* p. 5.

7. David Noble in *America by Design: Science, Technology, and the Rise of Corporate Capitalism* (Oxford University Press, 1977) inverts the hierarchy and sees this explosion not as state-centric manipulation, but as a "wholesale public subsidization of private enterprise" to serve the ends of technocratic corporate managers working as government contractors (p. 322). John Kenneth Galbraith in *The New Industrial State* (Boston, MA: Houghton Mifflin, 1967) envisions technocracy as having a decision-making mind of its own within a given institutional constellation, the "Technostructure," which operates autonomously from corporate or governmental intentions, often to the detriment of the public good. Don Price argues in *The*

The AEC and NASA are far and away the canonical American institutional examples of technocracy under this definition. The similarities on the surface are obvious. Both the AEC and NASA were characterized by geographically dispersed scientific research laboratories operating as scientific fiefdoms in a confederate framework.[8] Both consolidated to a great extent an entire realm of technology in federal, civilian agencies. Unlike other new technologies, such as the microcomputer or early aviation, both were handed over wholesale to civilian agencies created specifically to oversee them rather than entrusting progress to the military or private sector. In introducing the problem the framers of the Atomic Energy Act faced, AEC historians Richard G. Hewlett and Jack M. Holl noted: "How does one best go about introducing a new technology into society? A familiar problem for large manufacturers, the management of technological innovation was hardly a common function for federal officials, except in the area of regulation . . . in the case of nuclear power, the entire technology was confined within the government."[9] This fundamental historical similarity, domination and encapsulation of an entire area of technology by a civilian government agency, is the basis for the current argument.

This paper will examine the links between atomic energy and the processes in the executive and legislative branches that culminated in the signing into law of the National Aeronautics and Space Act on July 29, 1958. While a detailed comparative history of the roles, structures, and functions of NASA and the AEC would immensely contribute to the historical literature, the current analysis will focus more narrowly on the way in which the experience with atomic energy produced unspoken assumptions and shaped the very imagination of politicians of what the new NASA should and could become during the ten-month period from the launch of Sputnik to the passing of the National Aeronautics and Space Act. Specifically, it will be argued that NASA's rise in the 1960s as an engine of American international prestige was rooted in atomic diplomacy, and that certain debates in Congress about the new Agency

---

Scientific Estate (Belknap Press: 1965) that the fusion of political and economic power seen in the nuclear and Space Age has corrupted market principles by creating corporations solely dependent on government subsidies, resulting in a diffusion of political sovereignty that threatens the American constitutional order. Finally, no discussion of technocracy in America would be complete without mentioning Frederick Winslow Taylor's *Principles of Scientific Management* (New York, NY: Harper Brothers, 1911), which called for applying scientific principles to the training and management of workers to replace "rule of thumb" factory methods.

8. Peter J. Westwick in *The National Labs: Science in an American System, 1947-1974* (Cambridge, MA: Harvard University Press, 2003) perhaps borrowing from dialectical materialism stresses that the systemicity of the labs is central to an understanding of their operation. A single national lab cannot exist in isolation; classified journals and conferences and competition for personnel and research programs were central issues that defined the individual labs.

9. Richard G. Hewlett and Jack M. Holl *Atoms for Peace and War* (Berkeley, CA: University of California Press, 1989), p. 183.

were largely approached from within a framework of atomic energy, thereby limiting the range of discourse and influencing the shape of the new Agency.

While NASA grew by orders of magnitude in the 1960s, the features that specifically identified NASA as technocratic were frozen into the bureaucracy in this formative period. The sudden influx of money after Kennedy's famous decision to set NASA's sights on a Moon landing merely inflated NASA's existing latent potential.

THE ROLE OF PRESTIGE

A large debate in the historiography of NASA centers on the question of prestige. Is NASA's mission coincident with or even driven by American political imperialism? How did national prestige come to be measured by a cosmic yardstick? These questions are often posed in light of the two temporal sides of the Sputnik rupture. On the one hand, the Eisenhower administration was seemingly caught unawares of the worldwide impact the launch of Sputnik would have on public perceptions of American strength. On the other hand, John F. Kennedy would soon after catapult his career on the program to send humans to the Moon, a program that "transformed NASA from a scientific research agency into a goal-oriented bureaucracy."[10]

In the fall of 1957, high-level officials extrapolated the Sputnik launch into an across-the-board American deficiency in scientific ability. The Democratic majority under Senator Lyndon B. Johnson jumped on the opportunity to place blame on the Republican Eisenhower administration and relaunched hearings by the Preparedness Investigating Subcommittee of the Committee on Armed Services in the Senate in late November. General James H. Doolittle provided one of the early testimonies.[11] In his testimony, Doolittle felt convinced "that the rate of Russian progress is much more rapid than ours; that, in some areas, she has already passed us. If the rate continues, she will pass us in all."[12]

In a meeting of the Office of Defense Mobilization Science Advisory Committee (SAC) with President Eisenhower on October 15, Edward H. Land explained to the president the reasons for Soviet success:[13]

---

10. Giles Alston; Shirley Ann Warshaw, ed., Chap. "Eisenhower: Leadership in Space Policy" in *Reexamining the Eisenhower Presidency* (Westport, CT: Greenwood, 1993), p. 117.

11. Doolittle was already famous for his bombing raid on Tokyo shortly after the initiation of hostilities between the United States and Japan in 1942. He later went on to become Chairman of the NACA board, a position he held at the time of his testimony.

12. *Hearings before the Preparedness Investigating Subcommittee of the Committee on Armed Services, 85th Congress*, 1st and 2nd sessions, pt. 1, p. 111.

13. At the meeting, I. I. Rabi noted "most matters of policy coming before the President have a very strong scientific component" and "he didn't see around the President any personality who would help keep the President aware of this point of view." Eisenhower concurred and "said that he had felt the need for such assistance time and again." This discussion led to the suggestion by

The structure of Russian culture and thinking is such that
they are learning to live the life of science and its application
. . . Is there a way to tell the country that we should set out
on a scientific adventure in which all can participate? If this
can be done, with our concept of freedom and the independent, unfettered man, we can move far ahead. We need a
scientific community in the American tradition.[14]

Whether or not Land had accurately assessed the Soviet mentality towards
science or of the true implications of the Sputnik launch is of little importance.
The notable point is the reaction produced in the very highest echelons of
scientific and military advisory circles. Clearly, the hysteria and "fever"
that swept the country in the wake of the Sputnik launch was not limited
to an uninformed public. Indeed, the media and public were simultaneously
concerned with the integration crisis at Central High School in Little Rock,
Arkansas. For those in the government primarily concerned with national
security, Sputnik produced a larger effect than in the public at large.

The conception of Sputnik as a discontinuity is linked to the conception
of scientific prestige as a benchmark for national strength. Since Eisenhower
misjudged the impact Sputnik would have on the perception of the United
States, so the argument goes, only after the media frenzy and political attacks of
fall 1957 did the administration recognize the importance of science to national
prestige in the international sphere. Even in the face of Sputnik, Eisenhower
seemingly remained steadfast in his dislike of federal bureaucracy and shied
away from setting prestige as a goal of space research. On November 7, 1957,
Eisenhower announced the creation of the post of Special Assistant to the
President for Science and Technology in a televised address on national security.
The address summarized American nuclear assets while noting deficiencies in
science education in America. The speech concluded with a warning against
runaway spending:

It misses the whole point to say that we must now increase
our expenditures of all kinds on military hardware and
defense—as, for example, to heed demands recently made
that we restore all personnel cuts made in the armed forces.
Certainly, we need to feel a high sense of urgency. But this

James Killian for the creation of a scientific advisory panel to assist the proposed advisor. This
would become the President's Science Advisory Committee (PSAC), which began meeting in
November with Dr. Killian as its head. "Detailed (largely verbatim) notes on a meeting of the
ODM Science Advisory Committee with the President on October 15, 1957," folder 012401,
NASA Historical Reference Collection, NASA Headquarters, Washington, DC.

14. Ibid.

does not mean that we should mount our charger and try to ride off in all directions at once. We must clearly identify the exact and critical needs that have to be met. We must then apply our resources at that point as fully as the need demands. This means selectivity in national expenditures of all kinds.[15]

By analyzing metaphor in his speeches and press conferences, Linda T. Krug notes Eisenhower's "images created a vision of a nation of scientist-generals already hard at work planning how to unlock the secrets of the universe."[16] But the conclusion she draws that "Eisenhower was the only president who saw the space program as a viable entity in and of itself" is based on the assumption that Eisenhower never clothed hidden intentions in crowd-pleasing rhetoric.[17] Such sweeping conclusions about Eisenhower's personal views cannot be drawn from televised statements. All presidents must maintain a carefully groomed public persona. While Eisenhower's public proclamations often criticized big government, policy decisions and internal White House discourse did not match his rhetoric.

The National Security Council (NSC) engaged the question of prestige in relation to the planned American and Soviet satellite launches during the International Geophysical Year of 1957-1958. A Technological Capabilities Panel (TCP) was formed in 1954 under James Killian to investigate the satellite question and other technical issues deemed vital to national security.[18] The TCP issued its final report in February 1955 and the NSC, following the TCP's recommendation, concluded in May of that year that the U.S. effort (Project Vanguard) should be given high priority as "considerable prestige and psychological benefits will accrue to the nation which first is successful in launching a satellite."[19] The importance of such benefits was paramount to

---

15. Dwight D. Eisenhower, "Radio and Television Address to the American People on Science in National Security," November 7, 1957, http://www.eisenhowermemorial.org/speeches/19571113%20 Radio%20and%20Television%20Address%20on%20Our%20Future%20Security.htm.

16. Linda T. Krug, *Presidential Perspectives on Space Exploration: Guiding Metaphors from Eisenhower to Bush* (New York: Praeger Publishers, 1991), p. 29.

17. Ibid.

18. The TCP also drew the famous conclusion that establishing freedom of over-flight in space, i.e., sovereignty claims of airspace not extending beyond the atmosphere, was in the long-term interests of the U.S. This was motivated by the expectation that the U.S. would have a large lead over the U.S.S.R. in electronic satellite reconnaissance capability. For an overview of the TCP and its impact on the freedom of space, see McDougall . . . *the Heavens and the Earth,* ch. 5. Dwayne A. Day has recently uncovered documents tracing the origin of this principle to a CIA intelligence officer, Richard Bissell, and an Air Force aide working for the CIA. Dwayne Day, "The Central Intelligence Agency and Freedom of Space," paper presented at Remembering the Space Age: 50th Anniversary Conference, NASA History Office and National Air & Space Museum Division of Space History, Washington, DC, October 22, 2007.

19. "National Security Council Report 5520: Missile and Space Programs." See *A Guide to Documents of the National Security Council, 1947-1977* ed. Paul Kesaris, (University Publications of America, 1980).

U.S. foreign policy since "the inference of such a demonstration of advanced technology and its unmistakable relationship to inter-continental ballistic missile technology might have important repercussions on the political determination of free world countries to resist Communist threats, especially if the U.S.S.R. were to be the first to establish a satellite."[20]

The NSC concluded the U.S. scientific satellite effort should not hinder military missile developments and, therefore, should be vested in a separate, civilian-run program headed by the National Science Foundation. It is absolutely clear that the Eisenhower administration intended to use the satellite launch to reinforce American scientific prowess in the international arena.

The fact that prestige was an important element after that fateful October 4 and during the formative period of NASA is uncontroversial. In a PSAC meeting in March 1958, Hans Bethe commented, "it would be a great mistake for us to oppose popular enthusiasm even though misguided.[21] And in a recently declassified Office of Research and Intelligence Report issued just two weeks after Sputnik on October 17, 1957, it was concluded:

> The technologically less advanced—the audience most impressed and dazzled by the sputnik [sic]—are often the audience most vulnerable to the attractions of the Soviet system . . . It will generate myth, legend and enduring superstition of a kind peculiarly difficult to eradicate or modify, which the USSR can exploit to its advantage, among backward, ignorant, and apolitical audiences particularly difficult to reach.[22]

The report went even further in claiming the United States itself had fanned the flames of the fire in three ways: "first by fanfare of its own announcement of its satellite plans, second by creating the impression that we considered ourselves to have an invulnerable lead in this scientific and technological area, and third by the nature of the reaction within the U.S."

The importance of science to national prestige in the Eisenhower administration existed long before Sputnik; it originated in the experience with atomic energy. Eisenhower had long been an advocate of using atomic energy

---

20. Ibid.

21. PSAC Meeting, March 12, 1958. The transcribed notes of the PSAC are spotty at best, and the argumentative logic is nearly incomprehensible. They are reproduced in *The Papers of the President's Science Advisory Committee, 1957-1961,* microfilm, (University Publications of America, 1986).

22. Office of Research and Intelligence Report, "World Opinion and the Soviet Satellite: A Preliminary Evaluation," declassified 1993, folder 18106, (NASA Historical Reference Collection, NASA Headquarters, Washington, DC).

to further U.S. foreign policy, a fact exemplified by his personal championing of the Atoms-for-Peace program. ·

In his December 8, 1953 address to the U.N. General Assembly, President Eisenhower called for the establishment of an "international Atomic Energy Agency" to serve as a stockpile of nuclear materials for peaceful uses around the world. The proposal was "enunciated by the President almost as a personal hope," with few advisors and only one of the five Atomic Energy Commissioners, Lewis Strauss, aware of the proposal ahead of time.[23] The original proposal was devoid of details but is significant in that Eisenhower displayed a personal desire to use science and scientific prestige as a tool of international diplomacy. The policy was consciously constructed around the issue of prestige, e.g., the amount of uranium to be contributed by the United States was set at a high enough figure that the Soviet Union would not be able to match the American contribution.[24] While the implementation of the plan was slow in arriving, the middle of the decade saw tangible, albeit often ineffective, international cooperation in atomic technology with the U.S. as the international lynchpin and guarantor of atomic security. Science in the Eisenhower administration was part and parcel of foreign policy.

The tendency to employ science in the service of international prestige was expressed early on in the discussions concerning a new space agency. Coincidentally, Eisenhower asked James Killian (then president of MIT) to become his personal science advisor over breakfast on October 24, the purpose of the meeting being Killian's briefing of Eisenhower in preparation for the Atoms-for-Peace award being given to Neils Bohr later that day.[25]

In an Office of Defense Mobilization (ODM) Memorandum issued in January for Secretary of Health Arthur S. Fleming, the analogy to atomic energy was clearly enunciated: "In addition to the military importance of the scientific satellite one should not overlook the benefits of adequate emphasis on peaceful applications of rocketry just as the atoms-for-peace program has served to divert world attention from nuclear weapons."[26] And in a legislative leadership meeting on February 4, President Eisenhower cautioned against pouring "unlimited funds into these costly projects where there was nothing of early value to the nation's security. He recalled the great effort he had made for the Atomic Peace

23. Hewlett and Holl, *Atoms for Peace and War,* pp. 210-213.

24. John Krige, "Atoms for Peace, Scientific Internationalism, and Scientific Intelligence" *Osiris* 21 (1996): 164.

25. James R. Killian, Jr., *Sputnik, Scientists, and Eisenhower,* (Cambridge, MA: The MIT Press, 1977), p. 24.

26. Executive Office of the President ODM Memorandum to Arthur S. Fleming, "Scientific Satellites," January 23, 1957, folder 012401, NASA Historical Reference Collection, NASA Headquarters, Washington, DC.

Ship but Congress would not authorize it, even though in his opinion it would have been a very worthwhile project."[27]

The relation of prestige to spaceflight has trickled down to the present day. Political pundits still routinely call the value of human spaceflight into question. NASA is frequently attacked as a wolf in sheep's clothing; that is, NASA's stated peaceful exploratory goals are often argued to be merely a facade covering deeper political and military motives. The origins of this dichotomy can be traced directly back to the emphasis placed on prestige during the conception of NASA in the Eisenhower administration, which was in turn based on the experience of atomic foreign policy. By the time a man-in-space investigatory panel was commissioned in 1959 by George Kistiakowsky, then head of the PSAC, it was clear that putting humans in space was solely a prestige issue:

> In executive session of the panel, we talked about these things and I emphasized the need to spell out in our report what cannot be done in space without man. My opinion is that that area is relatively small and that, therefore, building bigger vehicles than Saturn B has to be thought of as mainly a political rather than a scientific enterprise.[28]

Indeed, it can be concluded that space represented a welcome new opportunity for Eisenhower's continuing desire to demonstrate American technological prowess because of a decline in the perception of atomic energy as a positive international technology, a decline spurred on by rising fears of global nuclear annihilation. Certainly the destructive element of nuclear technology had been publicly decried immediately after the Hiroshima and Nagasaki bombings, but the shift in scale from local (bomber-delivered atomic bombs) to global (intercontinental ballistic missile [ICBM]-delivered hydrogen bombs) damned any hope for an unproblematic public perception of nuclear technology. The first hydrogen bomb tests by the United States in 1952 and the Soviet Union in 1953 were followed by the irradiation of the Japanese fishing boat *Lucky Dragon 5* by the Castle Bravo test in March 1954, leading to a widespread public concern over the effects of nuclear radiation.

An illustrative example of the qualitative transformation of atomic energy in the public imagination can be drawn from science fiction. Isaac Asimov's Foundation trilogy, published between 1951 and 1953, portrayed humanity in the far future as a galactic empire in decline. The Foundation, created by a

---

27. Supplementary Notes, Legislative Leadership Meeting, February 4, 1958, folder 18106, NASA Historical Reference Collection, NASA Headquarters, Washington, DC.

28. George Kistiakowsky *A Scientist at the White House: The Private Diary of President Eisenhower's Special Assistant for Science and Technology* (Cambridge, MA: Harvard University Press, 1976), p. 409.

visionary scientist who foresaw the collapse of civilization using new historical-predictive methods, becomes the sole possessor of knowledge of atomic technology and hence the last hope for humanity's future.[29] But by the end of the 1950s, post-apocalyptic novels set in nuclear winter ruled the genre: Nevil Shute's *On the Beach* (1957), Pat Frank's *Alas, Babylon* (1959), and Walter M. Miller, Jr.'s *A Canticle for Leibowitz* (1959). Space, then, was a natural avenue into which the Eisenhower administration could expand its policy of scientific prestige in the service of the state while avoiding the stigmas becoming associated with nuclear technology.

The National Aeronautics and Space Act of 1958

Of special importance to the current analysis are the sections of the National Aeronautics and Space Act of 1958 that were inspired by the Atomic Energy Acts of 1946 and 1954. Specifically, these are: the relation of the Department of Defense to the new Agency, the role of international cooperation, and the apportionment of intellectual property.

When President-elect Eisenhower was briefed on AEC activities in November 1952, he took special exception to Gordon Dean's acquiescence to the Air Force's demand for atomic-powered plane research in the face of good evidence that such a program would not produce a viable aircraft. "Looking out the window he declared that this kind of reasoning was wrong. If a civilian agency like the Commission thought that a military requirement was untenable or wasteful in terms of existing technology, there was an obligation to oppose it."[30] This was a prescient moment for it foreshadowed the problem of divvying up responsibility between competing civilian and military institutions during the formation of NASA.

Analogies to the Atomic Energy Commission were widespread throughout the legislative creation of the new space agency. During the congressional hearings, Eilene Galloway, a national defense analyst at the Library of Congress, was invited by representative McCormack (the chair of the House committee) to write a report on the issues facing Congress in the drafting of the National Aeronautics and Space Act.[31] Her report was widely read and was reprinted in both the Senate and House proceedings and is notable for several reasons. First, Galloway drew the immediate conclusion that a comparison to the issues

---

29. Special thanks to Dan Bouk for pointing out this poignant example from a trilogy I have read four times yet somehow overlooked: Isaac Asimov's *Foundation* (Gnome Press, 1951), *Foundation and Empire* (Gnome Press, 1952), and *Second Foundation* (Gnome Press, 1953).

30. Hewlett and Holl, *Atoms for Peace and War,* p. 14.

31. Galloway also served as special consultant to Lyndon Johnson during the Senate hearings and has since become a noted aerospace historian.

facing the drafters of the Atomic Energy Act of 1946 (informally known as the McMahon Act) would be fruitful. To Galloway, the similarities were obvious:

> Atomic energy and outer space are alike in opening new frontiers which are indissolubly linked with the question of war and peace. They combine the possibility of peaceful uses for the benefit of man and of military uses which can destroy civilization. Both are national and international in their scope. They involve the relation of science and government, the issue of civilian or military control, and problems of organization for the executive branch and the Congress. If only their similarities are considered, the legislative task would appear to be the easy one of following the pattern of our present atomic energy legislation.[32]

According to Galloway, the dissimilarities between the two are centered around the problem of delineating military and civilian aspects of aerospace technology. While the boundaries are reasonably clear in the atomic case (bombs versus reactors), nearly every aspect of aerospace technology overlaps the two sides of the military-civilian divide. This is perhaps an oversimplification in that much effort had gone into the Atomic Energy Act of 1954 to allow the development of a civilian atomic energy industry and the civilian–military divide in practice was quite problematic. Still, it remains true that, in the case of atomic energy, a relatively clear boundary between civilian and military applications could be established through strict regulation of nuclear materials. In the case of NASA this was not true, yet still a formal divide was automatically assumed to be of paramount importance. In part this was due to concerns of needless duplication of effort and bureaucratic infighting over jurisdictional matters. However, previous experience with the AEC weighed heavily on lawmakers, particularly in the House of Representatives, who now saw science as intimately tied up with national security and felt a need for such a relationship to be codified in law. The administration favored a more informal relationship, as had been the case with the NACA. Both sides weighed heavily on precedent to reinforce their arguments.

The debate surrounding the obligations of the new space agency to the Department of Defense and vice versa has long been the center-point of the history of the National Aeronautics and Space Act of 1958. This is for the reason that the delineation of the role of military and civilian agencies has obvious current political implications, but it remains true that much of the contemporary debate also surrounded the issue. The wording of §102(b) of

---

32. Eilene Galloway, *The Problems of Congress in Formulating Outer-space Legislation*, (Washington, DC: U.S. Government Printing Office, March 1958).

the National Aeronautics and Space Act established the following criterion by which specific projects could be judged to be NASA- or Defense-centric:

> The Congress declares that the general welfare and security of the United States require that adequate provision be made for aeronautical and space activities. The Congress further declares that such activities shall be the responsibility of, and shall be directed by, a civilian agency exercising control over aeronautical and space activities sponsored by the United States, *except that activities peculiar or primarily associated with the development of weapons systems, military operations, or the defense of the United States . . . shall be the responsibility of, and shall be directed by, the Department of Defense . . .*[33]

The Act also established a National Aeronautics and Space Council headed by the President and including the Secretary of State, Secretary of Defense, NASA Administrator, and the Chairman of the AEC. The inclusion of the AEC chairman here is quite curious. In addition, any disputes between departments and agencies over jurisdictional matters were to be settled by the President under advisement of the council.

The original Bureau of the Budget draft bill was quite different from the arrangement in the AEC, which embodied communication with the Department of Defense in its Military Liaison Committee. In his official commentary sent to the Bureau of the Budget on the original bill, Strauss suggested "the act provide for inter-agency liaison similar to that which has operated so satisfactorily in the case of the Military Liaison Committee in the atomic energy program."[34] The House bill included such a liaison committee and, in addition, another for the AEC. The administration had favored informal cooperation in the form of uniformed seats on the advisory in the same style as the NACA had traditionally pursued. The Senate kept the administration's arrangement. In the final compromise bill, the military liaison committee was added, while the AEC liaison was dropped.

An internal Bureau of the Budget memo in May snidely remarked on the House bill that "among the trappings of the Atomic Energy Act inserted in this bill are sections establishing and prescribing the functions of a Military Liaison Committee and an Atomic Energy Liaison Committee. Both Committees are to be headed by chairmen appointed by the President . . . The Department of Defense as well as NACA has opposed this creation of statutory liaison committees, and

---

33. National Aeronautics and Space Act of 1958, Public Law 95-568, *http://history.nasa.gov/spaceact. html*. Emphasis and ellipses added.

34. Lewis Strauss to Maurice Stans, Director of the Bureau of the Budget, March 31, 1958, folder 012405, (NASA Historical Reference Collection, NASA Headquarters, Washington, DC).

every effort should be made to secure their elimination in the Senate."[35] The inclusion of the liaison committees in the House bill suggests a strong tendency to adopt portions of the AEC paradigm wholesale. It is particularly remarkable in this case because the civilian-military boundary proposed for NASA was quite different than the model in the AEC. That is, NASA would by default carry on the bulk of aerospace research, but the Department of Defense, by sufficiently justifying its need directly to the President, could develop its own aerospace projects. This is in stark contrast to the complete monopolization of basic atomic research by the AEC, which necessitated a reliable and clear avenue of communication to and from the military.

The differences between NASA's and the AEC's relationships with the military deserves elaboration. From the beginning, the AEC was to encompass all levels of nuclear research, nuclear materials production, reactor design, and bomb construction. This centralization was a result of the realities of atomic energy. First, the Manhattan District was already in place during the establishment of the AEC and maintaining its internal configuration was necessary for the uninterrupted production of atomic weapons. Second, and more important, atomic energy as a technology is unique for a material reason: the regulation of atomic technology is in large part the regulation of a single element and its derivatives. Indeed, the Atomic Energy Act categorically transferred "all right, title, and interest within or under the jurisdiction of the United States, in or to any fissionable material, now or hereafter produced" to the Commission. In effect, all atoms on U.S. territory with 92 or more protons were declared to be the property of the federal government. In addition, an entire new class of information was created. Termed "Restricted Data," this wide umbrella automatically "classified at birth" any and "all data concerning the manufacture or utilization of atomic weapons, the production of fissionable material, or the use of fissionable material in the production of power."[36] Regulation of fissionable material was also the assumed primary task of early atomic weapons nonproliferation efforts. Containment of atomic technology was seen as synonymous with ownership of nuclear materials.

From the inception of the AEC the production and control of nuclear materials was the prime directive of the organization. Fissionable material was simultaneously obviously dangerous, necessary for national defense, and could be relatively easily collected and controlled. The implication of this material reality was tremendous for the bureaucratization of atomic technology in a central governmental agency. In the case of aerospace technology, such a clear compartmentalization was not a natural outgrowth of the relevant technology.

---

35. Letter from Alan L. Dean to Wiliam Finan, June 2, 1958, folder 12400, NASA Historical Reference Collection, NASA Headquarters, Washington, DC.

36. Atomic Energy Act, 1946. Public Law 585, 79th Congress, *http://www.osti.gov/atomicenergyact.pdf.*

Still, the basic structure of the AEC was to provide a perceived "obvious model" for creating an aerospace agency.

§205 of the National Aeronautics and Space Act provided engagement in "a program of international cooperation . . . and in the peaceful application of the results thereof." The Senate Special Committee had noted in a report entitled *Reasons for Confusion over Outer Space Legislation and how to Dispel it* that "the main reason why we must have a civilian agency in the outer space field is because of the necessity of negotiating with other nations and the United Nations from some non–military posture."[37]

The Act specifically authorized the Administrator to grant access to NASA employees to AEC restricted data. This violated long-standing AEC policy, which based access on AEC classified status. Strauss thus raised the concern that the act would allow the President to "disseminate Restricted Data to foreign governments . . . We think that an extension of this existing authority to the proposed Agency would be undesirable and unworkable."[38] In his testimony before the Senate Special Committee, Strauss stressed his preference for limiting international agreements at the outset, and noted that "the history of these new agencies, if the Atomic Energy Commission is a prototype, has been that, in the course of time, the basic law is amended by spelling out in greater detail the extent to which cooperation with other nations may be carried on."[39] The strong ties to the AEC are evident.

The issue of intellectual property centered on the allocation of patents. The House bill patterned itself on the Atomic Energy Act, giving the government exclusive ownership of any intellectual property arrived at due to NASA-related work. The American Patent Law Association lobbied against such a provision, for the obvious reason that long-term profits from owning patents was a prime incentive for firms bidding on contracts.[40] In a letter to William F. Finan, Hans Adler (both were in the Bureau of the Budget) wrote in reference to the patent provision in H.R. 12575 (the bill that became the National Aeronautics and Space Act): "this provision is also based on the Atomic Energy Act. However, we doubt that the Atomic Energy Act should serve as the proper precedent, since inventions in the atomic area have peculiar defense and secrecy aspects

37. Senate Special Committee on Space and Astronautics Report, "Reasons for Confusion over Outer Space Legislation and how to Dispel it" May 11, 1958, folder 012389, NASA Historical Reference Collection, NASA Headquarters, Washington, DC.

38. Letter, Lewis Strauss, General Manager of AEC, to Maurice Stans, Director Bureau of the Budget. March 31, 1958, folder 012405, NASA Historical Reference Collection, NASA Headquarters, Washington, DC.

39. Hearings Before the Special Committee on Space and Astronautics, United States Senate, 85th Congress, 2nd session, p. 50.

40. Richard Hirsch and Joseph John Trento, *The National Aeronautics and Space Administration* (New York: Praeger Publishers, 1973), p. 26.

which make private ownership difficult."[41] Again, we have an example of the adoption of policies crafted for atomic energy without reasoned analysis of their relevance to an aerospace agency. The final language adopted assigned intellectual property to the government, with the Administrator having the right to waive this right if he so desired.

It cannot be overstated how formative the experience with atomic energy was on the psyche of those determining the shape of NASA. The belief that atomic energy would infuse all aspects of future technology was widely held in 1950s America, and rocketry was no exception. The realities of chemical reactive propulsion dictate a maximum theoretical efficiency (specific impulse) due to limited available chemical enthalpy, but the exit velocity of a thermal nuclear rocket is limited only by material failure at high temperatures. The AEC, for these reasons, launched just such a nuclear rocket research program (ROVER) in 1956. Stanislaus Ulam, testifying before the House Select Committee on Astronautics and Space Exploration, reaffirmed that "it is not a question of conjecture or optimism, but one might say it is mathematically certain that it will be the nuclearly powered vehicle which will hold the stage in the near future."[42] With historical actors like Ulam making such statements, it becomes clear that the birth of NASA as an institution must be historically analyzed through the lens of the atomic experience. The concept of the stewardship of the state over technological affairs had become ingrained in the imagination in the atomic era and was adopted without serious protest during the formation of NASA. Indeed, a sharp contrast can be drawn to the violent reaction by private interests to the original Atomic Energy Act and the relatively benign reception of the National Air and Space Act. A profound transformation had occurred in the intervening years.

CONCLUSION

Under the Atomic Energy Commission, technocracy had been introduced to America. Under NASA, it was wedded to the federal framework. There are fundamental differences to the two cases, as in the ability to control nuclear material and the need to enforce atomic secrecy through the curtailment of granting patents. But throughout the whole of the discussions in both the executive and legislative branches during 1957-1958, it remains clear that the framers of the new aerospace agency were profoundly affected by their experience with atomic energy, specifically the AEC. When conceiving of a new agency, bureaucrats and legislators actively reached into the past and

---

41. Hans Adler to William Finan, "Subject: HR 12575." June 4, 1958, folder 12400, (NASA Historical Reference Collection, NASA Headquarters, Washington, DC).

42. *Hearings Before the Select Committee on Astronautics and Space Exploration*, 85th Congress, 2nd session, p. 602.

cherry-picked elements from their prior experience with atomic energy while passively making unconscious assumptions based on the technological realities of atomic energy. Often the decisions they arrived at were not appropriate for the aerospace case.

NASA represented a form of technocracy that divorced military interests as completely as possible. In the 1960s, NASA would become an Agency mobilized for social change. Thomas Hughes argues in *American Genesis* that, during the Great Depression, the Tennessee Valley Authority (TVA) was a push for regional social development by progressive politicians via electrification and the management of water resources.[43] NASA followed in these footsteps. Perhaps not so coincidentally one of the original commissioners of the TVA, David Lilienthal, would later become the first chairman of the AEC.

But NASA was technocracy in an evolved form. It combined three trends that had not yet together existed in any American organization: 1) Big Science, i.e., the close cooperation of large numbers of scientists and engineers in a vertically integrated hierarchy organized for the production of massive projects; 2) a mandate that pushed science for social benefits and simultaneously minimized obligations to the military; and 3) science in the service of national prestige abroad.

The Atomic Energy Commission took over the operation of the entire American atomic machine, from enrichment to reactor design to bomb testing in the South Pacific. NASA, instead, was given a mandate to push the boundaries forward in aerospace technology only insofar as they could be peacefully used. This was, then, a pivotal transformation in the history of American technocratic institutions. Under the presidencies of Kennedy and Johnson, NASA was a juicy target to be expanded, but this was merely opportunism. NASA's form had already been cemented in 1958, a form which had atomic roots.

---

43. Thomas P. Hughes, *American Genesis: A Century of Invention and Technological Enthusiasm 1870-1970* (New York, NY: Viking Penguin, 1989), pp. 360-381.

CHAPTER 5

CREATING A MEMORY OF THE GERMAN ROCKET PROGRAM FOR THE COLD WAR

Michael J. Neufeld

In the middle of April 1945, as Allied armies swept into what little remained of the Third Reich, American newspapers carried horrifying reports, followed by photos of recently liberated concentration camps in central Germany. Prominent among them was a camp in the city of Nordhausen. Several thousand corpses and a few hundred emaciated survivors were found, along with a smaller number of dead and dying a few kilometers away at the Mittelbau–Dora main camp, which was located next to an amazing underground V-weapons plant known as the Mittelwerk. A couple of weeks later, a new wave of shock spread through Allied populations when official newsreels of the camp liberations reached movie theaters, including footage of Bergen–Belsen, Buchenwald, and Nordhausen. Some American newspapers explicitly made the connection between the horrors of the latter and V-2 missile production.[1]

Yet within a year or two, that connection had almost sunk without a trace. By the time the U.S. Army held a war crimes trial for Nordhausen in 1947, the U.S. press almost ignored it as yet another trial. When former project leaders Gen. Walter Dornberger and Dr. Wernher von Braun, both by then living in the U.S., came to give interviews and publish memoirs in the 1950s about the V-2 project and the Peenemünde rocket center, they were able to essentially omit the underground plant and its concentration–camp prisoners from their stories as there was little information in the public domain to challenge such a formulation. Other writers, notably Willy Ley—the former German spaceflight society member and refugee from the Nazis who more than anyone else founded space history in the English–speaking world—also said virtually nothing about

---

1. "Tunnel Factory: Yanks Seize V-2 Plant in Mountain," *Washington Post,* April 14, 1945; Ann Stringer, "Dead and Dying Litter Floor of Nazi Prison Barracks," *Los Angeles Times,* April 15, 1945; "Germans Forced to Bury Victims," *New York Times,* April 15, 1945; "Tribune Survey Bares Full Horror of German Atrocities," *Chicago Tribune,* April 25, 1945; "Waiting for Death" (photo), *Los Angeles Times*, April 26, 1945; Bosley Crowther, "The Solemn Facts: Our Screen Faces a Responsibility to Show Newsreels and Similar Films," *New York Times,* April 29, 1945; "Camp Horror Films are Exhibited Here," *New York Times,* May 2, 1945; "Mrs. Luce Tells Nazi Slave Policy, Aimed to Protect Secret Weapons," *New York Times,* May 4, 1945.

Two survivors in the Nordhausen-Boelcke Kaserne camp at the time of liberation by the U.S. Army in April 1945. The horrors of Nordhausen and the nearby underground V-weapons plant were briefly infamous in the Western press. (National Archives)

these atrocities. It appears likely that Ley knew little about them due to a deliberate policy of silence by the ex-Peenemünders and the U.S. government. The former clearly had strong motivations of self-interest, and the latter wished to protect the program of importing engineers, scientists, and technicians from Nazi Germany that became best known as Project Paperclip.[2]

Of course, those were not the primary reasons why Ley, von Braun, and Dornberger gave interviews and wrote books and articles. These pioneers wanted to tell their part in the exciting story of German rocket development

2. On Paperclip, see Clarence G. Lasby, *Project Paperclip: German Scientists and the Cold War* (New York: Atheneum, 1971); Linda Hunt, "U.S. Coverup of Nazi Scientists," *Bulletin of the Atomic Scientists* (April 1985), pp. 16-24, and *Secret Agenda: The United States Government, Nazi Scientists and Project Paperclip, 1945 to 1990* (New York: St. Martin's Press, 1991); Tom Bower, *The Paperclip Conspiracy: The Battle for the Spoils and Secrets of Nazi Germany* (London: Michael Joseph, 1987). My assertions about U.S. press coverage of Nordhausen between 1945 and the 1980s are based on keyword searches of Proquest Historical Newspapers. Smithsonian researchers have electronic access to seven papers: *New York Times, Washington Post, Boston Globe, Christian Science Monitor, Chicago Tribune, Atlanta Journal-Constitution,* and *Los Angeles Times.*

from the Weimar amateur groups through the creation of the V-2 and its export to the U.S. Ley and von Braun in particular were also trying to sell the public something they fervently believed in: spaceflight. However, in the process, they were compelled to provide a sanitized history of Nazi rocket activities palatable to Western audiences during the Cold War. Because von Braun's German-led engineering team played an important role in American missile development in the 1950s, they needed to justify the Germans' presence and the obvious continuities between Nazi and American rocketry, as did the U.S. government, which faced episodic Soviet-bloc denunciations over the issue. After Sputnik, when the space race with the Soviets became a central public concern, popular writers supplemented the pioneering efforts of Ley, von Braun, Dornberger, and others with books built on the foundation laid by the three former Germans.

Among the most noteworthy aspects of this early German rocket historiography as it developed in the 1960s are: 1) a romanticization of the Nazi rocket center at Peenemünde as fundamentally aimed at space travel, rather than weapons development for Hitler—although that was less the case for Dornberger, the military commander; 2) a corresponding depiction of the Peenemünders as apolitical or even anti-Nazi engineers driven by space dreams, which was both an exaggeration and a conflation of von Braun's experience with that of his group; and 3) a suppression of almost all information about concentration-camp labor and membership in Nazi organizations. These tendencies were bolstered by the deeper Cold War memory cultures of the United States and West Germany, which promoted an often selective view of World War II that neglected the Holocaust. As a result, the Mittelwerk and its attached Mittelbau-Dora camp virtually fell out of history—at least outside the Soviet bloc—until the 1970s, and in the United States, for the most part until 1984. This paper will examine the phases of the creation of this memory of the German rocket program and what social, cultural, and political factors allowed it to flourish relatively undisturbed for three decades.

The postwar history of the German rocket program—and the genre of space history in the English-speaking world—began largely with one book, Willy Ley's *Rockets*. It originally appeared in May 1944 before he had any knowledge of the V-2, but it was greatly expanded after the war in multiple editions such as *Rockets and Space Travel* (1947) and *Rockets, Missiles and Space Travel* (1951). From the outset, Ley included not only the origins of rocketry, early space travel ideas, and the history of military rockets, but also a memoir of his involvement with Weimar rocket activities and the VfR, the German spaceflight society (1927-1934). It was quite natural for him to add the history of the rocket programs of Nazi Germany, predominantly the Army program and its Peenemünde center that produced the V-2. His sources included various newspaper and magazine articles, notably in the 1947 edition in which he repeated a lot of wild rumors and nonsense from the press. However, over time he greatly improved his account, based on his personal contacts with Wernher von Braun and later with other

Peenemünders. In early December 1946, immediately after the U.S. government unveiled Project Paperclip to the American press and public, von Braun visited Ley at his home in Queens, New York, their first encounter since sometime in 1932 or 1933. They enthusiastically discussed the German project until 2:45 a.m. Ley told Herbert Schaefer, the only other Weimar rocketeer to emigrate during the 1930s, "that I found no reason to regard v.B. as an outspoken anti-Nazi. But just as little, if not even less, did I find him to be a Nazi. In my opinion the man simply wanted to build rockets. Period." While this judgment contained a lot of truth, it would not be the last time that von Braun received a free pass on his Third Reich activities from his fellow space enthusiasts.[3]

Ley had fled to the U.S. in 1935 to escape the Nazi crackdown on the private rocket groups and later wrote for the leftist New York tabloid *P.M.*, so this willingness to accept von Braun's account is intriguing and not entirely easy to explain. The end-of-war concentration camp revelations were not far in the past. Certainly a passionately shared absorption with space travel has everything to do with it, but it also seems likely that Ley willingly accepted the assumptions that Americans brought to the problem of the complicity of scientists, engineers, and doctors with Nazi crimes: that it was fairly straightforward to separate the few fanatical Nazis from the bulk of mere opportunists who only wanted to work in their specialty. Crimes against humanity were ascribed to the SS; technically trained people were given almost a free pass unless there was evidence of specific involvement and/or Nazi enthusiasm. In the case of the V-2 and its underground plant, those assumptions can be seen at work from an early stage in the reports of Major Robert Staver, who led U.S. Army Ordnance's technical intelligence team there; he described the rocketeers as "top-notch engineers" no different than Allied "scientists" in developing weapons of war. These assumptions also played out in Project Paperclip, where behind a veil of classification, U.S. military agencies screened engineers and scientists almost solely on the basis of membership in Nazi organizations while explaining away virtually all "problem cases" as opportunism. Even Wernher von Braun, who had been (admittedly somewhat reluctantly) an SS officer, was finally legalized as an immigrant in 1949 on those grounds. But his file, like those of the others, remained classified until the 1980s, so he was able to leave the potentially damaging fact of his SS membership out of his memoirs and the official biographies that the U.S. Army and later NASA distributed.[4]

3. Willy Ley, *Rockets: The Future of Travel Beyond the Stratosphere* (New York: Viking, 1944), *Rockets and Space Travel* (New York: Viking, 1947), *Rockets, Missiles and Space Travel* (New York: Viking, 1951). Compare the Peenemünde chapters in the latter two. For von Braun's visit and the quotation: Ley to Schaefer, December 8, 1946, in file 165, box 5, Ley Collection, National Air and Space Museum Archives (original in German, my translation).

4. Staver to Ordnance R&D, June 17, 1945, in Box 87, E.1039A, RG156, National Archives College Park; Hunt, *Secret Agenda,* chaps. 3, 4, 7; Michael J. Neufeld, *Von Braun: Dreamer of Space, Engineer of War* (New York: Alfred A. Knopf, 2007), pp. 120-122, 234-238, 245, 323-324, 347-348, 404-406, 428-429.

Willy Ley (right) and Wernher von Braun (middle) with Heinz Haber (left), c. 1954. These
three were the scientific advisors to Walt Disney's mid-1950s space television
series. (National Air and Space Museum, Smithsonian Institution)

Von Braun wrote his first memoir in 1950 for a British Interplanetary
Society book never came to pass. It eventually appeared in the society's journal
in 1956, somewhat rewritten and, in one case at least, bowdlerized. His original
manuscript made a rather bald statement of amoral opportunism regarding the
1932 discussions between his Berlin rocket group and the German Army, which
led to his working for the latter as a civilian: "We felt no moral scruples about
the possible future use of our brainchild. We were interested solely in exploring
outer space. It was simply a question with us of how the golden cow could be
milked most successfully."[5] That statement vanished in the published version,
but it had already appeared in print five years earlier in a lengthy and fascinating
profile of von Braun in the *New Yorker* magazine on April 21, 1951. Whether he

---

5. Wernher von Braun (hereinafter WvB), "Behind the Scenes of Rocket Development in
Germany 1928 through 1945," ms., 1950, in file 702-20, WvB Papers, U.S. Space and Rocket
Center (hereinafter USSRC); WvB, "Reminiscences of German Rocketry," *Journal of the British
Interplanetary Society* 15 (May-June 1956): 125–145.

actually said it to the writer, Daniel Lang, during the interview or Lang lifted it from the manuscript that von Braun lent him is unclear. But the memoir and the profile offered the same fundamental account: von Braun, seized with dreams of spaceflight since his teenage years in the 1920s, went along with the German Army as it offered money for rocketry, then Hitler came to power, which led to vastly increased resources and the building of Peenemünde and the V-2. Late in the war, the intervention of higher Nazi powers increased as these weapons became of interest to Hitler—who von Braun saw a few times— leading to Heinrich Himmler's attempt to take over the rocket program for the SS. After von Braun rebuffed Himmler's initiative, he was arrested by the Gestapo with two colleagues in early 1944 for drunken remarks in which they stated that they would rather go into space than build weapons. He was only rescued because of the intervention of his mentor, General Dornberger. When the Third Reich collapsed a year later, von Braun led his team away from the Soviets and surrendered to the Americans. He hoped that in the U.S. he would eventually realize his space dreams, albeit again in the employ of the military.[6]

As an account of the trajectory of his life to that point, the article was reasonably accurate; what he left out was that which not-so-subtly altered the story. For example, he did not mention joining the Nazi Party in 1937, when the party pressed him to do so, although Lang did quote one of von Braun's U.S. Army superiors, who dated it to 1940. In fact, von Braun himself told the Army in 1947 that he had joined the Party in 1939, so he himself consciously or unconsciously falsified the date. Over time, this key indicator of Nazi commitment, or the lack of it, drifted in popular accounts, such that his first biographer in English, Erik Bergaust, dated von Braun's entry to 1942; the latter made no attempt to correct him. Von Braun naturally also suppressed his brief membership in an SS cavalry unit and riding club in 1933-34 and his 1940 "readmittance" (as his SS record calls it) to the black corps as an officer. His memoir article did discuss the underground plant near Nordhausen briefly, but the brutal exploitation of concentration-camp workers was blamed solely on SS General Hans Kammler, thereby holding the whole matter at arm's length. Von Braun left the impression that the underground plant was completely separated from Peenemünde. The fact that SS prisoners had also worked at the rocket center and many other V-2 sites, and that he had been inside the Nordhausen facility at least a dozen times, he also suppressed. Given his intimate encounters with the Nazi elite, however, it was hard for him to deny that his prominent place in that regime, but his arrest by the Nazis allowed him to depict himself as ultimately more a victim of the regime than a perpetrator.[7]

6. Ibid.; Daniel Lang, "A Romantic Urge," in *From Hiroshima to the Moon: Chronicles of Life in the Atomic Age* (New York: Simon & Schuster, 1959), pp. 175-193, quote on p. 180, originally published in *New Yorker* (April 21, 1951): 69-70, 72, 74, 76-84.

7. "Affidavit of Membership in NSDAP of Prof. Dr. Wernher von Braun," June 18, 1947, Accession 70A4398, RG330, National Archives College Park; WvB NSDAP file card, former

In mid-1952, another memoir appeared under his name, "Why I Chose America," in a periodical aimed at women and families, the *American Magazine.* Ghostwritten by an interviewer with von Braun's superficial editing, this article came in the wake of a sudden increase in his fame. In March, he had finally made his space-advocacy breakthrough with the publication of his lead article in a space series in *Collier's* magazine, which had a circulation of millions. Although "Why I Chose America" was clearly written less in his voice than that of the ghostwriter, it is revealing for how much it makes transparent the context of that time, specifically, the era of McCarthyism and the Red Scare. It centered his alleged decisive moment at the end of the war, when he had to choose between East and West—in fact, he was basically in the power of General Kammler and scarcely in a position to do anything but follow his orders to evacuate southwest to get away from the Soviets. It was fortunate that Kammler's orders matched his own desires. "Why I Chose America" also makes much of his disillusionment with Nazism and with totalitarianism in general, notably as a result of his arrest, and it hammers on his Americanization, his conversion in El Paso to born-again Christianity, and his happiness with his new home in Huntsville. In short, this article made von Braun—a German who could not be naturalized until 1955 because of his delayed legal entry—into a patriotic Cold-War American.[8]

It is not at all clear how much "Why I Chose America" influenced the later literature on von Braun and Peenemünde. While certainly read by a much larger initial audience than his own memoir, which only came out in 1956 in an obscure space periodical, the latter was reprinted in a book and taken as a fundamental source by many later journalists and authors. The 1952 piece, on the other hand, probably faded away, especially in comparison to the Lang 1951 profile in a much more prominent magazine. In any case, the canonical von Braun stories of his rise, success at Peenemünde, arrest, and rescue by the U.S. Army were reinforced in the summer of 1958 when the Sunday newspaper supplement, the *American Weekly,* published his third and longest memoir, also ghostwritten, "Space Man—The Story of My Life." This three-part piece came in the wake of Sputnik, and the national hero status he achieved as a result of his prominent place in launching the first U.S. satellite, Explorer 1. The topic of Nordhausen and concentration-camp labor appear again only in the most marginal way. His Americanization was once again emphasized, a seemingly necessary strategy in view of his burdensome past. It is noteworthy that by this time von Braun's life story, at least for that concerning his past in Nazi Germany, had hardened into

BDC records, microfilm in National Archives College Park; WvB, "Behind the Scenes...," ms., 1950, in file 702-20, WvB Papers, USSRC; WvB, "Reminiscences"; Lang, "A Romantic Urge"; Erik Bergaust, *Reaching for the Stars* (Garden City, NY: Doubleday, 1960), p. 23.

8. WvB, "Why I Chose America," *The American Magazine* 154 (July 1952): 15, 111-112, 114-115.

a clichéd pattern of anecdotes visible in all media profiles and in the first book-length biographies that appeared in English and German in 1959 and 1960.[9]

Several years earlier, General Walter Dornberger published his book, *V-2*, which became the most influential account of the German rocket program aside from the specifics of von Braun's life. Judging by a manuscript in English now in the Deutsches Museum's archive in Munich, Germany, Dornberger originally tried writing it for an American audience in a language he then scarcely commanded, probably while working for the U.S. Air Force in Dayton, Ohio, from 1947 to 1950. (He then joined Bell Aircraft in Buffalo, New York, to work on rocket plane projects, ultimately becoming its vice president for engineering.) In 1951, von Braun pointed out his former boss's manuscript to his new German publisher, Otto Bechtle, who was arranging for von Braun's own bad science-fiction novel, *Mars Project*, to be rewritten in German by a popular aviation writer and former Nazi propagandist, Franz Ludwig Neher. Neher did the same, and much faster, for Dornberger's memoir, which appeared as *V-2: Der Schuss ins All* (*V-2: The Shot into Space*) in the fall of 1952. It would be nice to know who invented the subtitle, which so neatly captures the reinvention of a Nazi terror weapon as the space rocket it most certainly was not, at least before it was launched at White Sands, New Mexico, with scientific instruments.[10]

Although Dornberger was a space enthusiast as well, the book was a straight military account of the program, which only mentions the space aspects in passing. Neher's unacknowledged rewrite was a success; *V-2* became an instant classic. Translations appeared in Britain and America in 1954, the latter edited and introduced by Willy Ley.[11] It entrenched certain stories about the German Army rocket program, some of which have been almost impossible to dislodge in the popular media, such as the claim that the Reichswehr only began working on rockets because they were not banned in the Versailles Treaty. Noteworthy is Dornberger's account of the relationship between the rocket program and the Nazi leadership, above all Hitler. The former rocket general claimed that because the Führer's doubts early in the war, the program was delayed by two years, making it "too late"

---

9. WvB, "Space Man—The Story of My Life," *American Weekly* (July 20, 1958), 7-9, 22-25; (July 27, 1958), 10-13; (August 3, 1958), 12, 14-16; Heinz Gartmann, *Wernher von Braun* (Berlin: Colloquium, 1959); Bergaust, *Reaching* (see above).

10. On the story of von Braun's novel see Michael J. Neufeld, *Von Braun: Dreamer of Space, Engineer of War* (New York: Knopf, 2007); chaps. 10-11; WvB to Otto Bechtle, March 9, 1951, in German Corr. 1949-54, Box 43, WvB Papers, Library of Congress Manuscript Division; Walter Dornberger, Ms., "V 2: Around a Great Invention" (1948), NL165/010 and NL165/011, Deutsches Museum Archives, Munich; Walter Dornberger, *V-2: Der Schuss ins Weltall* (Esslingen: Bechtle, 1952), reprinted as *Peenemünde: Die Geschichte der V-Waffen* (Frankfurt/Main and Berlin: Ullstein, 1989). Von Braun's novel was recently published as *Project Mars: A Technical Tale* (Burlington, Canada: Apogee, 2006).

11. Walter Dornberger, *V-2* (New York, NY: Viking, 1954).

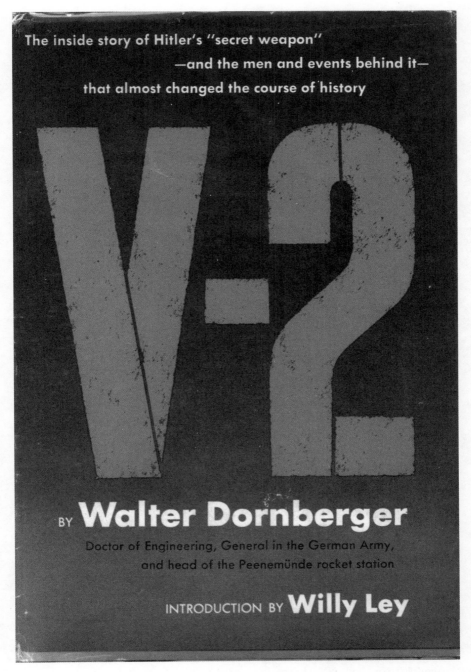

Walter Dornberger's 1954 memoir, along with the works of Ley and von Braun, fundamentally shaped the initial manner in which the German rocket program was remembered. (author's collection)

to affect the outcome—an argument much in line with the postwar memoirs of other German officers, who scapegoated Hitler for everything. In Dornberger's influential account—but also in von Braun's memoirs—their breakthrough with Hitler only comes on a visit to his Wolf's Lair in July 1943, when the dictator suddenly became a missile enthusiast. To emphasize the story, both Dornberger and von Braun omitted a visit they made to Hitler at the same place in August 1941 and underplayed the steps on the road to V-2 mass production made by Armaments Minister Speer with Hitler's approval in 1942. Blaming the Führer certainly fit the mood in the new West German Federal Republic, the population of which was inclined to focus on German suffering, while blaming a handful of leading Nazis for all crimes, above all for the crime of losing the war.[12]

With hindsight created by the revelations about the Mittelbau–Dora camp in the 1970s and 1980s, the most striking thing about Dornberger's book is that it barely mentions the underground plant and omits any reference to concentration-camp labor whatsoever. As someone intimately involved in decision-making about slave laborers, and as one who (like von Braun) encountered them personally on numerous occasions at Nordhausen, Peenemünde, and many other construction and production sites of the rocket program, Dornberger could only written it that way as a deliberate choice to suppress a central feature of the program that was just too dangerous to the reputation of the Peenemünders to discuss. As a result, he successfully falsified history by omission. But of course, in his depiction of himself, von Braun, and other leading rocket engineers, he also managed to make them all appear as non-Nazis, even as anti-Nazis, by laying emphasis on the meddling of Himmler's SS and other National Socialist organs late in the war. Dornberger's own Nazi enthusiasm, and that of several leading members of von Braun's team, like Arthur Rudolph, Ernst Steinhoff, and Rudolf Hermann, also vanished. Regarding a conversation that he, von Braun, and others had with Himmler at the Peenemünde officer's club, Dornberger states: "We engineers were not used to political talk and found it difficult." He claimed they were all repelled by Himmler's "inhuman policy of force." Later in the same chapter, he states: "We hardly ever discussed politics in Peenemünde. We were out of the world. Whenever two people met in the canteen or at mess, their conversation would

---

12. On Versailles, see Michael J. Neufeld, "The Reichswehr, the Rocket and the Versailles Treaty: A Popular Myth Reexamined." *Journal of the British Interplanetary Society* 53 (2000), 163-172. On Hitler and the V-2 priority battle, see Heinz Dieter Hölsken, *Die V-Waffen: Entstehung—Propaganda—Kriegseinsatz* (Stuttgart: Deutsche Verlags-Anstalt, 1984); Michael J. Neufeld, "Hitler, the V-2 and the Battle for Priority, 1939-1943," *Journal of Military History* 57 (July 1993), 511-538, and *The Rocket and the Reich: Peenemünde and the Coming of the Ballistic Missile Era* (New York, NY: The Free Press, 1995), chaps. 4-6. On West German memory, see Robert G. Moeller, *War Stories: The Search for a Usable Past in the Federal Republic of Germany* (Berkeley, CA: University of California Press, 2001) and Jeffrey Herf, *Divided Memory: The Nazi Past in the Two Germanys* (Cambridge, MA: Harvard University Press, 1997).

turn with five minutes to valves, relay contacts, mixers, . . . or some other technical detail that was giving us trouble." In short, he describes them as all just apolitical engineers serving their country, which certainly was how they wanted to see themselves after the war.[13]

Dornberger's book fed into the space-oriented narrative of German rocket-program history launched by Ley, but it also helped create a second genre: the military-oriented V-weapons literature. In the late 1950s and early 1960s, several books were published, mostly in Britain, on British intelligence and countermeasures and the V-1 and V-2 campaigns, including Air Marshal Sir Phillip Joubert de la Ferté's *Rocket* (1957), Basil Collier's *Battle of the V-Weapons 1944-45* (1964), and David Irving's *The Mare's Nest* (1965). Irving, who was already noticeably pro-German but not yet infamous as a Nazi apologist and Holocaust denier, provided the most complete account on both Allied and German sides of the V-weapons campaign in the last two years of the war, but it is noteworthy that, although he did much more original research than the others, he minimized the Mittelwerk/Nordhausen story about which he certainly knew more. Surprisingly, there was more information in a contemporaneous American book, James McGovern's *Crossbow and Overcast* (1964), which featured the transfer of the von Braun group to the U.S. Army. But even as McGovern reported the horrors discovered in and near Nordhausen in 1945, he followed von Braun's lead in holding the whole thing at arm's length from the German rocketeers by blaming it all on the SS—perhaps not surprisingly, as two of his key sources were Dornberger and von Braun.[14]

At this point, let's step back and look at the larger contexts of the American memory of National Socialism, the concentration camps, and the Holocaust as it took shape between 1945 and 1965. Although it is not easy to demonstrate that these contexts shaped the memory of the Peenemünde and the German rocket program that Ley, von Braun, and Dornberger created and popular writers extended, it is difficult to believe that they did not have some influence. It is particularly noteworthy in regard to Mittelbau-Dora that the Holocaust was little discussed between the end of the main Nuremberg trial in 1946 and the Eichmann trial in Israel in 1961. Other than the Anne Frank story, which was presented with an uplifting, universalistic message in the book and movie, the

13. Dornberger, *V-2*, 187, 192, 194. This self-image is demolished in a new book by Michael Petersen, *Missiles for the Fatherland*, forthcoming with Cambridge University Press. Petersen demonstrates the committed work of the engineers for the Nazi regime and the intimate relations between Peenemünde and the underground slave-labor Mittelwerk plant. On Dornberger's pro-Nazi political attitudes see his personal notes quoted in Neufeld, *The Rocket*, 182-183; on the others, see the evidence cited in ibid., in Neufeld, *Von Braun*, and in Hunt, *Secret Agenda*.

14. Phillip Joubert de la Ferté, *Rocket* (London: Hutchinson, 1957); Basil Collier, *Battle of the V-Weapons 1944-45* (London: Hodder and Stoughton, 1964); David Irving, *The Mare's Nest* (Boston and Toronto: Little, Brown, 1965); James McGovern, *Crossbow and Overcast* (New York, NY: William Morrow & Co., 1964), pp. 120-122.

topic was nearly taboo. The Jewish community in the U.S. spoke of the Shoah reluctantly, wishing to assimilate into Cold War America. Before Raul Hilberg published his groundbreaking *The Destruction of the European Jews* in 1961, he had a very difficult time finding a publishing house to take it; afterward, his book was either ignored or attacked. The American public apparently just was not ready to deal with the topic, and the same applied in Europe.[15]

However, the relationship between the Jewish Holocaust and Mittelbau–Dora is not straightforward, as few of the prisoners there were Jewish. The camp was filled with Soviet POWs and Polish forced laborers who had somehow ended up in SS hands, plus French and Belgian resistance fighters, German political prisoners, German criminals, gypsies, and several other groups. Jewish prisoners did not arrive in the camp until May 1944 and were rarely employed in V-2 production. But at the end of the war, thousands of starving, largely Jewish survivors of Auschwitz and Gross Rosen were dumped into the Mittelbau–Dora camp system and constituted a large fraction of the dead and dying discovered by the U.S. Army in 1945.[16] Although it can be posed only as a counterfactual hypothesis, it seems to me that if the consciousness and knowledge of the camps and the Holocaust that arose after the late sixties had existed in the fifties, it would have been much harder for Dornberger and von Braun to sweep the Nordhausen story under the rug. Indeed, in the 1970s, the rising attention to the Holocaust in the Western world did have an indirect effect on the attention paid to Mittelbau–Dora, eroding the received story of Peenemünde constructed in the 1950s.

Despite the lack of interest in the worst of Nazi crimes in the late 1940s and 1950s, reinforced by the Cold War alliance with the newly constructed West Germany, it cannot be said that the American environment was entirely friendly to the Peenemünders as they told their stories of the German rocket program. There were large number of veterans of the war and members of ethnic and religious groups who had no reason to like Germans. There were many false alarms in the media about the rise of neo-Nazism in the Federal Republic. In 1960-1961, the West German government became worried about an "anti-German wave" in the American public and media as the result of anti-Semitic incidents in German cities, as well as William Shirer's best-selling book *The Rise and Fall of the Third Reich*, the movie *Judgment at Nuremberg*, and the

---

15. Peter Novick, *The Holocaust in American Life* (Boston/New York: Houghton Mifflin, 1999), chaps. 4-6; Raul Hilberg, *The Politics of Memory: The Journey of a Holocaust Historian* (Chicago, IL: Ivan R. Dee, 1996).

16. On the history of Mittelbau-Dora, the definitive work is Jens-Christian Wagner, *Produktion des Todes: Das KZ Mittelbau-Dora* (Göttingen: Wallstein, 2001). See also André Sellier, *A History of the Dora Camp* (Chicago, IL: Ivan Dee, 2003).

revelations of the Eichmann trial. German crimes had scarcely been forgotten, although equated for years to Communism under the label totalitarianism.[17]

Anti-German prejudice bubbled up repeatedly in public comments about von Braun, who rapidly became by far the most famous of the rocketeers in the 1950s. It certainly explains the heavy handed stress on his Americanization and his supposed non- or anti-Nazi record in his two ghostwritten memoirs and in the first American biography written about him. It even surfaced in the heroic movie about him released in August 1960, *I Aim at the Stars,* an American-German coproduction. Von Braun's most trusted German producer wrote to him from Hollywood in June 1958 about the process of formulating a story treatment: "As you know, they are anxiously trying to show that you were no Nazi, although you were a member of the Party and built the V-2 for Hitler." In the end, the movie script incorporated a hectoring American character who pursues von Braun with questions like why he had not been hanged at Nuremberg. Apparently, the American script writers were just not comfortable making him the unalloyed hero of his own heroic "biopic." Even so, the movie opened to protest in Munich, London, Antwerp, and New York; but it bombed basically because it was tedious. Comic Mort Sahl's punch line became the most memorable thing about it: *I Aim at the Stars* should have been subtitled *But Sometimes I Hit London.*[18]

Such public doubts and media fiascos notwithstanding, Wernher von Braun had an enviable image in the American, and even more so, in the West German press in the late 1950s and early 1960s. Hero worship was everywhere, and was even prominent in less friendly counties like Britain and France. Von Braun was the vindicated prophet of spaceflight, instrumental in launching the first U.S. satellite and the first U.S. interplanetary probe, and the most visible symbol of the space race with the Soviets—at least until gradually displaced by the astronauts. He was cast in the mold of scientific hero, with his Nazi past neatly explained away based on the standard accounts. As von Braun and his group of Germans had become central to American space efforts (they were transferred to NASA in 1960 to become the core of Marshall Space Flight Center in Huntsville, Alabama) it was easy to conflate von Braun's biography, and that of a tiny band of space true-believers who came with him, with his entire group. Journalists and book authors simply glossed over the fact that most of his 120-odd engineers, scientists and technicians had been hired or recruited during the Third Reich and previously had nothing to do with rocketry or spaceflight. The set storyline was that the dream of spaceflight and landing on the Moon had arisen in the Weimar

---

17. Brian C. Etheridge, "*Die antideutsche Welle:* The Anti-German Wave in Cold War America and Its Implications for the Study of Cultural Diplomacy" in Jessica Gienow-Hecht, ed., *Decentering the United States: New Directions in Culture and International Relations* (Berghahn Books: 2007).

18. Bergaust, *Reaching*; Mainz to WvB, June 12, 1958, in file 208-7, WvB Papers, USSRC; Neufeld, *Von Braun*, 325–326, 346–353.

rocket and space societies, but Von Braun & Co. had to take a "detour" via military rocket development because that is where the money was. That detour continued in work for the U.S. Army, but then von Braun began campaigning for space travel in the 1950s, and the missiles he had developed became one of the foundations of the U.S. space program, leading ultimately to his group's central role in landing a human on the Moon in Apollo. Von Braun himself, together with close associates like Fred Ordway, went on to write space-history works in the 1960s and 1970s that fortified a spaceflight narrative privileging the Germans. Rip Bulkeley has rightly labeled this the "Huntsville school" of history.[19]

Detailing how the received version of the Peenemünde and von Braun story formulated in the 1950s was gradually undermined takes us beyond the scope of this paper, but it is instructive to look at a few key points. In the 1960s, the East German Communists tried several times to embarrass the United States and von Braun by outing his SS officer status and his involvement with Mittelbau-Dora. Julius Mader, a popular author who was a covert officer of the East German secret police, published *Geheimnis von Huntsville: Die wahre Karriere des Raketenbarons Wernher von Braun* (*Secret of Huntsville: The True Career of Rocket Baron Wernher von Braun*) in 1963, a book that was translated into Russian and other East-Bloc languages and circulated in nearly a half million copies. Out of it sprang a major feature film, *Die gefrorenen Blitze* (*Frozen Lighting*), that the East German official film studio released in 1967. But the Cold War divide was so strong that the book and the movie had very little impact in West Germany and none at all in the U.S., where they were almost unknown. Only slightly more effective was the East German involvement in the West German war-crimes trial in Essen from 1967-1970 of three SS men from the Mittelwerk. The chief East German lawyer succeeded in getting von Braun called as a witness, but NASA, seeking energetically to protect the rocket engineer, got the testimony moved to the German consulate in New Orleans in early 1969 and successfully kept most of the press away. During the Apollo 11 Moon landing in July of that year, the famous columnist Drew Pearson wrote that von Braun had been an SS member in the context of otherwise praising him, but offered no proof as to where he got this information. The rest of the American media completely

---

19. On hero-worship in West Germany and even in France, see Neufeld, *Von Braun,* pp. 323-324, 408-410. For typical products of the German-centered narrative see Ernst Klee and Otto Merk, *The Birth of the Missile: The Secrets of Peenemünde* (New York, NY: E. P. Dutton and Co., 1965) (translation of *Damals in Peenemünde*); Wernher von Braun and Frederick I. Ordway, III, *A History of Rocketry and Space Travel* (New York, NY: Thomas Y. Crowell, 1967); Ordway and Mitchell R. Sharpe, *The Rocket Team* (New York, NY: Thomas Y. Crowell, 1979). On the Huntsville school, see Rip Bulkeley, *The Sputniks Crisis and Early United States Space Policy* (Bloomington, IN: Indiana University Press, 1991), 204-205. The classic explication of the "detour" thesis, based on a reading of the earlier secondary literature and not one scrap of archival research, is William Sims Bainbridge's *The Spaceflight Revolution: A Sociological Study* (New York, NY: Wiley, 1976).

ignored it. So entrenched was the apologetic life story that when von Braun died eight years later of cancer, his voluminous obituaries never mentioned that fact; many did not even bring up his membership in the Nazi Party.[20]

The Essen trial, along with the publication of Albert Speer's memoirs in 1969 in German and 1970 in English, did contribute to gradually opening up the history the Mittelwerk and the Mittelbau-Dora camp system, notably in West Germany. A rising consciousness of the history of the Holocaust and the camp system worked in the background to make it harder as well to retail the old history of the German rocket program. When Ordway finally published *The Rocket Team* with Marshall Center writer Mitchell Sharpe in 1979—a narrative of the von Braun group dominated by the V-2 story—they could no longer leave out the underground plant, even if they did produce a rather one-sided and abbreviated treatment. That same year, *Dora,* the memoir of French resistance fighter Jean Michel, appeared in English translation, further opening up the topic, although the book had much less influence than the *Rocket Team*—or at least it did until it helped spark an investigation by the newly formed Office of Special Investigations (OSI) of the U.S. Department of Justice. In October 1984, it announced that one of von Braun's closest associates, Arthur Rudolph, had left the country and denounced his U.S. citizenship as part of a voluntary agreement to forestall a court trial over his denaturalization. He had to admit his early membership in the Nazi Party and his prominent role in the management of slave labor in the Mittelwerk. This announcement provoked a wave of headlines across the U.S. and around the world. Suddenly Nordhausen appeared in multiple American newspaper articles for the first time since April and May 1945. Shortly afterward, thanks to the Freedom of Information Act and the work of freelance journalist Linda Hunt, von Braun's Party and SS record came out when his Army security files were declassified. The old history of the German rocket program, although still entrenched in many quarters, would never be defensible again. When the Cold War ended only five years later, making the former East German sites of Peenemünde and Mittelbau-Dora accessible, it only reinforced the trend. It

---

20. Julius Mader, *Geheimnis von Huntsville: Die wahre Karriere des Raketenbarons Wernher von Braun* (Berlin-East: Deutscher Militärverlag, 1963; 2nd ed., 1965; 3rd ed. 1967); Paul Maddrell, "What We Discovered About the Cold War is What We Already Knew: Julius Mader and the Western Secret Services During the Cold War," *Cold War History* 5 (May 2005), 235-258, esp. 239-242; Thomas Heimann and Burghard Ciesla, "*Die gefrorenen Blitze:* Wahrheit und Dichtung: FilmGeschichte einer 'Wunderwaffe'," in *Apropos: Film 2002. Das Jahrbuch der DEFA-Stiftung* (Berlin: DEFA-Stiftung/Bertz Verlag, 2002), pp.158-180. Neufeld, *Von Braun,* pp. 404-408, 428-429, 473; Kaul Antrag, December 4, 1967, and Hueckel to WvB, November 6, 1968, in Ger. Rep. 299/160, Nordrhein-Westfälisches Hauptstaatsarchiv/Zweigarchiv Schloss Kalkum; Pearson, "Prime Moon Credit is Von Braun's," *Washington Post,* July 17, 1969; WvB to Gen. Julius Klein, August 2, 1969, copy provided by Eli Rosenbaum/OSI.

opened the way to a new, more complex and often contradictory public memory of the German rocket program in the U.S., Germany, and the Western world.[21]

Two things predominantly shaped how the V-2 and the Third Reich rocket project was remembered in the first few decades after World War II: the prominence of ex-German rocketeers in the United States and their value to the West in the Cold War. Willy Ley, an anti-Nazi refugee, rose to fame in the U.S. and elsewhere as a science writer in World War II and after, and he offered a space-oriented perspective on German rocket history. Then von Braun and Dornberger arrived under Project Paperclip and provided their technical expertise to the United States; von Braun in particular then became a national celebrity in the 1950s through space promotion in *Collier's* and Disney, followed by his central role in launching the first U.S. satellite, the first American deep space probe, the first American astronaut, and the Apollo expeditions to the Moon. Von Braun became a national and Western asset in the Cold War struggle with the Soviets, one that the media wanted to protect even without official U.S. government efforts to manage his image. Since the Nordhausen and Mittelbau-Dora story and von Braun's SS membership were virtually unknown, in large part due to government secrecy, the received story of the German rocket program held up, even in the face of East German attempts to undermine it. Nothing so clearly indicates the shaping influence of the Cold War than that fact; two competing narratives of von Braun and Peenemünde arose on either side of the "Iron Curtain," especially after Mader's 1963 book, yet even in the free press of the West, very little changed. It took the rising consciousness of the Holocaust and the history of the Nazi camp system to begin to erode the traditional narrative. Holocaust consciousness also led to the formation of the Nazi-hunting Office of Special Investigations in the U.S., which finally broke open the story.

In conclusion, I would like to appeal to space historians to begin to pay closer attention to their own history. Some good historiographic overview articles have been written, but not many attempts have been made to write the history of space history, notably in its origin phases. This history will tell us much about the constitution and mentality of spaceflight movements. More than that, the growth of the literature on public and collective memory provides another rich field for exploration: how space history, which has been written mostly by space enthusiasts and friendly journalists throughout much of its existence, has shaped popular memories of rocket development and space travel in the larger publics of the West and East—not to mention the rest of the world. Some pioneering work

21. Bernd Ruland, *Wernher von Braun: Mein Leben für die Raumfahrt* (Offenburg: Burda, 1969), 227-239; Albert Speer, *Erinnerungen* (Berlin: Propyläen, 1969), translated as *Inside the Third Reich* (New York: Macmillan, 1970); Manfred Bornemann and Martin Broszat, "Das KL Dora-Mittelbau," in *Studien zur Geschichte der Konzentrationslager* (Stuttgart: Deutsche Verlags-Anstalt, 1970) and Bornemann, *Geheimprojekt Mittelbau* (Munich: J. F. Lehmanns, 1971); Ordway and Sharpe, *Rocket Team*; Jean Michel with Louis Nucera, *Dora* (1975; New York, NY: Holt, Rinehart and Winston, 1979); Hunt, "U.S. Coverup" and *Secret Agenda*; Bower, *Paperclip Conspiracy*; Neufeld, *Von Braun,* 474-475; Eli Rosenbaum/OSI interview, July 26, 2006.

FRONT PAGE

New York Times

NEW YORK, THURSDAY, OCTOBER 18, 1984 · · · 60 cents beyond 75 t except o

German-Born NASA Expert Quits U.S. to Avoid a War Crimes Suit

By RALPH BLUMENTHAL

A German-born space official who developed the rocket that carried Americans to the moon has quietly left the United States and surrendered his citizenship rather than face Justice Department charges that he had brutalized slave laborers at a Nazi rocket factory during World War II.

Announcing the action yesterday in a brief statement, the Justice Department said that the official, Arthur Rudolph, as director for production of V-2 rockets at an underground factory attached to the Dora-Nordhausen camp from 1943 to 1945, "participated in the persecution of forced laborers, including concentration camp inmates, who were employed there under inhumane conditions." A third to one half of Dora's 60,000 prisoners died.

The National Aeronautics and Space Administration, which had awarded high awards to Mr. Rudolph for his work for the agency from 1962 from 1969, had no comment on the Justice Department announcement. [Page A13.]

The announcement on Mr. Rudolph, who was brought to the United States in 1945 with Wernher von Braun and more than a hundred other Nazi German technicians and scientists, was negotiated in advance with Mr. Rudolph.

The announcement did not mention his prominent role in the United States space and missile programs. Nor did it say where he had gone. Investigators said it was West Germany.

Officials said it was unlikely that Mr. Rudolph, who is 77 years old, would face prosecution in West Germany as the statute of limitations has expired. Mr. Rudolph was not carried on an Allied list of war criminals drawn up after the war. Other officials of the rocket factory were convicted of war crimes and jailed or executed.

Mr. Rudolph's whereabouts in West Germany has not been disclosed. His lawyer, George Main of San Jose, Calif., declined to take a reporter's call

Continued on Page A12, Column 3

More than anything else, the 1984 revelation of Arthur Rudolph's involvement in the abuse of concentration-camp labor undermined the traditional narrative of the German rocket program. (author's collection)

has been done, mostly on the United States, but a rich field of opportunities exists for those who are willing to use the tools of social and cultural history and collective memory to delve into the reception of space history, not only its generation.

CHAPTER 6

OPERATION PAPERCLIP IN HUNTSVILLE, ALABAMA

Monique Laney

In 1984, Arthur Rudolph renounced his American citizenship and moved back to Germany with his wife 30 years after becoming an American citizen. Previously, Rudolph had enjoyed more than a 20-year career with the U.S. Army and NASA from which he retired in 1969. Rudolph was one of the German rocket engineers who had been brought to the United States following World War II under the secret military project that would come to be known as "Operation Paperclip." He decided to leave the country after so many years because he was being investigated by the OSI, which alleged that Rudolph had been involved in the horrific and often deadly treatment of forced and slave laborers from the Dora concentration camp who had been used to produce V-2 rockets at Mittelwerk in the last years of the war. Before Rudolph passed away in 1996, several attempts were made to bring him back into the United States, and even today some of his former colleagues and friends want to see his name cleared.[1] According to one of his former coworkers, Huntsville responded to the government action against Rudolph with "unanimous disgust" and "did its best to try to fight it."[2]

---

1. The Rudolph case and attempts to have his name cleared have been covered extensively in local and national newspapers and magazines since 1984. This case has evoked strong emotions where terms such as "witch hunters" and "holocaust deniers" seem to be commonplace in an ongoing battle over some form of truth. Two authors who have interrogated the case at some length from opposite sides of this controversy, reflecting some of the accusatory and at times inflammatory rhetoric, are Thomas Franklin (pseudonym), *An American in Exile: The Story of Arthur Rudolph* (Huntsville, AL: Christopher Kaylor Company, 1987) and Linda Hunt, *Secret Agenda: The United States Government, Nazi Scientists, and Project Paperclip, 1945-1990* (New York, NY: St. Martin's Press, 1991).

2. Charles A. Lundquist interview, Huntsville, AL, July 12, 2007. Lundquist was born and raised in South Dakota. He was awarded his Ph.D. in astrophysics at the University of Kansas in 1954. Due to an earlier education deferment, he was then drafted to join the Army Ballistic Missile Agency (ABMA) at Redstone Arsenal near Huntsville, Alabama. Lundquist left Huntsville in 1962 to work for astronomer Fred Whipple until 1973 at the Smithsonian Astrophysical Observatory (SAO) in Cambridge, Massachusetts. He returned to Huntsville in 1973 as Director of the Space Science Lab at the Marshall Space Flight Center. In 1981, Lundquist took the job of Director and later Associate Vice President for Research at the University of Alabama in Huntsville (UAH). Although officially retired, he currently still helps out at the archives at UAH.

This remark about the unanimity of the community in regards to the so-called "Rudolph case" gives me pause. Not only does it imply that there were no opposing voices in Huntsville, but it also contradicts national and international discourses about the German rocket engineers that have scrutinized these scientists past work in Nazi Germany.[3] Such a remark exemplifies a counter-narrative, fighting to be heard in light of national narratives while simultaneously excluding narratives that contradict its own dominant stance in the local community. This "hegemonic counter-narrative" is the impetus for my dissertation, which explores the impact of Operation Paperclip on narratives of the first and second generation Germans, non-German Huntsville residents, and the local and national media, as well as on debates between laypersons and historians negotiating how to evaluate the engineers' past.

Huntsville, Alabama, has been home for most of the German rocket team members associated with Wernher von Braun who were brought to the United States under the secret military project known as Operation Paperclip. Those who arrived before 1950 were sent to Fort Bliss near El Paso, Texas, where they worked for the U.S. Army and shared their expertise in rocketry developed while designing and testing V-2 rockets in Germany during World War II. After one to two years in Fort Bliss, the men's dependents were allowed to join them. In 1950, the Army moved its rocket development program to Redstone Arsenal near Huntsville, Alabama. This meant that, with few exceptions, most members of the German team moved to Huntsville with their families, and their children would come to consider Huntsville their hometown. In 1960, NASA established the Marshall Space Flight Center on Redstone Arsenal where most of the Germans had been transferred under Wernher von Braun's direction.

The following analysis is an excerpt of a larger project that investigates the impact of Operation Paperclip on the German families and their Huntsville neighbors. The project is based primarily on oral histories because answers concerning impact are largely dependent on how and by whom the past is told. I have interviewed German and non-German Huntsville residents with different social and cultural backgrounds who lived in Huntsville in the 1950s and 1960s. By listening to the ways in which individuals recount the past and

---

3. I am referring primarily to the many national and international newspaper and magazine articles and documentary films reporting on and evaluating the German rocket engineers in the United States since their presence was made public in 1946. For some book length sources, see Tom Bower, *The Paperclip Conspiracy: The Hunt for the Nazi Scientists* (Boston: Little, Brown, 1987); Linda Hunt, *Secret Agenda: The United States Government, Nazi Scientists, and Project Paperclip, 1945-1990* (New York, NY: St. Martin's Press, 1991); John Gimbel, *Science, Technology, and Reparation: Exploitation and Plunder in Postwar Germany* (Stanford, CA: Stanford University Press, 1990); Clarence G. Lasby, *Project Paperclip: German Scientists and the Cold War* (New York, NY: Atheneum, 1971); and Michael J. Neufeld, *The Rocket and the Reich: Peenemünde and the Coming of the Ballistic Missile Era* (New York, NY: Free Press, 1995). For examples of sources that reflect a less scrutinizing approach to this history, see footnote 11 on page 94.

evaluate certain events from today's perspective, one can learn which narratives of the past have dominated and which have been marginalized and, with closer analysis, the societal discourses that influence the telling of the stories over time can be discerned.

There are many reasons for scholars to use oral histories in their research. Sometimes it is a way to humanize the subject matter and give it a sense of reality. In other cases, oral histories may be the only source available, thereby functioning as evidence in lieu of, or as a supplement to, written documents. I use oral history in a different manner. I am interested in the storytelling of oral history, (i.e., the struggle over memory). What parts of the past do people find important or significant to their lives, and how does that interact with the way they talk about events in the past and the present? How are these stories affected by the individual's position in relationship to dominant groups in society? What do discrepancies and errors tell us about "the work of desire and pain over time?"[4]

Because those on the margins of society are often affected very differently by changes at the center, listening to their stories allows us to question the logic and dynamics of dominant narratives. As pointed out earlier, knowing who is speaking from the margin or from the center is not always as clear for this project because it is determined by what grouping we focus on as the center. For example, while the German rocket engineers and their families were a minority group with unique histories and cultural backgrounds in Huntsville, their perceived "otherness" was based primarily on positive distinctions. Despite national and international scrutiny, they do not seem to have lost their status as a powerful minority that has adjusted to and blended in easily with the white Christian majority of Huntsville—making them part of "the center" in Huntsville. As members of a minority group, their positions were very different from those of other minority groups (e.g., members of the African American community). However, the idealization of the German rocket team by many Huntsvillians contrasts with national and international perceptions. So, while the German minority is part of the Huntsville majority, in regard to the German rocket engineers, Huntsville's majority perspective reflects that of a minority within the nation and internationally.[5] This distinction is particularly important for this project because the main issue I grapple with is how narrations of the past reflect changing power structures (who is marginalized in reference to whom), sometimes reinforcing old structures, and sometimes creating new and unexpected alliances.

---

4. Alessandro Portelli, *The Order Has Been Carried Out: History, Memory and Meaning of a Nazi Massacre in Rome*, 1st Palgrave Macmillan ed. (New York, NY: Palgrave Macmillan, 2003), p. 16.

5. The question of national and international perspectives needs clarification because it depends largely on definitions of who speaks for the national or international community. This is a weakness of this paper that will be addressed in detail in my larger work.

In addition to first and second generation Germans in Huntsville, I interviewed members of the African American and Jewish communities and World War II veterans, as well as former coworkers, neighbors, and friends of the German rocket engineers and their families. As the following exploration will illustrate, not surprisingly, those who historically wielded little power and had practically no voice in the Huntsville community have a very different perspective on the impact of the Germans on Huntsville than those who considered themselves to be equals. Perhaps less expected is how this marginalized narrative in Huntsville seems to echo larger national and international narratives.

When asked about the impact of the German rocket engineers and their families on Huntsville, practically every interviewee, regardless of personal background, mentions music, specifically the Huntsville Symphony Orchestra. After that, they usually attribute the first Lutheran Church in town and UAH, the U.S. Space & Rocket Center, Broadway Theatre League, and the Ballet Company to efforts of the German team. The word "culture" is prominent among most of the responses, and those who grew up alongside the children of the rocket engineers emphasize how "smart" the children were, often outdoing locals in school, which was typically linked directly to the father's reputation as "rocket scientists."[6]

With the space program came significant economic development for the formerly small cotton town in Northern Alabama that many Huntsville residents link directly to the arrival of the German rocket experts from Fort Bliss, Texas, in 1950, despite the simultaneous arrival of many American engineers, scientists, and technicians and their families.[7] The town's population increased

---

6. The term "rocket scientist" is a misnomer used by the media and in popular culture and applied to a majority of engineers and technicians who worked on the development of rockets with von Braun. It reflects a cultural evaluation of the immense accomplishments of the team but is nevertheless incorrect. For an explanation of why a distinction should be made between scientists and these engineers, see Michael J. Neufeld's latest publication, *Von Braun: Dreamer of Space, Engineer of War* (New York, NY: Alfred A. Knopf in association with the National Air and Space Museum, Smithsonian Institution, 2007): "A Note On The Name And On Terms."

7. The notion that the Germans were the main cause for the town's rapid development is a popular myth in Huntsville, most likely derived from the German team's later success and prominence. It distorts the fact that the approximately 120 German rocket engineers were part of a larger Army transfer that included the move of approximately 500 military personnel, 65 civilian personnel, and 102 General Electric (GE) employees in addition to the German specialists. By 1955, employment at Redstone Arsenal had increased from 699 in June 1949 to 6,442. For more information, see "Fort Bliss, Texas, Rocket Office to Be Moved to Redstone Arsenal," *Huntsville Times*, November 4, 1949. "Quarters: Of Guided Missile Area Set up at Redstone Arsenal," *Huntsville Times*, April 16, 1950, "150 Redstone Families Here, Others Coming," *Huntsville Times*, July 9, 1950. "Move Scheduled by GE Employes [sic]," *Huntsville Times*, April 3, 1950. Helen Brents Joiner and Elizabeth C. Jolliff, *The Redstone Arsenal Complex in Its Second Decade, 1950-1960*, ed. Historical Division (Redstone Arsenal, AL: U.S. Army Missile Command, 1969).

almost tenfold, from 16,437 to 137,802, over the following two decades, adding people from other regions of the United States as well as from other countries.[8] Formerly a small cotton mill town that prided itself in being the "watercress capital of the world," Huntsville soon became known as "Rocket City," to some even "Space Capital of the Universe," for its new space- and missile-related industry. Today, real estate companies and travel agents like to advertise that Huntsville has the "highest percentage of engineers and more Ph.D.'s per capita than any city in the country."[9] Huntsville's main newspaper perceives the town as hugely indebted to the German rocket specialists, especially Wernher von Braun, who was titled "Huntsville's first citizen" in an article describing the ceremonies at Huntsville's courthouse to send him and his family off to work for NASA in Washington, DC, in 1970.[10] Enforcing the notion that the city owes much of its prosperity to the arrival of these immigrants, the Von Braun Civic Center was named after the most prominent member of the rocket team—a constant reminder of the Germans' presence in town while simultaneously showcasing Huntsville's defiant position towards more critical national and international viewpoints.

While the casual observer or newcomer to Huntsville is not likely to hear any contradictions to this impressive and positive portrayal, the people in the town were not then, nor are they now, unanimous in their assessment of the newcomers. For this excerpt of my research, I focus on interviews with members of the African-American community. I intend to illustrate the significance of social positioning based on racial categorizations in a Jim Crow environment for perceptions of the German families' impact on Huntsville. In the larger project, I propose that some dominant narratives in Huntsville stem from power structures based in the history of slavery and Jim Crow, as well as in certain forms of anti-Semitism and elitism among those with economic power in this once small cotton mill town. The German group clearly fit well into these preexisting structures and, as we will see, were sometimes perceived as reinforcing them.

Naturally, the perspectives of individuals evaluating this event are based on personal backgrounds and level of contact with members of the German group. The most prominent public voices typically heard on this subject are either members of the German community, former coworkers, or close

8. These population numbers are for Huntsville proper. W. Craig Remington and Thomas J. Kallsen, eds., *Historical Atlas of Alabama: Historical Locations by County*, vol. 1 (Tuscaloosa, AL: Department of Geography, College of Arts and Sciences, University of Alabama, 1997).

9. Holly McDonald, "Home page: Welcome to Huntsville," Keller Williams Realty Web site, *http:// www.hollymcdonald.com/* (accessed August 22, 2007), and "Welcome to Huntsville: It's a Great Place to Live," *Inspired Living: Greater Huntsville Relocation Guide* (Huntsville, AL: Price-Witt Publications LLC, Fall/Winter 2005): 12.

10. Bill Sloat, "Rocket City Launches von Braun," *Huntsville News* (February 25, 1970): 1–2.

friends of some of the German families.[11] My approach counters these mostly monolithic accounts and acknowledges British social historian Paul Thompson's statement that the aim of oral historians is to make history "more democratic" by "shifting the focus and opening new areas of enquiry . . . by bringing recognition to substantial groups of people who had been ignored" and by radically "questioning . . . the fundamental relationship between history and the community" because "the self-selected group will rarely be fully representative of a community."[12]

When talking about the impact of the Germans on Huntsville, the lack of voices from the African American community in a southern town is hard to ignore. I first became aware of this when reviewing a public forum titled "Creating Rocket City," recorded on video in 2003 as a contribution to the public library's celebration of the centennial of the Wright brothers' first flight.[13] As the title implies, this panel was intended to discuss the city's development since the 1950s. It included one African American speaker, Hanson Howard, who was, according to the organizer, included "for diversity."[14] A camera scan over the audience reveals that Howard was apparently also the only African American person in the room. When I asked him about that, he noted simply, "I figured that's the way it would be . . . You know, it wasn't a surprise."[15]

11. The following are examples of longer articles or book-length accounts illustrating to what I am referring and do not include numerous local newspaper articles. Erik Bergaust, *Rocket City, U.S.A.; from Huntsville, Alabama to the Moon* (New York: Macmillan, 1963); Placide D. Nicaise, *Huntsville and the Von Braun Rocket Team: The Real Story*, ed. Scientists and Friends (Monterey, California: Martin Hollmann, 2003); Frederick Ira Ordway, III and Mitchell R. Sharpe, *The Rocket Team*, 1st MIT Press pbk. ed. (Cambridge, MA: MIT Press, 1982); Ruth G. von Saurma and Walter Wiesman, "The German Rocket Team: A Chronology of Events and Accomplishments," *The Huntsville Historical Review* 23, no. 1 (1996), 20-29; Ruth G. von Saurma, "Personal Recollections of Huntsville's Rocket and Space Highlights, 1949-1980," *The Huntsville Historical Review* 27, no. 1 (2000), 37-52; Ruth G. von Saurma, "Growing up in Huntsville," *The Huntsville Historical Review* 23, no. 1 (1996), 13-19; Ernst Stuhlinger, "German Rocketeers Find a New Home in Huntsville," *The Huntsville Historical Review* 23, no. 1 (1996), 3-12; Ernst Stuhlinger, "Sputnik 1957—Memories of an Old-Timer," *The Huntsville Historical Review* 26, no. 1 (1999), 26-31; Ernst Stuhlinger and Frederick Ira Ordway, III, *Wernher Von Braun, Crusader for Space: A Biographical Memoir*, Original ed. (Malabar, FL: Krieger Pub., 1994); The Marshall Retiree's Association, "Reminiscence of Space Exploration History Fireside Chats, February 17 2000," video available at *http://media.eb.uah.edu/NASA_ARCHIVES/farside_chats/index.htm*; Bob Ward, *Dr. Space: The Life of Wernher Von Braun* (Annapolis, MD: Naval Institute Press, 2005).

12. Paul Thompson, "The Voice of the Past: Oral History," in *The Oral History Reader*, ed. Alistair Thomson, Robert Perks (London, New York: Routledge, 1988), 26.

13. David Lilly, *A Century of Flight: 'Creating Rocket City'* (Huntsville, AL: Huntsville Public Library, 2003).

14. David Lilly, e-mail correspondence, October 10, 2005.

15. Hanson Howard interview, Huntsville, AL, May 10, 2006. Hanson Howard is a retired warrant officer who moved to Huntsville from Maryland with the Army in 1960. He was stationed in Germany twice, as well as to White Sands, New Mexico, and Vietnam. He currently works as Business Counselor at Northeast Alabama Region Small Business Development Center and is executive director of the Service Corps of Retired Executives. Howard requested to be

Why is this not a surprise? Why is the African American community not more represented in a public forum that discusses the enormous economic, cultural, and societal development of Huntsville since the Army moved its rocketry development program to town? This paper intends to shed some light on how we might answer these questions.

While the percentage of African Americans in Huntsville's population is lower than in many areas in Alabama, it has been and is again at about 30 percent.[16] Segregation officially ended in the early 1960s, but the visitor today can easily observe that the town still appears to be de facto segregated into predominantly black neighborhoods in the northwest and white neighborhoods in the southeast parts of the city. However, the dominant perception is that Huntsville integrated its public facilities rather quickly in comparison to other towns in Alabama and before the signing of the 1964 Civil Rights Act.[17]

Charles Ray, a retired employee of the Army's Equal Employment Opportunity (EEO) office at Redstone Arsenal and the owner of Nelms Funeral Home in Huntsville was born in Madison County, Alabama, in 1936. He is an Alabama A&M alumni and has lived in the Huntsville area most of his life. Ray explains that racism in the area was less "rabid" due to the relatively

---

sent back to Redstone Arsenal in late 1966 after his tour in Vietnam. He is an Alabama A&M alum and active in Huntsville's community as a member of a ballet association, a United Way volunteer, and president of the board of the American Red Cross.

16. While in 1950 African Americans made up 32 percent of Huntsville's population, in 1960 and 1970 the percentage had declined to 14 percent and 12 percent respectively. This decline has been explained by the "lack of in-migration" of African Americans from the surrounding farmland into the city despite the decline in agriculture in the 1950s and 1960s. While the city's population expanded, it was mainly the white population that grew larger. According to the U.S. Census for 2000, the percentage of African Americans in Huntsville has risen back to 30 percent. U.S. Census Bureau (*http://www.census.gov/*), "Huntsville, Alabama: Factsheet," (2000), Huntsville-Madison County Public Library Archives, "Comparative Growth Rates: Huntsville, Madison County, Birmingham, Alabama; Source: US Census of Population 1920-1960," (file "Huntsville Population"), Huntsville-Madison County Public Library Archives, "General Population Characteristics—1970: Huntsville and Madison County Alabama, Source: U.S. Department of Commerce/Bureau of the Census," (Huntsville, AL: file "Huntsville Population"). See also Andrew J. Dunar and Stephen P. Waring, *Power to Explore: A History of Marshall Space Flight Center, 1960-1990, NASA Historical Series* (Washington, DC: National Aeronautics and Space Administration, NASA History Office, Office of Policy and Plans, 1999), p. 126.

17. According to the documentary film, *A Civil Rights Journey*, one of the town's historical markers received the addition "1962—First City in Alabama to begin segregation" to mark Huntsville's uniqueness in respect to civil rights in the state of Alabama. For information on the Civil Rights Movement in Huntsville, see Dunar and Waring, *Power to Explore: A History of Marshall Space Flight Center, 1960-1990, NASA Historical Series* (Washington, DC: National Aeronautics and Space Administration, NASA History Office, Office of Policy and Plans, 1999), chapter 4, and Sonnie Hereford, III, M.D., *A Civil Rights Journey*, DVD (United States: Sonnie Hereford, III, 1999). For a description of the Jim Crow system in context of Alabama 20th century history, see Wayne Flynt, *Alabama in the Twentieth Century, The Modern South* (Tuscaloosa, AL: University of Alabama Press, 2004), chapter 7.

low percentage of African Americans in the community, which meant that "we posed no political or economic threat to whites." In addition, the rural community was not as clearly segregated.

> We . . . lived in an area where there were large farms (that) had absentee owners, and they had sharecroppers as well on those farms. So, we lived together, had fun together and it was awfully inconvenient sometimes . . . for the kids, because we had to separate to go to school.[18]

While the private sphere of the surrounding rural areas was apparently not segregated in the first place, the urban community of Huntsville had an additional extraordinary incentive to integrate quickly—federal contracts with attached mandates to demonstrate equal opportunity employment practices.[19] Michael Smith, a retired professor of political science who was born and raised in Huntsville and left the area after attending Alabama A&M at the age of 20 in 1968, returned in 1985 to teach at Calhoun Community College in nearby Decatur. He explains: "That's how you got . . . racial integration in the town, because we got word from . . . the Kennedys . . . that if they didn't straighten the stuff out, then the government might have to look at these contracts."[20]

While all interviewees note the significant economic impact that came with the arrival of German immigrants, the relationship of my African American interviewees to the Germans is characterized mostly by memories of segregation and its implications. In short, being German meant being white, especially in the 1950s when the rocket engineers and their families were trying to establish themselves in town. "When Dr. Von Braun . . . and his group (came) . . . there were celebrations and welcoming committees and so forth, but none of *our* people were invited to come to participate."[21]

Sonnie Hereford, III, a retired family doctor and former civil rights activist, was born in Madison County in 1931. He attended Alabama A&M before moving to Nashville, Tennessee, for medical school. After his residency, he returned to Huntsville to practice medicine until 1993. Since then he has been teaching premed and prenursing students at Calhoun Community College.

---

18. Charles Ray, Jr. interview, Huntsville, AL, July 17, 2007.

19. For a brief history of the efforts to implement civil rights reforms at the Marshall Space Flight Center and associated businesses in the Huntsville area, see Andrew J. Dunar and Stephen P. Waring, *Power to Explore: A History of Marshall Space Flight Center, 1960-1990*, NASA Historical Series (Washington, DC: National Aeronautics and Space Administration, NASA History Office, Office of Policy and Plans, 1999), chapter 4.

20. Michael Smith interview, Huntsville, AL, July 29, 2007.

21. Sonnie Hereford, III interview, Huntsville, AL, July 19, 2007. Italics indicate emphasis by the speaker.

Hereford's above remark most likely refers to the barbecue party organized by the town's Chamber of Commerce to welcome the newcomers from Fort Bliss, Texas, as "Special Civic Guests." The event was announced repeatedly in the town's main newspaper in 1950,[22] and it was clearly a well-planned and large-scale undertaking:

> Local civic clubs will be asked to cancel their meetings dur-
> ing that week, to meet during the municipal party at the Big
> Spring Park . . . Elaborate entertainment and reception com-
> mittees will also be established, to make sure that everyone
> is introduced, and has the opportunity of knowing others
> who will be present . . . Sponsoring Chamber officials are
> hopeful that the barbecue and fellowship will create contin-
> ued harmonious relations between the various segments of
> the Huntsville population.[23]

Obviously, the creation of "harmonious relations" did not apply to those between the white and black communities of Huntsville. When Ray explains, "We just did not move in the same circles," he is responding to such insults by referring to them as a matter of choice by the ones being excluded—not just by those who were actively doing the excluding. In this way, the insult is transformed and therefore rendered nonexistent. The need for such a transformation speaks to the level of pain these acts of exclusion caused.

In light of these segregated circumstances that were not unusual during the 1950s in the United States, it does not surprise that interest in and knowledge of the group of Germans was and is relatively low among members of the African American community. Ray adds,

> *I* did not know of their . . . direct participation in the affairs
> of Huntsville. I'm sure, you know, the financial uplift that
> they brought to Huntsville with the program was controlled
> not by Germans, but by the white management at NASA.

---

22. "2,700 Attend Newcomer Fete Despite Rains," *The Huntsville Times*, August 10, 1950, "Invitations Sent for Civic Patry [*sic*]," *Huntsville Times*, July 17, 1950, "Newcomer Party of Welcome Set," *Huntsville Times*, July 16, 1950, "Newcomers' Civic Party Rescheduled for Aug. 9," *Huntsville Times*, July 26, 1950, "Rain Will Not Halt Newcomer Outing Today," *Huntsville Times*, August 9, 1950.

23. "Chamber to Hold Barbecue Party," *Huntsville Times*, July 23, 1950. The guest list included not only the German rocket engineers and their families. In fact, they were in the minority. Invited were "Incoming scientific personnel at Redstone Arsenal, the contract companies, enlisted men and officers and their families." Ibid.

And that's essentially what the community dealt with, was the white power structure at NASA.[24]

The Germans' commonality with other white citizens of the area expressed itself in other ways as well. Just like most white Southerners of Huntsville, the newly arrived Germans did not appear to be openly opposed to the system of racial segregation. The fact that they had been members of the privileged majority in Nazi Germany in a system that segregated and persecuted Jews and other minorities made their silence towards the Jim Crow system appear to be a continuation of the same tragic callousness towards those constructed as racially or otherwise inferior.[25] The commonality based on beliefs in being of the same race seemed tightly linked to common experiences and histories of racial privilege.[26]

Smith points towards this important relationship and connects the use of slave labor for the production of the V-2 rockets during World War II with the slave labor system of the United States in the seemingly not so distant past, offering one explanation for why the German rocket engineers' past seems to be overlooked by many in Huntsville:

And, so these people love von Braun and will not hear about the Mittelwerks . . . (in) Huntsville . . . I don't hear any

---

24. Ray interview, July 17, 2007.

25. Germans who were adults during the Nazi period in Germany may not have all known to what extent the persecution of Jews was being carried out, but there is no doubt they saw and would have often been part of the treatment of Jews and other minorities as second class citizens or worse. Very few tried to intervene. Robert Gellately, *Backing Hitler: Consent and Coercion in Nazi Germany* (Oxford and New York: Oxford University Press, 2001).

26. The term "race" is a highly contested and socially constructed term. Which race a person is associated with or applies to himself or herself depends on factors such as location and historical context. As in many other countries, racism and racialization have a long standing history in the United States and in Germany. For the relationship of racism and racialization between Europe and the United States, see, for example, David Theo Goldberg, *Racist Culture: Philosophy and the Politics of Meaning* (Cambridge, MA, Oxford, UK: Blackwell Publishers Inc., 1993) and Stefan Kuhl, *The Nazi Connection: Eugenics, American Racism, and German National Socialism* (New York, NY: Oxford University Press, 1994). For the United States specifically, see: Michael Omi and Howard Winant, *Racial Formation in the United States* (London: Routledge, 1994); Edward W. Said, *Orientalism* (London:Penguin, 1978); Kimberlé Crenshaw, Neil Gotanda, Gary Peller, Kendall Thomas (eds.), *Critical Race Theory* (New York, NY: The New Press, 1995); Lipsitz, *The Possessive Investment In Whiteness*, (Philadelphia, PA: Temple University Press, 1998). For Germany specifically, see: Tina Campt, *Other Germans: Black Germans and the Politics of Race, Gender, and Memory in the Third Reich* (Ann Arbor, MI: University of Michigan Press, 2004); Heide Fehrenbach, *Race After Hitler: Black Occupation Children in Postwar Germany and America* (Princeton, NJ: Princeton University Press, 2005); Friedrichsmeyer, Sarah, Lennox, Sara, Zantop, Susanne, *The Imperialist Imagination: German Colonialism and Its Legacy* (Ann Arbor, MI: University of Michigan Press, 1998); Uli Linke, *German Bodies: Race and Representation after Hitler* (New York, NY: Routledge, 1999).

> particular outrage over the fact that it might be true . . . Again,
> Huntsville is in no position . . . because . . . you had slavery
> right here . . . and many of the same people are still here today
> . . . And even if they're not, they're from other parts of the
> South . . . and you have the same immorality.[27]

Smith emphasizes the slave labor system of the South, even though the system was not limited to this area of the United States. This may be his interpretation of the Jim Crow system that was implemented only in the Southern States and is generally seen as having continued the racist ideals of the earlier slave labor system. The mentality of those who enforced and complied with the Jim Crow system is, therefore, similar to that of those who enforced and complied with the previous slave labor system. This implies that white Northerners had potentially acquired a different mindset since the slave labor system ended while white Southerners had not. Whether that is a fair assertion or not, Smith's comment is significant because of the connection he makes between German and American histories of cruelties committed against those perceived as racial minorities.

Sonnie Hereford describes how white people in Huntsville dealt with segregation, which seems remarkably similar to attitudes of the majority of Germans towards the plight of Jewish people and other minorities during the Third Reich.

> Many Caucasians here . . . were just nonchalant. I mean with
> some of them, they knew that the black people were being
> mistreated, but they weren't trying to do anything about it,
> and then there were some of them *doing* the mistreating. You
> know what I mean. And there were a few that wanted to
> work with us to try to change it.[28]

Of course, what made the Germans different from other white people in town was their newcomer status and foreignness. That status apparently came with expectations on behalf of African American residents that were quickly disappointed.

> Well, we were hoping that . . . they might join us in our fight
> for freedom . . . But I guess we were naïve . . . We were
> thinking . . . since they were encountering some resistance . . .
> maybe they'll join with us and all of us will fight . . . But . . . I

---

27. Smith interview, July 29, 2007.

28. Hereford interview, July 19, 2007.

don't recall any . . . of the people from the German community actually helping us.[29]

The resistance some of the Germans may have encountered when they
first arrived in Huntsville obviously did not make them feel as much solidarity
with the African American minority as apparently expected.[30] The expectation on behalf of African American residents in Huntsville may have been the
result of reports from Germany during the first years of the allied occupation.[31]
While the economic situation was still very dire for most Germans in Germany,
and the Allied Occupation was still in full force immediately after the war,
many Germans reportedly treated African American soldiers quite cordially.
They were presumably responding to experiences with African American soldiers who were generous and friendly despite their position of relative power
as occupying soldiers. The Germans' friendliness was surprising for many
African American soldiers, who had heard horrific stories about German racism and who were still segregated in their military units as well as at home. The
German response was therefore reported as an unusual experience in African
American magazines and newspapers around the country, which may have left
the impression that Germans generally have a more favorable attitude towards
African Americans and, therefore, more sympathy for their plight.[32]

---

29. Ibid.

30. Most of my German interviewees noted that they found very little antagonism, let alone resistance
to their arrival in Huntsville. This perception has been confirmed by other interviewees who
were residents in Huntsville at the time. I was told of a few incidents of outright animosity but
also that those sentiments seemed to dissipate quickly. I will address this phenomenon in more
detail in the larger project.

31. Sentiments towards black soldiers in Germany were indeed very different after World War
II, in contrast to World War I when nationalists interpreted France's use of black soldiers
to occupy Germany as an added insult. After World War II, German attitudes towards
African Americans varied and again became the focus of negative attention with the rise of
interracial children born to African American soldiers and German women. However, the
general impression seemed to be that Germany was less racist towards African American
soldiers. For a discussion of the impact of the American occupation on sentiments towards
American soldiers in the years immediately following the Second World War, see, John
Gimbel, *A German Community under American Occupation: Marburg 1945-52* (Stanford, CA:
Stanford University Press, 1961). Maria Höhn describes changing German attitudes towards
African-American soldiers following the war in *GIs and Fräuleins: The German-American
Encounter in 1950s West Germany* (Chapel Hill, NC: University of North Carolina Press,
2002). Heide Fehrenbach and Yara-Colette Lemke Muniz de Faria have both published
important research on the treatment of interracial children born in postwar Germany.
Heide Fehrenbach, *Race after Hitler: Black occupation children in postwar Germany and America*
(Princeton, NJ: Princeton University Press, 2005); Yara-Colette Lemke Muniz de Faria,
Zwischen Fürsorge und Ausgrenzung: Afrodeutsche "Besatzungskinder" im Nachkriegsdeutschland
(Berlin: Metropol Verlag, 2002).

32. This perception was apparently still prevalent in the late 1950s. Recalling his service in
Germany in 1958, General Colin Powell once stated that "[for] black GIs, especially those

The disappointed expectations of solidarity and active support from Germans for the causes of African Americans in Huntsville points to an important fact: being German in the United States had a very different meaning than that of being German in Germany, especially after World War II.[33] The Germans who came to Huntsville in 1950 and, therefore, to the United States very shortly after the war would not have experienced the German occupation of Germany to the same extent and have had the opportunity to bond with African American soldiers. In addition, instead of experiencing the Allied Occupation that had provoked strong feelings of humiliation by many Germans in the postwar years, the German rocket engineers and their families encountered very little difficulty in Huntsville, blended well with the white majority, and were generally welcomed with open arms by a community that appreciated the prosperity and cultural influence they brought to town.

In some ways, the German families seemed to make matters even worse for the African American community. Even though strangers to the town, they had more privileges than some of the town's longstanding residents. Hereford explains:

> I think some people in my community were maybe jealous . . . because . . . *they* were permitted to go to the theatres and the concerts and to the sports arena, and *we* were not . . . And they were permitted to go to the restaurants and the hotels and motels and what have you, and we were not . . . I think there was maybe some animosity . . . [34]

---

out of the South, Germany was a breath of freedom—they could go where they wanted, eat where they wanted, and date whom they wanted, just like other people." Maria Höhn, *GIs and Fräuleins: The German-American Encounter in 1950s West* Germany (Chapel Hill, NC: University of North Carolina Press, 2002), p. 13.

33. For the meaning of being German in the United States, I refer to works on German immigrants to the United States, such as Wolfgang J. Helbich and Walter D. Kamphoefner, *German-American Immigration and Ethnicity in Comparative Perspective*, (Madison, WI: Max Kade Institute for German-American Studies, University of Wisconsin, 2004) and Russell A. Kazal, *Becoming Old Stock: The Paradox of German-American Identity* (Princeton, NJ: Princeton University Press, 2004). The post-World War II period of German immigration has not yet been analyzed thoroughly, with an important exception that focuses on the postwar emigration from Germany to the United States and Canada. For examples, see Alexander Freund, *Aufbrüche nach dem Zusammenbruch: Die Deutsche Nordamerika-Auswanderung Nach dem Zweiten Weltkrieg* (Göttingen: V & R Unipress, 2004), two articles by the same author: "Dealing with the Past Abroad: German Immigrants' Vergangenheitsbewältigung and their Relations with Jews in North America since 1945," *GHI Bulletin*, Fall 2002, 31:51-63, and "German immigrants and the Nazi past," *Inroads*, Summer 2004, 15:106 (12), as well as an unpublished dissertation by Helmut Buehler, "The Invisible German Immigrants of the 21st Century: Assimilation, Acculturation, Americanization" (Ed. Dissertation, University of San Francisco, 2005).

34. Hereford interview, July 19, 2007.

The Germans were not only perceived as not supportive of African American causes or simply privileged in comparison; as Smith explains, they also appeared to actively engage in undermining African-American institutions, such as Alabama A&M, by supporting the Jim Crow system of dual education:

> This is one of the negative things they've done—they were (the ones) who helped found the University of Alabama here in Huntsville. And I say negative because there was already a state supported school in Huntsville and it's called Alabama A&M. And so you now have this clash, this friction, this tension, between the new white school and the old black school, both state supported. So, that's one of the things the Germans also did. Von Braun. So, in other words, von Braun may have brought his European ethnocentrism . . . from Germany to Huntsville. And it was nothing out of the ordinary for him to . . . advocate the opening of a Jim Crow school. So, the Germans were *not* advocates of racial integration, as far as I know.[35]

Von Braun was indeed instrumental in getting substantial funding for the University of Alabama, Huntsville, which was founded before desegregation in 1961.[36] Instead of pointing to von Braun, Clyde Foster offers another explanation

35. Smith interview, July 29, 2007. For more information about the dual system of education in Alabama, see the higher education desegregation case known as John F. Knight, Jr., and Alease S. Sims, et al. vs. the State of Alabama, et al., Civil Action No. CV 83-M-1676, "which began in Montgomery in 1981." The case was "concerned with eliminating vestiges of historical, state enforced, racial segregation and other forms of official racial discrimination against African Americans in Alabama's system of public universities." Both Alabama A&M and UAH among others were defendants in this case. For a detailed historical overview of the "history of discrimination against African Americans in higher education" in Alabama, see Opinion 1 Fed. Supp., Vols. 781-835 787, F. Supp. 1030, Knight v. State of Ala., (N.D. AL 1991). After approving multiple remedial decrees, the court ordered the case closed in December 2006. Information about the case and full-text PDF files of the opinions are available at: *http://knightsims.com/index.html* (accessed January 5, 2008).

36. UAH existed as an extension of the University of Alabama in varying forms since 1950. In 1961, von Braun intervened in the town's education politics by addressing the Alabama legislature, requesting funds to build and equip a research institute on the UAH campus. The Alabama legislature approved $3 million in revenue bonds for the University. See "Alabama A&M University: Historical Sketch," Office of Information & Public Relations, *http://www.aamu.edu/portal/page/portal/images/AAMUHistory.pdf* (accessed December 10, 2007) and "If you really investigate UAH's history, how it all started and when and why, you might decide that the whole thing goes back to 1943 and the day Pat Richardson was hit in the neck by a softball," Phillip Gentry, University Relations, *http://urnet.uah.edu/News/pdf/UAHhistory.pdf* (accessed December 11, 2007). For accounts of von Braun's role in the expansion of UAH, see Ben Graves, "Panelist #7," in *A Century of Flight: 'Creating Rocket City'* by David Lilly (Huntsville, AL: 2003) and Bob Ward, *Dr. Space: The Life of Wernher Von Braun*, pp. 170-171.

for what happened in this case. Foster was born in 1931 and moved to the Huntsville area in 1950 from Birmingham, Alabama, to attend Alabama A&M. He worked for ABMA since 1957 and transitioned to NASA in 1960, along with most of the German rocket engineers. As the former mayor of nearby Triana, an all-black community west of Redstone Arsenal, Foster was instrumental in getting running water and a sewage system to the community, as well as building houses and creating job opportunities for community members.

Similar to Smith, Foster was appalled that Huntsville already had a university that was "A hundred and seventy-five years of age (yet) they come in and start a UAH."[37] However, he also describes the difficulties in convincing the existing institution to implement an apparently much-needed engineering degree. He says,

> That's one of the things I tried to get A&M at that particular time to take advantage of . . . I couldn't get (th)em interested. They (were) hard to convince back in the sixties, to get them to understand what's at hand, and what would be available. I guess they d(id)n't wanna rock the boat. A lot of it . . . had been caused by what segregation had done . . . [38]

What Foster is referring to are the effects of systemic racism that often led to what he calls "acting like Uncle Tom." The effects of racism had put African Americans at an immense disadvantage long before they could even think about attending college, let alone try to compete for the new jobs coming to town with the Germans. Hereford describes the school he attended: "We had no library. We had . . . no lunch room . . . no chemistry lab or biology lab, and *I* wanted to be a *physician*. And so I went to school (in Huntsville) for twelve years and we had none of those things."[39]

Foster describes the effects of this lack of educational opportunities on income opportunities available to the black community. Referring to the technical skills needed to be part of the booming space industry he states,

> . . . we didn't have a population with the prerequisites that would be needed to do this type of work . . . At that time (we) only . . . had barbershop(s), funeral home(s), beauty salon (s), (and) café(s). Depending on what city, there might have been some small type . . . hotel.

---

37. According to its Web site, Alabama A&M University opened in May 1875, which means that it had existed for 86 years when von Braun lobbied for UAH.

38. Clyde Foster interview, Triana, AL, July 19, 2007.

39. Hereford interview, July 19, 2007.

So, while the Germans seemed to have brought a lot of jobs to the town, being able to take advantage of that was not a matter of equal opportunity. Foster summarizes his experience as an African American man in 1950s Huntsville as follows:

> To have been born . . . in one of the most difficult times to be born and to compete, here in Alabama, confronted with all of the segregation, George Wallace, and on the other hand you've got Dr. Wernher von Braun with the team from Germany from the University of Berlin! And how do you wanna compete?[40]

Concluding Remarks

In my larger project, the impact of the German families on Huntsville's African American and other communities will be addressed in more detail. In this brief exploration, however, I hope to have provided some possible explanations for why Hanson Howard was not surprised to be the only African American in the room for a forum about the "creation of rocket city." Perhaps more importantly, I hope to have illustrated the significance of who is telling stories of the past, which includes the individual's relationship to dominant and normative groups in the past and the present.

It should be clear now that the assertion that Huntsville is "unanimous" on anything related to its German members' past is an overstatement. While sentiments towards the Germans in Huntsville were apparently always more positive than people not familiar with the town might expect, especially for the time shortly after World War II, they were and are certainly not unanimous. The notion of unanimity is not unique, but it creates a distorted picture, perpetuating injustices of the past. Troublesome and uncomfortable as it might be for those involved, it seems plausible that common experiences and histories of racial privilege have allowed the incoming Germans and white Huntsville residents to form a bond of complicity. As most of the Germans remained mainly silent about racial segregation in the South, most white Huntsville residents did not, and do not, raise questions about the Germans' past in Nazi Germany. Many members of both groups see themselves as victims of systems they view as beyond their control and seem to find explanations declaring the inevitability of their compliance satisfactory. In this way, the individuals are cleared from responsibility.[41]

---

40. Foster interview, July 19, 2007.

41. This is clearly a very complicated issue with many more aspects than I can discuss here. For the longer project, I will analyze individual's comments on this subject in detail, placing them in their appropriate historical and national context.

As Germans in the United States, the German rocket engineers and their families did not have to contend with the same issues as Germans in postwar Germany who were forced to confront the past as a nation for decades. However, in the United States, these immigrants were and are confronted with the histories of other immigrants, including refugees from Germany who had fled Nazi Germany during and after the war, as well as Germans of the following generations, like myself, who came to the United States at a later point in time and who are accustomed to a more honest approach to dealing with the German past.[42] When discussing the rocket engineers' significance to the nation, we cannot afford to distort history to such an extent that we ignore the histories of other immigrants or of those who were already part of the American fabric. The controversies provoked by the Rudolph case alluded to at the beginning of this chapter make that abundantly clear.

By discussing the impact of historical events on members of the Huntsville community not typically consulted, I aim to disturb the notion of unanimity and create a more complex and complete picture. I believe this is as important for narratives about the "creation of the rocket city" as it is to the "history of rocketry" and, therefore, "space history" that should include the impact of the changes on different social groups of the Huntsville community just as much as the use and abuse of forced laborers from concentration camps for the production of Saturn V's forerunners and its implications.

This is a first attempt to listen to those often unheard voices within the Huntsville community as they tell the story of their German neighbors. However, listening to these voices alone would be a distortion as well, which is why I will continue the interview-based research weaving these and other voices into the fabric of my dissertation.

As stated earlier, the main issue with which I grapple is how narrations of the past reflect changing power structures and sometimes create new and unexpected alliances. In this case, it seems that despite their marginalization in Huntsville, voices from the African-American community may be more reflective of the larger national community, which often contrasts with the common idealization of the German rocket engineers in Huntsville. How these and other voices in Huntsville contest each other and interact with national narratives will be at the center of my dissertation where I take the concept of

---

42. For references on how Germans in Germany have been grappling with the Nazi past, see, for example, Michael Kohlstruck, *Zwischen Erinnerung und Geschichte: Der Nationalsozialismus und die jungen Deutschen* (Berlin: Metropol Verlag, 1997); Claudia Fröhlich and Michael Kohlstruck, *Engagierte Demokraten: Vergangenheitspolitik in kritischer Absicht* (Münster: Westfälisches Dampfboot, 1999); Charles S. Maier, *The Unmasterable Past: History, Holocaust, and German National Identity* (Cambridge, MA: Harvard University Press 1997); Jens Fabian Pyper. *"Uns Hat Keiner Gefragt": Positionen der dritten Generation zur Bedeutung des Holocaust* (Berlin: Philo, 2002); and Philipp Gassert and Alan E. Steinweis, *Coping With the Nazi Past: West German Debates on Nazism and Generational Conflict, 1955-1975* (New York, NY: Berghahn Books, 2006).

impact a step further. I argue that national histories, memories, and identities that immigrants bring with them become intertwined with those of their host society, inevitably leading to a struggle over meaning. For the Germans in Huntsville, this struggle was very different than for immigrants considered racially and culturally different or economically less desirable from those in positions of power in the 1950s in the United States.[43]

There are, of course, other reasons for this group's very unique experience, which I will continue to interrogate as I pursue this project. I will close with the possibly most poignant comment on the interrelated nature of the histories of Germany and the United States I have encountered at this point. As he describes his visits of concentration camps in Germany, Charles Ray notes: "(H)ell, that could have been *here*, 'cause we have a tendency not to question power."[44]

ON SOURCES

The video *A Century of Flight: Creating Rocket City* (2003) referred to in this paper is available at the Huntsville-Madison County library in Huntsville, Alabama. The transcript is in my possession.

The oral histories from which excerpts are used in this paper were collected based on snowball sampling during the summer of 2007 when I had the opportunity to speak to eight members of the African-American community of Huntsville. I have written permission to use the interviewees' names for research reports and to store the transcripts for future researchers at an interested and reputable archive or academic library when my project is completed. Currently, the audio files and transcripts are in my possession.

The lack of women's voices in this paper is an obvious weakness. I have made a deliberate effort to talk to African American women but have so far been unsuccessful. I am aware that this may have multiple historical reasons that challenge my project as I continue to seek interviewees.

---

43. This is not to say that this aspect has changed drastically for immigrants today. The spectrum of literature on immigration to the United States is vast. Here are just a few examples describing experiences of other immigrants to the United States in the postwar period. This selection illustrates the significance of professional occupation to the immigrant experience, which I will address for the German rocket engineers in my dissertation. Chaterine Cheiza Choy, *Empire of Care: Nursing and Migration in Filipino American History* (Durham and London: Duke University Press, 2003), Evelyn Nakano Glenn, *Issei, Nissei, War Bride: Three Generations of Japanese American Women in Domestic Service* (Philadelphia, PA: Temple University Press, 1986), David Guitiérrez, *Walls and Mirrors: Mexican Americans, Mexican Immigrants, and the Politics of Ethnicity* (Berkeley, CA: University of California Press, 1995), Pierrette Hondagneu-Sotelo, *Doméstica: Immigrant Workers Cleaning and Caring in the Shadows of Affluence* (Berkeley, CA: University of California Press, 2001), Matthew Frye Jacobson, *Whiteness of a Different Color: European Immigrants and the Alchemy of Race* (Cambridge, MA: Harvard University Press, 1998), and Reed Ueda, *Postwar Immigrant America: A Social History* (Boston, MA: Bedford Books of St. Martin's Press, 1994).

44. Smith interview, July 29, 2007.

I also asked my German and African American interview partners specifically about women and men who had worked for the German families as nannies, domestics, gardeners, or in any other service capacity around the house but, unfortunately, most seem to have passed away by now.

CHAPTER 7

THE GREAT LEAP UPWARD:
CHINA'S HUMAN SPACEFLIGHT PROGRAM AND CHINESE NATIONAL IDENTITY

James R. Hansen

In late September 2007, the Chinese National Tourism Administration posted an article on its Web site entitled, "China to build largest aeronautics theme park." To be located near the new Wenchang Satellite Launching Center on the island province of Hainan, the announced aeronautics theme park will cover a total area of 61 mu (over 1,000 acres), and, in addition to a "space gate" (*Tai Kong Zhi*), "simulated space hall," and aeronautics museum, will feature "a vacation center for aeronautics experts," an "entertainment zone," a "commercial zone where people can buy souvenirs," and an "aeronautics leisure center." In the center of the park will stand a giant viewing tower from which the thousands of expected visitors may view what officials of the China Space Technology Group (the PRC organization that is working jointly with the Hainan provincial government to build the park) call China's "spectacular satellite launching process." Those spectators who cannot all fit onto the tower may watch the launches from bleachers on a huge floating platform just off the island in the South China Sea capable of holding over 3,000 people.

At the end of its press release, China's National Tourism Administration (CNTA) declared that Wenchang will be the world's largest aeronautics theme park, significantly larger than the visitor's center at the U.S. Kennedy Space Center. Its operation will "promote international technological cooperation in the space field and greatly boost the tourist industry in Hainan." Picking up on the story, one American news service ran the headline, "Watch Out Disneyland, China May Be Jumping Ahead of You"—a valid comparison considering that the original Disneyland in California was built on 320 acres (compared to 1,004 for the Hainan park) and the original plot of Epcot at Disneyworld in Florida only 600 acres.

The same day of the article about Wenchang, the CNTA released another story whose subject was seemingly so different from the first that it might have confused many Western analysts trying to understand the conflicting forces of modernity and traditionalism at work in today's China. Entitled "Sacrificial ceremony to Confucius opens in Shandong," the story covered a grand sacrificial ceremony marking the 2,558th birthday of Confucius, which was taking place in the hometown of "China's most honored ancient philosopher." At the local

Confucian temple, "a descendant of the great man" lit a fire signifying the root of Chinese culture, and a 3,000-year-old sacrificial dance—one recently honored by the Beijing government as part of China's "National Intangible Heritage"—was "nobly performed." With 3,000 people in attendance (the same number who will be able to watch a rocket launch at Wenchang from the floating platform), including dozens of representatives from Hong Kong, Macao, and Taiwan, 500 middle-school students delighted the crowd by reading aloud from Confucius's *Analects*.

These two stories placed together in one's mind offer a curious starting point for an exploration into the nature of Chinese national identity in the early 21st century and how that nascent identity may be bolstering today's burgeoning Chinese human spaceflight program. A principal question addressed at "Remembering the Space Age: 50th Anniversary Conference," held in Washington, DC, in October 2007 was, *"Has the Space Age fostered a new global identity, or has it reinforced distinct national identities?"* The argument made by this author at that conference was that anyone in the West trying to make sense of the Chinese "nation," or even what it means to be "Chinese," must begin by working to sort out all the ethnic, racial, and national self-identities that exist within this vast and vastly complicated land.

On its Web site, the CNTA rushes to point out that China is a "happy family" composed of 56 different "nationalities": the Han, Manchu, Mongol, Uygur, Zhuang, Tibetan, and so on, and that in today's China at least four different Chinese "nations" coexist. The first is composed of all People's Republic of China citizens. The second is the "Han" nation, as Han peoples account for more than 90 percent of the country's population—and are always the first "nationality" to appear on any Chinese list. The third consists of the PRC plus Hong Kong, Macao, and contested Taiwan. The fourth consists of overseas Chinese who retain some feeling of dual nationality. Understanding what these different national self-identities represent, where they came from, how they have interacted with communist ideology and doctrine, and how they might still connect to Confucianism or other aspects of the "National Intangible Heritage"—not to mention how they overlap, harmonize, or rub each other the wrong way, let alone relate to China's ambitions in space—more than merits, it requires, some very significant expertise in the social, cultural, and political history of China.

Any investigation into the character of the national identity in China today must begin by becoming familiar with the ideas and interpretations expressed in three recent books on the subject: *China's Quest for National Identity* (Cornell University Press, 1993), an anthology edited by Lowell Dittmer and Samuel S. Kim; *Discovering Chinese Nationalism in China: Modernization, Identity, and International Relations* (Cambridge University Press, 1999), by Yongnian Zheng; and *A Nation-State by Construction: Dynamics of Modern Chinese Nationalism* (Stanford University Press, 2004), by Suisheng Zhao.

China's Quest for National Identity offers ten quality articles from 1993 on the meanings—and lack of meanings—of "national identity" in China from the early dynasties of ancient times through the imperial and colonial periods to the present day. In the book's final chapter, its editors Lowell Dittmer (a professor of political science at Berkeley) and Samuel Kim (a Senior Research Scholar at the East Asian Institute of Columbia University) explore the question: "Whither China's Quest for National Identity?" In a nutshell, Dittmer and Kim conclude that post-Tiananmen China faced an "unprecedented national identity crisis," the basic dilemma for which was what to do about its "apparent inability to completely embrace or reject socialism."[1] Not a word in the book, published ten years before the first Chinese astronaut went into space in *Shenzhou V,* mentioned Chinese missile development or its fledgling human spaceflight program. Doubtless to say, such a book would do so today.

The 1999 book *Discovering Chinese Nationalism in China: Modernization, Identity, and International Relations,* by Yongnian Zheng, a Research Fellow in the East Asian Institute of the National University of Singapore, presents a very different picture of China's New Nationalism" than that portrayed by most Western intelligence analysts.[2] Zheng, following the lead of Edward W. Said's classic 1978 study *Orientalism,* a masterpiece of comparative literature studies and deconstruction, places his emphasis on how fundamental misperceptions occur when Westerners attempt to understand non-Western cultures.[3] It was in order to emphasize the internal forces of nationalism in China, rather than those forces imperfectly or inappropriately perceived in the West, that Zheng entitled his book, *Discovering Chinese Nationalism in China.*

Western perceptions deeply distort what the recent rise of nationalist feeling in China is all about, Zheng argues. Many of the misperceptions tie into such geopolitical and strategic notions as the "China threat" and "China containment," which derive from the West observing rapid economic growth, an increase in military spending, military modernization, growing anti-West sentiment, assertiveness in foreign behavior, and the rise of the New Nationalism, and then interpreting these developments solely from the "outside," relying on analysts who have never even visited China. Following Zheng's thesis, such misperceptions come directly into play when Western observers consider an

---

1. Samuel S. Jim and Lowell Ditmer, "Whither China's Quest for National Identity?" in *China's Quest for National Identity* (Ithaca, NY and London: Cornell University Press, 1993), p. 287.

2. A good look into U.S. intelligence approaches to understanding what has been going on in the Chinese space program can be made on the Web site *http://GlobalSecurity.org/space/world/ china.index.html.* In particular, see the following: Office of the State Council, "White Paper on China's Space Activities," November 2000; Mark A. Stokes, "China's Strategic Modernization Implications for the United States," U.S. Army Strategic Studies Institute, September 1999; and J. Barry Patterson, "China's Space Program and Its Implications for the United States," Air University, Maxwell AFB, AL, April 19, 1995.

3. Edward W. Said, *Orientalism* (New York, NY: Pantheon Books, 1978).

event like China's January 2007 ground-to-space missile destruction of one of its weather satellites. Many in Western intelligence saw the act as clear enough proof that Chinese nationalism is aggressive and a destabilizing force for international peace and security, though not seeing a similar American missile strike (from a U.S. warship) in mid-February 2008 to destroy an American satellite loaded with toxic fuel as anything that need greatly bother the Chinese.[4] Though it is outside the scope of this essay to cover the arguments made by the United States and China for and against the two anti-satellite (ASAT) events, a few of the ironies associated with them may be registered in the form of the following questions: How could China warn against and then strongly criticize the U.S. missile attack of February 2008 without mentioning its own anti-satellite missile test of the previous year? Conversely, why didn't Beijing use the U.S. interception to justify ex post facto the unannounced destruction of its own defunct satellite in January 2007? By what truly legitimate arguments can China (or Russia, for that matter) call for a complete ban on space weapons and then be involved in testing such weapons? How can the United States expect other countries such as China (and Russia) to stay away from the development of space weapons technology while simultaneously opposing treaties and other measures to restrict space weapons?

Professor Zheng's insights into China's "New Nationalism" may offer some help in promoting a better understanding of what is going on inside China today, technologically and otherwise. Understood from within China's society and culture rather than from without, China's New Nationalism should be seen, in Zheng's view, not as aggressive but as an understandable voice of

---

4. The Chinese ASAT missile in 2007 was a medium-range ballistic missile launched from Xichang Satellite Launch Center, China's major launch complex, located in Sichuan province in south central China. The target destroyed was an eight-year-old Chinese weather satellite in orbit some 535 miles above Earth. As a number of Western commentators emphasized at the time, this apparently successful ASAT test was a major space "first"—for the first time in history a missile launched from the ground destroyed a satellite, suggesting that the Chinese could now, at least theoretically, shoot down spy satellites operated by the United States or other nations. (In a 1985 test, the U.S. shot down one of its satellites with a missile fired from a fighter aircraft.) For representative U.S. and British immediate reactions to the Chinese anti-satellite missile test in January 2007, see Marc Kaufman and Dafna Linzer, "China Criticized for Anti-Satellite Missile Test; Destruction of Aging Satellite Illustrates Vulnerability of U.S. Space Assets," *Washington Post,* January 19, 2007, A01; Jon Kyl, "China's Anti-Satellite Weapons and American National Security," Heritage Foundation Lecture No. 99, January 29, 2007, accessed at *http://www.heritage. org/Research/NationalSecurity/hl990.cfm*; and "Chinese missile destroys satellite in space," January 21, 2007, accessed at *http://www.telegraph.co.uk/news*. For Chinese reaction to the U.S. destruction of its satellite in February 2008, see David Byers and Jane Macartney, "China and Russia cry foul over satellite," February 21, 2008, accessed at *http://www.timesonline.co.uk/tol/news/world/ us_and_americas/article3408155.ece*; "China Warns U.S. on Satellite Missile Test," February 26, 2008, accessed at *http://www.redorbit.com/modules/news/tools.php* and Thom Shanker, "Satellite is destroyed but questions remain," *International Herald Tribune,* February 21, 2008, accessed at *http://www.iht.com/articles/2008/02/21/america/satellite.php*.

frustration over an "unjustified" international order, with many in China seeing Western interests around the world as anything but benign and Western states as anything but innocent in their intentions toward China. In the past decade Chinese leadership has worked hard to better integrate their country into the international community. Admittedly, China's January 2007 ASAT test appeared to contradict Beijing's oft-stated opposition to the "weaponization" of space, but, following Zheng's thesis, one should not leap to the conclusion that China's insistence on testing a missile defense system is a "reckless move" driven predominantly by China's "traditional Sino-centrism" or by its "dangerously aggressive aspirations" for great power status.[5] For the past several years, the Chinese have chafed at what they have seen as efforts by the United States to exclude it from full membership in the world's elite space club. So by early 2007, Beijing set out to establish a club of its own—as the primary "space benefactor" to the developing world.

Some of this came quickly to fruition. In May 2007, the Chinese launched a communications satellite for Nigeria.[6] For the central African country, Beijing not only designed, built, and launched the satellite but also provided a large loan to help pay the bill. China also signed a satellite contract with another big oil producer, Venezuela, and also began to develop an Earth-observation satellite system—and alternative to GPS—in association with Bangladesh, Indonesia, Pakistan, Peru, Thailand, and Iran.[7] In the next several years, most observers of the PRC believe that China could launch as many as 100 satellites, not only to create a digital navigational system but also to help deliver television to rural areas, improve mapping and weather monitoring, and facilitate scientific research.

The Chinese have also chafed at not being allowed to participate in the U.S.-led ISS, a feeling of prejudice against China that became abundantly clear to NASA Administrator Michael Griffin and other members of his NASA entourage when they made an official visit to China in September 2006. Following an agreement between President Hu Jintao and President George W. Bush and U.S. acceptance of a special invitation from Dr. Laiyun Sun,

---

5. Initially declining to confirm or deny that any ASAT test had happened, Beijing eventually made an official declaration that "The test was not directed at any country and does not constitute a threat to any country. . . . China has always advocated the peaceful use of space, opposes the weaponization of space . . ., and has never participated and will never participate in any arms race in outer space." Naturally, Western observers refused to accept that China's destruction of its weather satellite did not have any military associations driven by strategic objectives.

6. Associated Press, "China Launches Satellite for Nigeria," May 14, 2007, accessed at *http://www. space.com/missionlaunches/070514_china_nigcomsat1.html.*

7. See Zhao Huanxin (*China Daily*), "China to develop Venezuela satellite," November 3, 2005, accessed at *http://www.chinadaily.com.cn/english/doc/2005-11/03/content_490429.htm;* "Venezuela to increase oil sales to China," August 17, 2006, accessed at *http://www.chinadaily.com.cn/english/ doc/2006-08/17/content_666707.htm;* and "China signs 16 international space cooperation agreements, memorandums in five years," October 12, 2006, accessed at *http://english.peopledaily. com.cn/200610/12/eng20061012_311154.html.*

Administrator of China's National Space Administration, Administrator Griffin and a small group of NASA officials traveled to China for a historic 5-day visit. Nothing concrete resulted from the exchange of pleasantries, but the visit was nonetheless a milestone in the history of space diplomacy as Griffin was the most senior U.S. space official ever to go to China.[8]

In Beijing and Shanghai, the NASA delegation met with officials of the China National Space Administration (CNSA), Chinese Academy of Space Technology, Chinese Academy of Sciences, Center for Space Science and Applied Research, Shangai Institute of Technical Physics, and China Meteorological Administration. They met for talks with China's Chief Minister for Science and Technology, who asked, very frankly, if China could participate in the ISS program, an overture the NASA folks were ready always to politely but quickly sidestep. "The tone of our meetings in China was at all times very cordial, very polite, very welcoming," recalls former astronaut and NASA chief scientist Shannon Lucid, who made the trip. "But there was always the undercurrent that China really wanted to be part of the ISS. At just about every press conference, we would be asked why China couldn't be part of ISS. Administrator Griffin always handled that very well, explaining that you could not have cooperation on something like the ISS unless everything was open and above-board. You absolutely needed that for safety reasons."[9]

The Chinese space programs officials who asked that question were clearly not part of the Chinese military, which is in charge of the country's human spaceflight program. In fact, upon arrival in China, the NASA party was informed by its hosts that it would be able to visit the Jiuquan Satellite Launch Center but while there would be able to tour a few launching pads but would not be given a tour of any of the buildings where spacecraft were tested and prepared for launching. Given the time and trouble of traveling all the way out to the remote site in the high Gobi Desert just to see launch pads, Griffin informed the Chinese that the NASA delegation did not care to make the trip. What he and the NASA group hoped to see, Griffin later told Western reporters, were engineering facilities and to be in a position to have "eye-level" discussions with fellow engineers.[10]

Nonetheless, the NASA officials returned home with a much enhanced appreciation for China's commitment to space exploration. According to former astronaut and NASA chief scientist Shannon Lucid, the enthusiasm of the Chinese people for space exploration, experienced up close and personally, is

---

8. See "Transcript, NASA Administrator Michael Griffin Press Conference, Shanghai, China, September 27, 2006," accessed at http://www.nasa.gov/about/highlights/griffin_china.html.

9. Shannon Lucid, Houston, TX, telephone interview with author, June 13, 2007, transcript, 1.

10. See Warren E. Leary, "NASA Chief, on First China Trip, Says Joint Spaceflight Is Unlikely," *New York Times*, September 28, 2006, accessed at *http://www.nytimes.com/2006/09/28/science/space/28nasa.html.*

highly impressive. Their enthusiasm seems authentic—and no mere invention of the communist state. In Shannon Lucid's view, space exploration has developed into "the foremost symbol" of what the Chinese wish for their society to become. "Right now space exploration is probably more important symbolically to the Chinese than it is to the American people," says Lucid. "Of course, it was symbolically very important to us back in the 1950s and 1960s, because of the Cold War. It will become more important to Americans symbolically in the future if some other country starts to prove that it is better at it than us."[11]

Lucid, who was born in Shanghai in 1943 to a Christian missionary couple, has returned to China as an adult no less than three times and expresses her thoughts about the feelings of the Chinese people about their space program with greater familiarity of Chinese culture and history than most Westerners: "The Chinese are very proud of their space program. As a people, they are very connected to their long history and feel that being a leader in space is part of their legacy, as the Chinese were the ones to invent rockets centuries ago." In 1997, Lucid and fellow U.S. astronaut Jerry Ross accepted a Chinese invitation to attend to an international congress on science fiction literature held in Beijing. "The Chinese were really into science fiction," Lucid recalls. "I thought at the time, this is what it must have been like in Germany in the 1920s when so many of their people got caught up in science fiction and an enthusiasm for rockets and space, or in the United States and the Soviet Union back in the 1950s." In Lucid's view, China is a dynamic and dramatically changing society whose people are very much looking forward to its future: "When you see Shanghai, wow! It is so modern with all its shops and big new buildings! It is so modern—absolutely amazing! You think to yourself, 'This is a communist country?! There is all this modern building going on, such a great hustle and bustle. There is such vitality among the Chinese.'"[12]

NASA Administrator Griffin returned from the China trip so impressed by what the Chinese were doing that before long he issued what arguably will become his most memorable, and certainly controversial, statement. At a Washington, DC, luncheon at the Mayflower Hotel on September 17, 2007, Griffin remarked that China would likely be on the Moon with human explorers before the U.S. ever manages it again. What Griffin said precisely—or not very precisely, at least in terms of his first sentence—was:

> I personally believe that China will be back on the Moon
> before we are. I think when that happens, Americans will not
> like it, but they will just have to not like it. I think we will see,
> as we have seen with China's introductory manned spaceflights

---

11. Lucid to author, transcript, p. 2.

12. Ibid, transcript, pp. 1–3.

so far, we will see again that nations look up to other nations that appear to be at the top of the technical pyramid, and they want to do deals with those nations. It's one of the things that made us the world's greatest economic power. So I think we'll be instructed in that lesson in the coming years.

Griffin concluded, "I hope that Americans will take that instruction privately and react to it by investing in those things that are the leading edge of what's possible."[13]

Whether there was a political design behind Griffin's seemingly impromptu comment was immediately debated. Some commentators thought what the NASA Administrator said was a good thing, because it could possibly spur the United States into a more aggressive stance on space exploration—read "Mars." Others felt Griffin's comment to be highly lamentable. One representative of the U.S.'s burgeoning private space industry offered a strongly negative opinion: "Those who have given up have already failed. It's clearly time for new leadership if (Griffin) believes what he said. To suggest that a program that plans to outspend China by more than 10 to 1 can't beat their space program hands-down practically defines pathetic."[14] More considered opinions regarded Griffin's assessment to be totally genuine, not cynical, and stemming not just from the early momentum of Chinese achievements in space or its grand statements of space ambitions but from what Griffin and company personally heard and saw on their historic visit to China a year earlier.

A third book fundamental to understanding the evolution of the Chinese national identity, and placing the contemporary Chinese space program into a deeper and richer cultural perspective, is *A Nation-State by Construction: Dynamics of Modern Chinese Nationalism* (2004), by Suisheng Zhao, the Executive Director at the Center for China-U.S. Cooperation at the University of Denver's Graduate School of International Studies. Zhao's thesis is that Chinese leadership in the 1990s abandoned Marxism for "pragmatic nationalism," which author Zhao

---

13. Griffin's speech was reproduced in its entirety under the title "America Will Not Like It," *New Atlantis* No. 18 (Fall 2007): 128-30.

14. See comment from Stephen Metschan from September 17, 2007, published at *http://www. spacepolitics.com/2007/09/17/griffin-china-will-beat-us-to-the-moon/*. Metschan is associated with DIRECT v2.0, an alternative approach to launching missions planned under NASA's VSE program. The idea behind the DIRECT is to replace the separate Ares-I Crew Launch Vehicle (CLV) and Ares-V Cargo Launch Vehicle (CaLV) with one single "Jupiter" launcher capable of performing both roles. Metschan is also founder and president of *TeamVision* Corp, developer of the *FrameworkCT*, a new class of business intelligence software focused on improving the early decision-making process in large and complex organizations. Prior to founding TeamVision, Metschan worked for Boeing on Advanced Engineering projects for NASA for over ten years. His primary focus was on the integration of analysis, design, manufacturing, finance, and marketing teams into a cohesive team framework to enhance the understanding of problems and their solutions for advanced space vehicle systems. He earned a B.S in mechanical engineering in 1989 from the University of Portland.

defines as a commitment to "avoid dogmatic constraints and [rather] adopt whatever approach proves most effective in making China strong."[15]

Specifically, Zhao demonstrates how the policies of the past two Chinese presidents, Ziang Zemin, and Hu Jintao, employed nationalism with "great diplomatic prudence," not just as an instrument by which the Chinese Communist Party could preserve its rule but also as the most effective way to assemble domestic support from the many disparate divisions in Chinese necessary prerequisite to the building of an effective modern state.[16] Zhao's book endorses "pragmatic nationalism" as an effective approach to Chinese governance and foreign policy, one that promotes economic development and is bringing a better life to the masses of Chinese people while at the same time avoiding major confrontations with the West.

Not that there are no alternative "nationalisms" in play in today's China. Zhao also identifies "liberal nationalism," whose spokesmen since the end of the Cold War have been pressing for greater public participation in the political process, challenging authoritarian rule (such as at Tiananmen Square), and explicitly calling for the adoption of liberal democratic ideas as the best means of promoting China's renewal. Zhao also shows how there is a still a strong strand of "nativism" in China, which calls for a return to self-reliance and Chinese tradition and traces the roots of China's weakness to the impact of imperialism on China's self-esteem as well as to the subversion of indigenous Chinese values, such as Confucian ethics. Finally, there is "anti-traditionalism," a very different sense of nationalism that holds that China's very traditions, such as a Confucian hierarchy and an inward-looking culture, constitute the main source of its weakness. Anti-traditionalists call for the complete rejection of these backward traditions and the rapid adoption of foreign culture and Western models of economic and political development. The anti-traditionalist strand also calls for China to accommodate a "progressive" internationalist system. Starting in the 1980s, anti-traditionalists called on the Chinese people to rejuvenate their nation by assimilating Western culture, adopting Western models of modernization, and adjusting to the capitalist world system. To achieve this goal, they demanded a fundamental change in the Chinese mindset, toward one supporting "the spirit of science and technology."[17]

What can Western observers of the Chinese space program really learn from this literature on Chinese national identity? How might it help us comprehend the conjunction between China today building the world's largest aeronautics theme park at Wenchang while simultaneously celebrating the 2,558th birthday of Confucius at Shandong?

---

15. Suisheng Zhao, *A Nation-State by Construction: Dynamics of Modern Chinese Nationalism* (Stanford University Press, 2004), p. xxx.

16. Ibid, p. xxx.

17. Ibid. p. xxx.

Regrettably, none of the books examined in this essay said much of anything about China's space program. Given that Zhao's book was published in 2004, one might think that at least his book would have done so. In an e-mail to Professor "Sam" Zhao at the University of Denver, I asked how he would apply the thesis of his book to the recent history of China's space program, particularly its recent achievements and present ambitions for human spaceflight. Zhao answered my inquiry: "China develops its space program because it sees the program as reflecting its comprehensive national strength and is an indication of China's growing levels of science and technology. The human spaceflight program will enhance China's international prestige and status and increase the *ninjiu li*, or 'cohesiveness,' of the Chinese people and the nationalist credential of the government domestically. Also, it will position China in the future competition for outer space resources."[18]

Although interesting and concise, Zhao's response was not very satisfying, as there must be more to deconstruct about the political, social, and cultural meanings of the Chinese human spaceflight program than his response suggested—certainly much more given how space exploration has been capturing the popular imagination in China since their first human spaceflight in *Shenzhou V* in 2003.

In a provocative 1997 book entitled *Space and the American Imagination* (Smithsonian Institution Press), American University political science professor Howard McCurdy delved deeply into the relationship of American space exploration to the larger U.S. popular culture. Adopting a cultural studies type approach, McCurdy showed how visions of the inevitability of human space exploration arising from ideas and imagery in popular science and science fiction connected in powerful ways to preexisting mythologies in American culture, most notably the myth of "the frontier." This public perception of space exploration as it boiled and bubbled in the 1950s and 1960s influenced the decisions of American policymakers to pursue exploration, and to pursue it via human spaceflight, which was the predominant vision to capture the popular imagination, rather than going with the often cheaper, perhaps more scientifically justifiable and technologically sophisticated unmanned programs. Without this tight linkage between reality and imagination, McCurdy concluded, the Apollo lunar landings would surely not have even been tried, let alone accomplished so quickly.

What was true for the American experience in terms of dynamic linkages between reality and imagination must also be true, in similar yet distinct fashions, for other national cultures. Another outstanding example of such a study into the culture of spaceflight rests in the scholarship of historian Michael Neufeld, particularly his 1990 article in the journal *Technology and Culture*, "Weimar Culture and Futuristic Technology: The Rocketry and Spaceflight Fad in Germany, 1923-1933," which analyzed how the enthusiasms of German

---

18. E-mail, Suisheng "Sam" Zhao to author, June 2, 2007.

science fiction writing, notably that of Max Valier (someone who did not truly even understand the principles of rocketry) ultimately led to the rocket being forged into a weapon of war.[19] Just previous to Neufeld's work, historian Walter McDougall in his 1985 book Pulitzer Prize-winning book . . . *the Heavens and the Earth: The Politics of the Space Age* (New York: Alfred A. Knopf) explored the critically important associations between literary and social movements in Tsarist Russia and the Soviet Union and subsequent technological develop-ments related to rockets and spaceflight. Recent excellent studies of the cultural context of the roots of the Russian space program have also been made by James Andrews[20] and Slava Gerovitch.[21] Given the fertility of such studies for our understanding of the history of spaceflight within the national cultures of the Soviet Union, Germany, and America, one must ask, how would a simi-lar analysis of Chinese culture inform our understanding of the meaning of space exploration to the Chinese? To date, very few books and articles on the Chinese space program offer anything like the penetrating insights that have been provided for the United States, Germany, and Russia.[22]

Similarly, just as there is no way to fathom what the U.S. space program has meant to American society over the past half century without understanding what Americans have wanted from their heroes—"space" heroes and otherwise—there is also no way to understand what the Chinese are after in space without understanding the iconography that has developed around their *y'uhángyuán* or "universe navigators." If the particular types of heroic iconography that have come to surround China's first space traveler, *Shenzhou V*'s Yang Liwei, is any sort of reliable indicator, Chinese society by 2003 was well on its way toward successfully mixing a rising sense of pragmatic nationalism, communist ideology, traditional Confucian values, and drive for economic and high-tech industrial competitiveness into an effective recipe for an expansive program of human spaceflight.[23] Evidently the Chinese space program has been tapping into

---

19. *Technology and Culture*, Vol. 31, No. 4 (October 1990): 725-52.

20. See James T. Andrews, *The Bolshevik State, Public Science, and the Popular Imagination in Soviet Russia, 1917–1934,* and *Visions of Space Flight: K. E. Tsiolkovskii, Russian Popular Culture, and the Roots of Soviet Cosmonautics 1857-1957,* published in 2003 and 2007, respectively, both by Texas A&M University Press.

21. See Slava Gerovitch, "'New Soviet Man' Inside Machine: Human Engineering, Spacecraft Design, and the Construction of Communism," *OSIRIS*, vol. 22 (2007): 135-5, and "Love-Hate for Man-Machine Metaphors in Soviet Physiology: From Pavlov to 'Physiological Cybernetics,'" *Science in Context*, vol. 15, no. 2 (2002): 339-374.

22. The most complete treatment of the Chinese space program can be found in Brian Harvey, *China's Space Program: From Conception to Manned Spaceflight* (Chichester, U.K.: Springer, 2004), but the book suffers from its lack of historical and cultural perspective.

23. On the Chinese enthusiasm for their first countryman to make a spaceflight, see my article, "The Taikonaut as Icon: The Cultural and Political Significance of Yang Liwei, China's First Space Traveler," in *The Societal Impact of Spaceflight* (NASA Special Publication-2007-4801, 2007), eds.

sources of popular support for human space exploration that belong not just to the predominant strand described by Suisheng Zhao as "pragmatic nationalism" but also into the other strands of nationalism with currency in China today, including liberal nationalism, anti-traditionalism, and even nativism.

If true, it will be no wonder if many thousands of Chinese tourists will soon be sitting on floating bleachers out in the South China Sea watching rockets lift off from their new Chinese Disneyland, perhaps taking Yang Liwei or his fellow "taikonauts" to humankind's next landing on the Moon, as NASA Administrator Michael Griffin has warned.

Unfortunately, many Western observers—especially in the defense intelligence community—persist in understanding Chinese developments from without instead of probing deeper into China, as today's top scholars on the modern Chinese identity would have them, from within the remarkably rich and infinitely complicated character of Chinese society—past, present, and future.

POSTSCRIPT

August 6, 2008: "Spaceman Yang to launch torch relay in Beijing," by Chen Jia, China Daily, accessed on September 11, 2008, at *http://www.chinadaily.com. cn/olympics/torch/2008-08/06/content_6907463.htm.*

"Yang Liwei, China's first astronaut, will run the opening leg of the Olympic torch relay in Beijing, which starts at 8 am today at the Meridian Gate of the Forbidden City. Basketball star Yao Ming, who some media said would run the first leg, will be the ninth torchbearer. "Yang helped China realize its dream to travel in space, and now we are living another dream of hosting the Games," Sun Xuecai, deputy director of the Beijing sports administration, told a news conference Tuesday.

September 9, 2008: "The New Red Scare—Avoiding A Space Race With China," by Loretta Hidalgo Whitesides, in *Wired*, accessed on September 11, 2008, at *http://blog.wired.com/wiredscience/2008/09/the-new-red-sca.html.*

"In the wake of the pageantry and sheer enormity of the Beijing Olympics, China is getting ready for its next beautifully scripted display of power and prestige: Its first space walk will be televised live by mid-October. The mission will carry three crew members, two of whom will move into the newly created EVA (extra-vehicular activity) airlock at the top of the Soyuz-like vehicle. One of these crew members will wear a newly designed Chinese EVA space suit, of which the country is very proud."

Steve J. Dick and Roger D. Launius, pp. 103–117. A slightly different version of my essay appeared as "Great Hero Yang," in *Air/Space Smithsonian* (Feb/March 2007), and can be accessed at *http:// www.airspacemag.com/issues/2007/february-march/great_hero_yang.php?page=1.*

CHAPTER 8

"The 'Right' Stuff: The Reagan Revolution and the U.S. Space Program"

Andrew J. Butrica

This paper addresses two questions related to the overall theme of "National and Global Dimensions of the Space Age": 1) has the Space Age fostered a new global identity or has it reinforced distinct national identities? and 2) how does space history connect with national histories and the histories of transnational or global phenomena? The evolution of the U.S. space program, I argue, is a direct outgrowth of the impact of ideology, specifically the conservative ideology of the so-called New Right. Because of the connection to this ideological agenda, space history has become linked with national history.

The intellectual origins of this paper began many years ago as an investigation into the influence of ideology on technology.[1] The ideology was the internationalist, pacifist, feminist, religious, and other beliefs of the Saint-Simonians. Named after its founder, Claude Henri de Rouvroy, le comte de Saint-Simon, the Saint-Simonians belonged to a French movement that flourished from roughly 1830 to 1870. Many of its members were engineering graduates of the prestigious École Polytechnique employed by the French state in a number of technical positions. They and the Saint-Simonian bankers sought to use transport technologies—such as canals and railways—to achieve a number of their ideological goals, one of which was to bridge the divide between the Christian and Moslem worlds using, among other means, a canal linking the Mediterranean and Red seas.[2]

In this study of the conservative space agenda, I suggest how the Reagan administration—as the triumph of the New Right—projected into space the conservative political agenda that elected it into office. America's turn to the right took place over several decades, and its intellectual origins can be traced back to the 1950s at the start of the Cold War. As historian George H. Nash

---

1. Recently, this question was taken up by Paul Forman in his "The Primacy of Science in Modernity, of Technology in Postmodernity, and of Ideology in the History of Technology," *History and Technology* 23, 1 (March 2007): 1-152.

2. This research was the subject of the paper "Saint-Simonian Engineers: An Aspect of French Engineering History" presented at the History of Science Society meeting in Philadelphia, Pennsylvania in October 1982.

wrote in 1976: "In 1945 no articulate, coordinated, self-consciously conservative intellectual force existed in the United States. There were, at best, scattered voices of protest, profoundly pessimistic about the future of the country."[3]

The emergence of the so-called New Right began in earnest during the 1960s, in parallel with—and to a large degree in response to—the rise of the New Left. The presidential candidacies of Barry Goldwater and George Wallace embodied the movement,[4] which was propelled by a zealous, if not obsessive, anticommunism belief; support for business and defense interests over social issues; and downright antipathy for the Great Society program—considered the epitome of the "welfare state" and "big government"—and the so-called "rights revolution," which sought equal protection under the law for African and Hispanic Americans, women, gays, and the disabled. Rather than address a range of social concerns, the New Right wanted to deregulate commerce, cut the size of government, and reduce corporate and individual taxes.[5] They believed in the positive benefits of technological progress and scorned the prevalent notion of limits to growth.[6]

The impact of ideology on the space program's evolution already has been taken up by Roger Launius and Howard McCurdy in their milestone work on presidential leadership, *Spaceflight and the Myth of Presidential Leadership* in which they wrote that, of all the factors influencing the space program, "ideology (was) the most important." Indeed, "From the beginning of the space age in 1957, the ideological debate over the program has revolved around the expense

---

3. Nash, *The Conservative Intellectual Movement in America: Since 1945*, 1st ed. (New York: Basic Books, 1976), p. xv.

4. Phyllis Schlafly, "A Choice, Not an Echo," in *Conservatism in America since 1930: A Reader* by Gregory L. Schneider, (New York: New York University Press, 2003), pp. 231-237. Schlafly discusses Goldwater's candidacy as the embodiment of conservative philosophy from the perspective of 1964.

5. For my discussion of these political events, I have relied mainly on William C. Berman's *America's Right Turn*, 2nd ed. (Baltimore: The Johns Hopkins University Press, 1998), pp. 2-3, 6-8, 21, & 39, as well as Mary C. Brennan's *Turning Right in the Sixties: The Conservative Capture of the GOP* (Chapel Hill: University of North Carolina Press, 1995); Dan T. Carter, *The Politics of Rage: George Wallace, the Origins of the New Conservatism, and the Transformation of American Politics* (New York: Simon and Schuster, 1995); Carter, *From George Wallace to Newt Gingrich: Race in the Conservative Counterrevolution* (Baton Rouge: Louisiana State University Press, 1997); and Godfrey Hodgson, *The World Turned Right Side Up: A History of the Conservative Ascendancy in America* (Boston: Houghton Mifflin, 1996).

6. Paul Neurath, *From Malthus to the Club of Rome and Back: Problems of Limits to Growth, Population Control, and Migrations* (Armonk, New York: M. E. Sharpe, 1994) reviews the limits to growth debate from the 18th century to the present, including the Club of Rome, and has a bibliography of the literature. Robert McCutcheon, *Limits to a Modern World: A Study of the Limits to Growth Debate* (London: Butterworths, 1979) provides a contemporary overview of the debate.

and direction of the enterprise, particularly the emphasis placed on human spaceflight initiatives as opposed to scientific objectives."[7]

According to Launius and McCurdy, during the 1950s conservatives endorsed a limited civilian space program focused on scientific research. With the launch of Sputnik, liberals clamored for an aggressive space program featuring human spaceflight with sizable federal expenditures and management in the form of NASA. Liberals also pushed the need to garner national prestige— something eschewed by conservatives—by taking on the Soviet Union in a space race. These ideological divisions began to shift as Richard Nixon entered the White House. By approving the Space Shuttle project, Nixon accepted the liberal space agenda of expensive spaceflight; however, he and his cabinet also saw the Shuttle's potential for conducting various military missions consistent with the conservative agenda.[8]

In addition, I argue, conservatives during the first years of the Reagan presidency envisioned the Shuttle as the principal technology for realizing their goals in space, namely, the commercialization and militarization of space. In many ways, it was tailor-made for their purposes. The Shuttle's technological limitations as an Earth-orbiting vehicle suited it ideally for an agenda that emphasized exploiting space rather than exploring it, especially regarding space applications (business and defense) in near-Earth space. All of the conservative space policy initiatives focused more on space applications than on exploration. Furthermore, the decision to place military and intelligence payloads on the Shuttle blurred the line between civilian and military missions, raising the question as to whether NASA—after a brief hiatus in fulfillment of Nixon's strategy of détente—was again in the service of the Cold War.

As conservative support shifted toward the space program, Launius and McCurdy explain, liberal support moved away from it. This "sea change in ideological attitudes toward space . . . drew its strength from the confluence of . . . the changing nature of American liberalism and the conservative embrace of frontier mythology." President John Kennedy made liberal use of the frontier analogy in his speeches, especially as a rationale for the ambitious Apollo project and the space race with the Soviet Union.[9] Once liberal interest in the Cold War

---

7. Launius and McCurdy, "Epilogue," in Launius and McCurdy, eds., *Spaceflight and the Myth of Presidential Leadership* (Urbana: University of Illinois Press, 1997), p. 235. James A. M. Muncy also made the point that, while space has been a partisan issue, "space has always risen and fallen on the waves of *ideology*." Muncy, "After the Deluge: What the GOP Takeover Could Mean for Space," *Space News* 4, 51 (December 19-25, 1994): 4.

8. Launius and McCurdy, "Epilogue," pp. 235-238.

9. One must not forget, too, the extensive references to "frontiers" and pioneering in Kennedy's July 15, 1960, acceptance speech to the Democratic National Convention in Los Angeles. The "New Frontier" slogan morphed into a label for his administration's domestic and foreign programs. "Address of Senator John F. Kennedy Accepting the Democratic Party Nomination for the Presidency of the United States," Memorial Coliseum, Los Angeles, July

waned, so did the necessity of dominating "this new sea." Liberals also increasingly rejected the frontier myth and its implied associations with exploitation and oppression. Conservatives lacked these misgivings about the frontier and embraced the economic benefits and material progress associated with the frontier myth.[10]

However, the impact of conservative ideology on the space program was far more pervasive than that described by Launius and McCurdy. The conservatives' own comparison of the space program under the Reagan administration with that of the Kennedy years was not without grounds. At the very least, the numerous new space initiatives undertaken by the Reagan administration made this a major turning point in U.S. space history at least on a par with that of the Kennedy-Johnson era.

Perhaps the most unforgettable Reagan space program was the Strategic Defense Initiative (SDI), a space-based antiballistic missile defense system. With homage to President Kennedy and the Apollo effort, Reagan committed the nation to building a space station by the decade's end. Less memorable was the National Aero-Space Plane (NASP), commonly confused with the Orient Express. The Orient Express would have been the nation's fastest aircraft capable of flying from Washington to Tokyo in two hours, while the NASP would have been the world's first single-stage-to-orbit spaceship. The most influential and lasting of the Reagan space initiatives was the formulation of the first national policy to foster the commercial use of space. As a result, the role of the private sector in space grew tremendously following the end of the Cold War, providing the aerospace industry with a respite from the defense cuts that came in the immediate ending of formal hostilities.

In addition to President Kennedy's space and frontier rhetoric, conservatives also embraced the Kennedy era's enthusiasm for large-scale space ventures overseen by NASA. One of the principal prophets of this conservative space agenda was Newt Gingrich. Elected to the House of Representatives from Georgia for the first time in 1978, Gingrich began formulating his ideas about the future, space, and technology in late 1982 and early 1983 as the economy began to turn around and a mood of optimism spread among conservatives and the public in general.[11] His ideas about space and technology are less important as reflections of his personal thinking than as a mirror held up to reflect the thoughts of a number of like-minded individuals who also viewed space as a new frontier for planting the flag of conservative ideas. Like the fabled frontier

15, 1960, John F. Kennedy Presidential Library and Museum, Historical Resources, *http:// www.jfklibrary.org/Historical+Resources/Archives/Reference+Desk/Speeches/JFK/JFK+PrePres/Ad dress+of+Senator+John+F.+Kennedy+Accepting+the+Democratic+Party+Nomination+for+the+Pre sidency+of+t.htm* (accessed November 12, 2007).

10. Launius and McCurdy, "Epilogue," pp. 238.

11. James A. M. Muncy interview, Washington, DC, January 12, 1999, tape recording and transcript, NASA Historical Reference Collection, NASA Headquarters, Washington, DC.

of the Old West, space was where new resources and new business opportunities abounded, and where there were no limits save those of the imagination. Another spokesperson for the conservative space agenda, James A. M. Muncy, Chairman of the Space Frontier Foundation, explained that "space is a natural extension of the Earth's frontiers, and that opening space to human enterprise and settlement is a unique American response to some liberals' calls for limits to growth as a rationale for ever-more-powerful statism."[12]

Implicit in Gingrich's writings is an enthusiasm for technological progress that went hand in hand with an intrinsic disdain for the idea of limits to growth and the associated notions of a future of lowered expectations and the need for state control and planning, all of which Gingrich attributed to liberalism.[13] In Gingrich's mind, which drew upon both futurology and science fiction, technology would take the lead in solving certain social issues. "Breakthroughs in computers, biology, and space," he declared, "make possible new jobs, new opportunities, and new hope on a scale unimagined since Christopher Columbus discovered a new world."[14] Technology and space were a fundamental part of the American ethos, the frontier spirit. In this future world driven by the frontier spirit and technological progress, the handicapped would no longer depend on welfare, having found gainful (tax revenue-generating) employment thanks to new technologies—"compassionate high tech"—and scientific discoveries. Essentially, the compassionate high tech position held that the benefit of investing in commercial and military space technology (in fulfillment of the conservative space agenda) would "trickle down" to Earth and lighten, if not resolve, the need for social welfare in a technology-oriented version of what came to be known as "trickle down economics."[15]

In order to turn this futuristic vision into reality, Gingrich proposed raising NASA's budget to its historic Apollo-era high and endorsed (as President Reagan did) both the Space Shuttle and the International Space Station. NASA's budget was way too small, he argued; the Agency's annual budget would run the Defense Department for only 11 days or Health and Human Services for only 8 days. Over 30 corporations—including such NASA contractors as RCA, General Electric, IBM, Westinghouse, and Western Electric—were larger than NASA. Gingrich saw nothing inconsistent with being a conservative and

12. Muncy, "After the Deluge: What the GOP Takeover Could Mean for Space," opinion piece written for *Space News*, published as Muncy, "After the Republican Deluge," *Space News*, 4, 51 (December 19-25, 1994): 4, fax copy, folder 644, box 22, X-33 Archive, Record Group 255, National Archives and Records Administration, Suitland, Maryland (hereafter, X-33 Archive).

13. Newt Gingrich and James A. M. Muncy, "Space: The New Frontier," in *Future 21: Directions for America in the 21st Century*, eds. Paul M. Weyrich and Connaught Marshner, (Greenwich, CN: Devin-Adair, 1984), p. 61.

14. *Gingrich and Muncy*, 62; Gingrich, *Window of Opportunity: A Blueprint for the Future* (New York, NY: Tom Doherty Associates, Inc., 1984), ix; Muncy, interview, 67.

15. Gingrich, 1, 7-9, 10, 27, 46, 49-50, 52, & 65-66.

favoring such large-scale federal expenditures. "Conservatives are not against a strong Government," he explained. "Conservatives are against big, bureaucratic welfare states."[16]

The large, expensive, bureaucratic programs started by the Reagan administration were consistent with the conservative space agenda, the two main pillars of which were the commercialization and the militarization of space. All of the Reagan administration's major space initiatives—from SDI to the Space Station Freedom and the NASP, with the exception of the commercialization of space per se—exemplified the expensive, large-scale, long-term projects that characterized the Cold War era. Furthermore, the commercialization and militarization of space were intimately interrelated in conservative thinking, creating a space-based mirror-image twin of the writings of Alfred T. Mahan (1840-1914)—a famous naval strategist and professor at the Naval War College.

Mahan stressed the interconnection between the commercial exploitation of the oceans and the military advantages of dominating the seas. He based his beliefs on studies of the role played by control of the sea, or the absence thereof, in the course of history up to the Napoleonic wars. He concluded that control of the seas was the chief basis of "the power and prosperity of nations." As with the New Right and traditional Republican thinking, encouraging commerce was a fundamental priority. In addition, Mahan saw no difference between "national interest" and "national commerce."[17] The use of private security forces in Iraq is yet another step in conservatives' continuing linkage of military and commercial interests.

Mahan's ideas appeared in print as Turner's frontier was closing and as the seas (and overseas interests) promised to serve the United States as a new imperial frontier. Mahan sought to extend the commercial and military influence of the United States in the Gulf of Mexico and the Caribbean Sea, especially at the Isthmus of Panama where a French company was planning to build a canal. Mahan's message that military dominance of the seas was essential to assuring both the nation's military and commercial strength became the foundational philosophy on which the conservative space military agenda was built. Maxwell W. Hunter, II and Lt. Gen. Daniel O. Graham, two of the principal architects and proponents of SDI, consciously followed in Mahan's intellectual foot steps by arguing for the construction of a space-based global defense system to bring

---

16. Gingrich, 53-54; Gingrich and Muncy, 62.

17. On Mahan and his theories, see Robert Seager, *Alfred Thayer Mahan: The Man and His Letters* (Annapolis, MD: Naval Institute Press, 1977); William Edmund Livezey, *Mahan on Sea Power*, rev. ed. (Norman, OK: University of Oklahoma Press, 1980); and Mahan, *The Influence of Sea Power on History, 1660-1783* (Boston, MA: Little, Brown, 1897), reprinted (New York, NY: Dover Publications, 1987).

about a Pax Americana similar to the oceanic Pax Britannica, in which national security and commercial interests were intertwined and mutually serving.[18]

The conservative military strategy in space also took as its starting point the rejection of the conduct of the Cold War instituted by the administration of President Richard Nixon. The two cardinal facets of Nixon's Cold War policy that Reagan and the conservatives rejected were détente and the 1972 Antiballistic Missile Treaty (SALT I). Instead, President Reagan chose to heighten the struggle against what he termed "the evil empire." Reagan also spoke against the reigning defense philosophy known as mutual assured destruction (MAD). The MAD strategy, simply stated, was that each party would be able to wreak destruction on the other, even if an initial strike substantially reduced the missile and nuclear forces of one side. Essentially, each side became the hostage of the other. Reagan's stance against MAD combined with ongoing studies of high-energy lasers and satellite weaponry to become SDI.

The military use of space was not new but rather a constant over the course of the Cold War. Starting in the 1960s, the U.S. military relied on satellites for reconnaissance (photographic, electronic, and oceanic), early warning of offensive missile launches, detection of nuclear explosions, communication, navigation, weather, and geodetic information. One might even say that, at least at one point in the conflict, all space efforts served a Cold War agenda. The Apollo program represented the Cold War at one level. Following the flight of Soviet cosmonaut Yuri Gagarin, on May 25, 1961, President Kennedy initiated the space race, a new Cold War battleground. He warned that the United States had to challenge the Soviet Union's space feats "if we are to win the battle that is now going on around the world between freedom and tyranny."[19] The Cold War and the nation's civilian space programs were now joined. This was truly total warfare that conscripted civilians and civilian agencies into a global struggle.

During this war, the construction of defensive systems served as a bargaining chip in treaty negotiations with treaties helping to limit the seemingly boundless search for, and construction of, new weapons. SDI was no exception; however, it took military space policy in a new direction by proposing to place defensive weapons in space. In contrast, earlier space strategies had

18. Erik K. Pratt, *Selling Strategic Defense: Interests, Ideologies, and the Arms Race* (Boulder, CO: Lynne Rienner Publishers, 1990), p. 96. See also, Graham, *High Frontier: A New National Strategy* (Washington: The Heritage Foundation, 1982); Graham, *The Non-Nuclear Defense of Cities: The High Frontier Space-Based Defense Against ICBM Attack* (Cambridge, MA: Abt Books, 1983).

19. *Public Papers of the Presidents of the United States: John F. Kennedy, 1961* (Washington, DC: Government Printing Office, 1962), 403-404. Asif A. Siddiqi, *Challenge to Apollo: The Soviet Union and the Space Race, 1945-1974,* NASA Special Publication-4408 (Washington: NASA, 2000), shows that Soviet military officers soon lost interest in civilian space projects following Sputnik. They felt that civilian projects hurt their attempts to fund military rocketry programs essential to the Cold War.

positioned defensive weapons on the ground until their use in space to "kill" enemy satellites. Another new direction taken by military space policy was the idea that, instead of protecting just the country's military defenses—as had been the case of earlier space weaponry—SDI would protect the entire population, both military and civilian. The first ground-based ASAT systems, which dated back to 1958,[20] involved launching a killer satellite atop a booster rocket to match the orbit of the target, then track it and detonate the killer satellite near the target. ASAT was not a designation for a single weapon system, but rather a generic term covering anything that could be used to attack, disable, or destroy a satellite from Earth or (in the case of SDI) from space. The 1972 Anti-Ballistic Missile (ABM) Treaty did not ban ASAT systems as neither side wanted to give up a space weapon that both sides were developing.

After a hiatus resulting from a combination of budgetary, political, and technical factors, the Soviet Union resumed ASAT testing in February 1976.[21] The resumption of testing galvanized the Ford administration into authorizing the development of an ASAT system, and President Jimmy Carter continued the ASAT project while seeking to revive existing arms control negotiations.[22] The U.S. and the Soviet Union once more were engaged in a space race with the Soviet Union again in the lead. SDI functioned as a continuation of this space race.

Part of the concern over the Soviet anti-satellite program was that country's progress in developing directed energy weapons using lasers and particle beams, which potentially could serve to arm ASAT weapons.[23] The United States was not without its own particle-beam and laser weapon research, which started under ARPA's Project Defender virtually from the time of the agency's creation. Laser weapons received increased interest following the invention of the gas-dynamic laser in the late 1960s.[24]

These developments in laser weapons and antiballistic missile systems, critics of MAD, and opponents of the 1972 ABM Treaty all came together under the rubric of SDI. According to historian Donald Baucom, the first appearance of the space-based battle station concept in the open literature was in a 1978 issue of *Aviation Week*.[25] The most likely source was Lockheed Corporation's

20. Paul B. Stares, *The Militarization of Space: U.S. Policy, 1945-1984* (Ithaca, NY: Cornell University Press, 1985), pp. 107, 109–110, 117–131, 135–136 & 145–146.

21. Stares, pp. 107, 109–110, 117–131, 135–136 & 145–146.

22. Pratt, 53; Donald R. Baucom, *The Origins of SDI, 1944-1983* (Lawrence, KS: University Press of Kansas, 1992), p. 76.

23. Clarence A. Robinson, Jr., "Soviets Push for Beam Weapon," *Aviation Week & Space Technology* 106, 18 (May 2, 1977): 16–23.

24. J. London and H. Pike, "Fire in the Sky: U.S. Space Laser Development from 1968," IAA-97-IAA.2.3.06, pp. 1–3, paper read at the 48th International Astronautical Congress, October 6-10, 1997, Turin, photocopy, folder 40, box 2, X-33 Archive; Pratt, 16–18; Baucom, 15–17.

25. Robinson, 42–43, 45, 48–49, 51–52; Baucom, 118.

Maxwell W. Hunter, II.[26] Hunter was a key figure in promoting what became SDI; indeed, he, along with three others, formed the so-called Gang of Four that pushed the concept in Congress.

The other major figure was retired Lt. Gen. Danny Graham, who had advised Ronald Reagan on national security matters during his gubernatorial and presidential campaigns. After Reagan's election, using Project High Frontier, help from members of the President's kitchen cabinet, and funding from the conservative Heritage Foundation,[27] Graham pushed his own version of a space-based defense system. Graham felt that by redirecting the arms race to space—where he believed the United States held the technological advantage—the country would achieve a "technological end run" around the Soviets and once again establish U.S. strategic superiority.[28] Once more, a belief in the positive benefits of technological progress drove the conservative agenda in space.

Reagan was disposed favorably toward antiballistic missile defense and against MAD, as he made clear several times, even as early as his 1976 bid for the Republican nomination.[29] The process that led to Reagan's call for creation of a space-based defense was slow and took many turns over the year and a half between the initial September 1981 meeting in Meese's office and Reagan's so-called Star Wars speech. That story has been told in some detail elsewhere.[30]

Despite the number of unconventional facets of SDI, it served as a bargaining chip in arms negotiations, namely in regards to the Nuclear and Space Talks (NST) in Geneva, not unlike the role of Nixon's Safeguard in SALT I talks. Through this diplomatic dialogue, which started in March 1985, the United States hoped to legitimize SDI and push its claims that the Soviet Union had violated the 1972 ABM Treaty. For its part, the U.S.S.R. denounced SDI as an impediment to arms control, and at the Reykjavik October 1986 summit talks, the Soviet Union proposed that both sides observe the ABM Treaty for another ten years, including the restriction on testing space-based ballistic missile defense systems outside the laboratory. The United States refused. In 1987, the Soviets "decoupled" the SDI from treaty negotiations; ending the program was no longer a prerequisite to an agreement. During September 1987 talks in

26. Baucom, 119; Hunter, "Strategic Dynamics and Space-Laser Weaponry," manuscript, October 31, 1977, file 338, box 13, X-33 Archive.

27. Graham, *Confessions of a Cold Warrior: An Autobiography* (Fairfax, VA: Preview Press, 1995), 118-120; Baucom, 145-146 & 150; Berman, 67-68; David Vogel, *Fluctuating Fortunes: The Political Power of Business in America* (New York, NY: Basic Books, 1989), 224-225; Dilys M. Hill and Phil Williams, "The Reagan Presidency: Style and Substance," 11 in Hill, Raymond A. Moore, and Williams, eds., *The Reagan Presidency: An Incomplete Revolution?* (New York, NY: St. Martin's Press, 1990).

28. Pratt, 96; Baucom, 164.

29. Pratt, 102, 103 & 104; Baucom, 130.

30. See, for example, Baucom; Graham, *Confessions*; Pratt; Stares; and Edward Reiss, *The Strategic Defense Initiative* (New York, NY: Cambridge University Press, 1992).

Geneva, the U.S.S.R. further modified its position to allow some antiballistic missile research in space. Talks later that year in Washington, DC, cemented a new relationship between the two countries, and on January 15, 1988, the Soviet Union presented a draft Strategic Arms Reduction Treaty (START) protocol, which committed both countries to abide by the 1972 ABM Treaty for ten years and froze the number of launchers.[31]

The 1972 ABM Treaty lasted for more than ten years until President George W. Bush, who was always critical of the treaty, took the next step and announced in December 2001 that the country was withdrawing from the treaty—a major goal of the conservative space agenda—effectively terminating the treaty on June 13, 2002.[32] Additionally, in recognition of the national priority that Bush gave to missile defense, Defense Secretary Donald Rumsfeld announced the elevation of the effort to agency status and its new designation, the Missile Defense Agency, on January 4, 2002.[33] With that bureaucratic boost, the conservative space agenda seemed alive and well.

The other major element of the conservative space agenda was the commercialization of space. As political scientist W. D. Kay has pointed out: "for the first several months of his presidency, Ronald Reagan did not appear to even have a science policy of any sort, let alone a plan for the U.S. space program."[34] That changed after the first flight of the Space Shuttle *Columbia* in April 1981, when "the general feeling within the White House after *Columbia* was that anything was possible."[35] The Space Shuttle, Kay added, "appeared to provide the Reagan White House with the final ingredient—the requisite technology—that it needed to integrate the U.S. space program into its larger political and economic goals."[36]

Indeed, the Space Shuttle stoked the Reagan administration's fires of enthusiasm for commercializing space, among other projects. The commercialization of space under the Reagan administration was an entirely new space initiative and was one of the two key pillars of the conservative space agenda along with the militarization of space. Reagan's commercial space policy grew out of an examination of military space policy carried out at the highest level, the

---

31. John C. Lonnquest and David F. Winkler, *To Defend and Deter: The Legacy of the United States Cold War Missile Program*, USACERL Special Report 97/01 (Champaign, IL: U.S. Army Construction Engineering Research Laboratories, 1996), 129-130; Reiss, 89-90.

32. U.S. Department of State, Fact Sheet, "ABM Treaty Fact Sheet," December 13, 2001, *http://www.state.gov/t/ac/rls/fs/2001/6848.htm* (accessed November 13, 2007).

33. "BMDO's Name Changed to Missile Defense Agency," *Aerospace Daily*, January 7, 2002, article 196406, [electronic edition].

34. W. D. Kay, *Defining NASA: The Historical Debate over the Agency's Mission* (Albany, NY: State University of New York Press, 2005), p. 125.

35. Kay, 128-129.

36. Kay, 127.

National Security Council, by order of the President in August 1981. A Senior Interagency Group, known as SIG (Space), came together under the direction of the President's Science Advisor, George Keyworth, and the National Security Council. It addressed a range of issues, such as launch vehicle needs, the adequacy of existing space policy for national security requirements, Space Shuttle responsibilities and capabilities, and potential new legislation.[37] The study led to the issuance of the National Space Policy (National Security Decision Directive 42) on July 4, 1982, which for the first time ever included business in space policy and marked the start of a national policy on space commerce.

The economic benefits of space (such as telecommunications, weather forecasting, remote sensing, and navigation) were not new; however, this was the first time in the history of the U.S. space program that a high-level official document made a direct reference to the American business community. The new National Space Policy thus marked a dramatic redefinition of space policy not seen since the launch of Sputnik in 1957.[38] Specifically, it laid out four goals to be accomplished in space; the third and fourth of which called for "obtain[ing] economic and scientific benefits through the exploitation of space" and for "expand[ing] United States private-sector investment and involvement in civil space and space-related activities."[39]

The release of the 1982 National Space Policy revealed its indebtedness to the Space Shuttle. National Security Decision Directive (NSDD) 42 called for making the Space Shuttle available to all commercial users, provided only that national security conflicts did not result. On July 4, 1982, the same date as the new space policy, President Reagan spoke before an audience of some fifty thousand people at Edwards Air Force Base, with American flags flying in the background, as the Space Shuttle *Columbia* landed.[40] This was the Space Shuttle's final test mission and the beginning of its operational status. It also was the first mission to carry a Pentagon payload and the first "Get Away Special" experiments conducted for a NASA business customer.[41] The Space Shuttle was now fully in the service of the conservative space agenda.

---

37. "National Space Policy," July 4, 1982, folder 386, box 15, X-33 Archive.

38. Kay, 127.

39. Christopher Simpson, *National Security Directives of the Reagan and Bush Administrations: The Declassified History of US Political and Military Policy, 1981-1991* (Boulder, CO: Westview Press, 1995), 136-143 (classified version) and 144-150 (unclassified version); Kay, 128.

40. Lyn Ragsdale, "Politics Not Science: The U.S. Space Program in the Reagan and Bush Years," Launius and McCurdy, eds., *Spaceflight and the Myth of Presidential Leadership* (Urbana, IL: University of Illinois Press, 1997), p. 133.

41. Judy A. Rumerman and Stephen J. Garber, *Chronology of Space Shuttle Flights, 1981-2000*, HHR-70 (Washington, DC: NASA, October 2000), p. 5.

The Space Shuttle's dual commercial-military purpose was renewed by a subsequent National Security Decision Directive issued on May 16, 1983,[42] with the central objective of encouraging the U.S. commercial launch industry. That policy made the Space Shuttle available to all domestic and foreign users, whether governmental or commercial, for "routine, cost–effective access to space." It also promoted the commercial use of expendable rockets by making government ranges available for commercial launches at prices "consistent with the goal of encouraging" commercial launches and by encouraging competition "within the U.S. private sector by providing equitable treatment for all commercial launch operators."[43]

The special National Policy on the Commercial Use of Space released on July 20, 1984, reflected the opinions that White House senior officials had heard from representatives from a range of companies interested in conducting business in space, such as Federal Express, McDonnell Douglas Astronautics, Grumman Aerospace, General Dynamics, and Rockwell International. It set out a series of initiatives that included research and development tax credits, a ten percent investment tax credit, accelerated cost recovery, timely assignment of radio frequencies, and protection of proprietary information.[44]

On November 18, 1983, President Reagan designated the Department of Transportation (DOT) as the lead agency to "promote and encourage commercial ELV [expendable launch vehicle] operations in the same manner that other private United States commercial enterprises are promoted by United States agencies." Rather than emulate the regulatory agencies scorned by the New Right, hampering commerce and inflating consumer prices, the DOT would "make recommendations . . . concerning administrative measures to streamline federal government procedures for licensing of commercial" launches (by the DOT). The agency also would "identify federal statutes, treaties, regulations and policies which may have an adverse impact on ELV commercialization efforts and recommend appropriate changes to affected agencies and, as appropriate, to the President."[45] Here was a regulatory mandate to encourage industry. Space commercialization was becoming a model of the Reagan Revolution, and the conservative space agenda.

---

42. Letter, Rosalind A. Knapp to David A. Stockman, December 12, 1983, folder 696, box 23, X-33 Archive.

43. Draft National Security Decision Directive, April 22, 1983, folder 696, box 23, X-33 Archive.

44. Craig L. Fuller to Richard G. Darman et al., note and attachment, "Space Commercialization," August 2, 1983, and Agenda, Space Commercialization Meeting, August 3, 1983, folder 696, box 23, X-33 Archive.

45. "Executive Order: Commercial Expendable Launch Vehicle Activities," attached to Michael J. Horowitz to Robert Kimmitt, December 12, 1983, and Rosalind A. Knapp to David A. Stockman, December 12, 1983, folder 696, box 23, X-33 Archive.

Congress subsequently gave the DOT's new role a legal basis with the passage of the Space Launch Commercialization Act, H.R. 3942 (Senate bill S.560), better known as the Commercial Space Launch Act of 1984. Members of Congress felt that the designation of a lead agency was insufficient because it lacked "legislative authority. The result could inhibit decision-making and interagency coordination and allow the present inefficient approaches to commercial launch approvals to persist."[46] Acting on the authority of both the Act and a Presidential Executive Order, Secretary of Transportation Elizabeth Hanford Dole established the Office of Commercial Space Transportation (OCST), which issued launch licenses and in general regulated the new launch-for-hire industry. NASA itself recapitulated the Reagan administration's evolving commercial space policy by issuing its own Commercial Space Policy[47] and creating its own Office of Commercial Programs in 1984.[48]

As a result of the new projects that the Reagan presidency started in conscious fulfillment of the conservative agenda, the United States ended up with a space program that was, at least in the eyes of the New Right, "politically correct." The malfunctioning Hubble Space Telescope and other issues suggested fundamental flaws in the way NASA operated, while the *Challenger* accident (coupled with the military's grounded expendable launchers) and the subsequent reevaluation of the space program brought on by the end of the Cold War signaled a turning point in U.S. space history. The conservative space agenda shifted accordingly. Although from society's perspective the changes that followed *Challenger* were neither as profound nor as pervasive as those wrought by Sputnik, the space program was never the same.

Conservatives now abandoned the Shuttle, which they held up as a symbol of everything wrong with NASA, and called for basic changes to NASA management. The chief institutional voice for these changes was Vice President Dan Quayle, who acted as head of the recently (1988) reestablished National Space Council. Quayle wanted to "shake up" NASA, which he believed was "to a great extent, still living off the glory it had earned in the 1960s." He complained that NASA projects were "too unimaginative, too expensive, too

---

46 U.S. House of Representatives, *Commercial Space Launch Act*, 98th Congress, 2d session, Report 98-816 (Washington, DC: GPO, 1984), 9.

47. "NASA Commercial Space Policy," October 1984, ii & v, "Summary of Policy Initiatives," and "Research and Development Initiatives," folder 386, box 15, X-33 Archive.

48. NASA Special Announcement, "Establishment of the Office of Commercial Programs," September 11, 1984; NASA News Press Release 87-126, "Assistant Administrator Gillam to Retire from NASA," August 19, 1987; "NASA Commercial Space Policy," October 1984, "Summary of Policy Initiatives;" and Isaac T. Gillam IV, "Encouraging the Commercial Use of Space and NASA's Office of Commercial Programs," *NASA Tech Briefs*, n.v. (Spring 1985): 14-15, all in folder 383, box 15, X-33 Archive; John M. Cassanto, "CCDS Shock Waves," *Space News*, January 24-30, 1994, 21.

big, and too slow."[49] He, like many other NASA reformers, wanted the Agency
to undertake "faster, cheaper, smaller" projects. If NASA shifted from large,
prolonged, expensive projects to smaller, faster, cheaper projects, critics argued,
the Agency would be able to accomplish more science for less money. Quayle
pushed NASA to undertake "faster, cheaper, smaller" projects in imitation of the
management style of the Strategic Defense Initiative Organization. The favored
management style of the New Right became the *de rigueur* management style of
NASA under Quayle and the Agency's new Administrator, Dan Goldin.[50]

The Space Shuttle was a programmatic relic of the Cold War; it embodied
the expensive, large-scale, long-term projects that characterized the Cold War era,
which conservatives had embraced unabashedly. Now it was out of place in the
fiscally conservative post-Cold War environment that favored cheaper, smaller,
short-term projects. Although NASA was expected to conform to the management
style born in the "black" world of national security secrecy, conservatives persisted
in burdening the country with expensive, large-scale, long-term projects, the
embodiment of which now became the Space Exploration Initiative (SEI), a
grandiose plan to return to the Moon and then land astronauts on Mars.[51] With
the return to power of conservatives under George W. Bush, the Space Exploration
Initiative returned from the dead as the "Vision for Space Exploration."[52]

As we consider the "National and Global Dimensions of the Space Age,"
we need to keep in mind how ideology—and in particular the conservative
space agenda—has so profoundly shaped the U.S. space program and how we
think about it. The changes brought about may appear to be the outcome of a
rational policymaking process, but are laden with the values of the New Right.
General acceptance of this conservative space agenda, of course, is assured by the
nation's ongoing turn to the Right. This ideological agenda, therefore, reflects
the country's own turn to the Right, and that conservative bent has shaped and
molded the distinct national identity of the United States and its space program.

49. Dan Quayle, *Standing Firm: A Vice-Presidential Memoir* (New York, NY: HarperCollins Publishers,
    1994), pp. 179 & 180.

50. Butrica, *Single Stage to Orbit: Politics, Space Technology, and the Quest for Reusable Rocketry* (Baltimore:
    Johns Hopkins University Press, 2003), pp. 134-137 & 150-151; and the general discussion
    in Howard E. McCurdy, *Faster, Better, Cheaper: Low-Cost Innovation in the U.S. Space Program*
    (Baltimore, MD: Johns Hopkins University Press, 2001), passim.

51. The most recent and complete scholarly treatment of the Space Exploration Initiative is
    Thor Hogan, *Mars Wars: The Rise and Fall of the Space Exploration Initiative* (Washington, DC:
    NASA, August 2007).

52. White House, Office of the Press Secretary, "President Bush Announces New Vision for Space
    Exploration Program," January 14, 2004, *http://www.whitehouse.gov/news/releases/2004/01/
    20040114-3.html* (accessed November 13, 2007).

CHAPTER 9

GREAT (UNFULFILLED) EXPECTATIONS: TO BOLDLY GO WHERE NO SOCIAL SCIENTIST OR HISTORIAN HAS GONE BEFORE[1]

Jonathan Coopersmith

The start of the Space Age, its morphing into the space race, and President John F. Kennedy's launch of Project Apollo excited not only engineers and scientists but also social scientists and historians. Neil Armstrong's words, "One small step for man, one giant leap for mankind," embodied not only the justified pride of a spectacular technological accomplishment but also the bold hopes of the American Academy of Arts and Sciences (AAAS) to harness the space program to apply the social sciences for the benefit of society and government.

For the AAAS, its "ultimate goal . . . would be to develop a system for the continuing monitoring of important effects of space efforts, together with a reporting of these effects in appropriate terms to the appropriate agency."[2] The participating historians had goals no less impressive. MIT professor Bruce Mazlish declared "In short, we are really attempting to set up a new branch of comparative history: the study of comparative or analogous social inventions and their impact on society."[3]

This paper examines this NASA-funded AAAS project in the mid-1960s to understand why such lofty goals existed, what the project accomplished, and where the humanities and social sciences stand in relation to the space program some four decades later.

AAAS PROJECT

Like the American President who set the Apollo program in motion, this effort had a Massachusetts origin. In February 1962, the AAAS established the Committee on Space Efforts and Society, which bid on and received a $181,000

---

1 I would like to thank Steven Dick, Roger Launius, and Peter Stearns for looking at early versions of this paper.

2. Earl P. Stevenson, "Report of the Committee on Space," *Records of the Academy (American Academy of Arts and Sciences)*, 1963/1964, 151.

3. Bruce Mazlish, "Historical Analogy: The Railroad and the Space Program and Their Impact on Society", in Bruce Mazlish, ed. *The Railroad and the Space Program. An Exploration in Historical Analogy* (Cambridge, MA: MIT Press, 1965), p. 12.

NASA grant for the "Conduct of a study of long-range national problems related to the development of the NASA program"[4] in April 1962.

According to its charter, the Boston-based AAAS was established in 1780, "to cultivate every art and science which may tend to advance the interest, honour, dignity, and happiness of a free, independent, and virtuous people."[5] Conducting such a project with significant societal implications fit fully with its activities. Indeed, its president Paul Freund considered this project "a major Academy study."[6]

The AAAS proposal to NASA asked

> From the standpoint of NASA objectives how can the resources of the nation be mobilized for the achievement of national goals developing out of advances in scientific knowledge and engineering capabilities, and what will be the predictable impact of enterprises so conceived on various sectors of our society? What will be the reciprocal impact back on NASA? Basically, the effort will be to develop a system by which the feedback indicators to NASA may be improved and to assist in making the NASA experience and achievements most meaningful in the public interest.[7]

Based upon an original proposal of two years, the AAAS project ultimately consumed four years and $45,000 more than expected.[8] This was one of several efforts funded by NASA as part of its 1958 mandate to study the "potential benefits to be gained from, the opportunities for, and the problems involved" in the space program.[9] Overall, in its first decade NASA spent nearly $35 million on 365 contracts to study the impact of the space program. Most of these contracts studied technology transfer and economic impacts.[10]

4. Earl P. Stevenson, "Report of the Committee on Space Efforts and Society," *Records of the Academy (American Academy of Arts and Sciences)*, 1962/1963, 141.

5. Accessed at *http://www.amacad.org/about.aspx* (downloaded August 7, 2007).

6. Paul A. Freund, "President's Report," *Records of the Academy (American Academy of Arts and Sciences)*, 1964/1965, 7.

7. Stevenson, op. cit., p. 141.

8. Earl P. Stevenson, "Report of the Committee on Space," *Records of the Academy (American Academy of Arts and Sciences)*, 1965/1966, 22.

9. "Introduction," in Raymond A. Bauer with Richard S. Rosenbloom and Laure Sharp and the Assistance of Others, *Second-Order Consequences. A Methodological Essay on the Impact of Technology* (Cambridge, MA: MIT Press, 1969), p. 2. Among other efforts were Lincoln P. Bloomfield, ed., *Outer Space: Prospects for Man and Society* (Englewood Cliffs, NJ,: The American Assembly, 1962) and Lillian Levy, ed., *Space: Its Impact on Man and Society* (New York, NY: W.W. Norton & Co., 1965). Levy was a journalist who joined the NASA Office of Public Affairs.

10. Mary A. Holman, *The Political Economy of the Space Program* (Palo Alto: Pacific Books, 1974), pp. 171-74. See also, T. Stephen Cheston, "Space Social Science," in Johnson Space Center's *Space*

Bruce Mazlish circa 1974 (Courtesy Calvin Campbell/MIT)

These contracts reflected the goal of NASA Administrator James E. Webb to harness and maximize the results of space spending to benefit all aspects of American society, including regional economic development and education. Webb's interest, however, was not universally shared within NASA, whose managers and engineers saw these goals as unnecessary externalities deflecting them from spaceflight.[11]

The first major study was a one-year project completed by the Brookings Institute in November 1960. To gather information, Brookings organized a two-day conference that included one historian, Melvin Kranzberg, a leading force in the creation of the history of technology as an academic discipline and a strong advocate of institutionalizing history in NASA.[12] In the summer of 1962, NASA funded an eight-week summer study of fifteen areas of space research at the State University of Iowa. If order of appearance indicated priority, then the lowest area was the social implications of the space program.[13]

These previous studies were more predictions, estimates, and recommendations than actual research.[14] The Brookings report called on NASA to establish an in-house capability of at least three senior social scientists. Their responsibilities would range from selecting research priorities and assessing ongoing projects to distributing the findings and assisting in their application at NASA. The report stated "one of the most pressing and continuing research challenges" would be to "develop effective methods to detect incipient implications of space activities and to insure that their consequences are understood."[15] What made the AAAS project different were its underlying goals and three publications. While the primary goal "briefly stated, is to examine the impact of space science and technology on American life," there was another motive:

Educators' Handbook, OMB/NASA Report Number S677. January 1983, *http://www1.jsc.nasa. gov/er/seh/social.html* (downloaded August 2, 2007).

11. W. Henry Lambright, *Powering Apollo. James E. Webb of NASA* (Baltimore, MD: Johns Hopkins University Press, 1995), pp. 99-100.

12. Donald N. Michael, "Proposed studies on the implications of peaceful space activities for human affairs" (Washington, DC: Brookings Institution, 1960), viii. Reprinted as a Report of the Committee on Science and Astronautics of the U.S. House of Representatives, 87th Congress, 1st Session, March 24, 1961. For Kranzberg's role, see Roger D. Launius, "NASA History and the Challenge of Keeping the Contemporary Past," *Public Historian* 21, 3 (Summer 1999), pp. 63-64.

13. "Some Social Implications of the Space Program," in National Academy of Sciences-National Research Council, *A Review of Space Research* (Washington, DC: National Academy of Sciences, 1962), 16-1-32.

14. Committee on Space Efforts and Society, "Space Efforts and Society: A Statement of Mission and Work," (Boston: AAAS, January 1963), reprinted in Raymond A. Bauer with Richard S. Rosenbloom and Laure Sharp and the assistance of others, *Second-Order Consequences. A Methodological Essay on the Impact of Technology* (Cambridge, MA: MIT Press, 1969), p. 211.

15. Michael, op. cit., pp. 3-4.

to encourage an enterprise of the size and importance of
NASA to incorporate in it a mechanism that would enable it
to guide its actions with respect to optimizing its second-order
social effects. To this end, our program has been designed to
demonstrate that effective and meaningful behavioral science
research could be done in this complex area.[16]

The chair of the AAAS committee was Earl P. Stevenson, the recently
retired president and chairman of Arthur D. Little, but he played only a nominal
role.[17] The real driving force was Raymond A. Bauer. Bauer (1916-1977),
described by the *New York Times* as "a pioneer in the application of behavioral
sciences," was a prolific and widely enquiring social psychologist at the Harvard
Graduate School of Business Administration who wrote and edited over 20
books, ranging from interviews with Soviet refugees in the 1950s to analyses of
advertising in the 1970s.[18]

Bauer's interest in the space program began with a 1960 survey of the
opinions of business executives about the space program and collaboration with
the Brookings report.[19] In a 1964 talk, his interest grew because

The point to be made is that the space program because of
its highly visible nature, and the developing concern for its
second-order consequences, has played a unique and valuable role that has turned our attention to problems we ought
to have been studying in any event. It seems to me highly
probable that just as the program of space exploration is the
leading edge of the advance of much of the new technology
(or at least serves as the symbol of this advance), in the same
way it may serve a very valuable catalytic function in getting
us to run our affairs better.[20]

---

16. "Report of the Committee on Space," *Records of the Academy (American Academy of Arts and Sciences)*, 1963/1964, 149, 155.

17. "Earl P. Stevenson, 84; Ex-Director and Head Of Arthur Little, Inc.," *New York Times*, July 5, 1978, B2.

18. "Raymond Bauer, 60; Business Professor Taught at Harvard," *New York Times*, July 11, 1977, 22; see also Ithiel de Sola Pool, "In Memoriam," *PS* 10, 4 (Autumn, 1977), 516-518, and Florence Bartoshesky, "Raymond A. Bauer: A list of his works," *Accounting, Organizations and Society* 6, 3 (1981), 263-270.

19. Raymond A. Bauer, "Executives Probe Space," *Harvard Business Review* 38 (Sep-Oct 1960), 6-14.

20. "Space Programs: The Joint Responsibility of Business and Government," 27, April 9, 1964. Box 5, file 42, Series II-B. HBS Research and Writing Records – Writings, 1941-1978. Raymond A. Bauer Papers, Harvard Business School Archives, Baker Library, Harvard Business School.

Raymond A. Bauer (Courtesy Harvard College Library)

In the summer of 1962, the committee held a summer conference to determine the most significant areas of study, which the AAAS and NASA then approved. Throughout the contract, the AAAS remained in "continuous conferring" with NASA.[21] In March 1963, the committee dissolved itself, and in April the AAAS Council created a smaller Committee on Space to supervise a study group that would conduct and organize the actual research.[22] The members were almost all from the Boston area with Harvard and MIT faculty monopolizing the committee and study group.

The committee discovered an unexpected challenge in convincing academics to conduct space-oriented research. Its 1962 request to sociologists for research proposals to "apply social science insight and imagination to the problem of massive technological innovation and the space program" received a poor response. Describing projects in language and concepts familiar to potential researchers seemed a necessary step.[23] The problem was not just faculty: four years later, Administrator Webb would turn "almost bitter about the response of the nation's university presidents" to NASA's Sustaining University Program, his effort to remake higher education into a more service-oriented, interdisciplinary enterprise.[24]

By 1965, the Committee considered its work "substantially completed." Another goal—stimulating research—had been accomplished with the Harvard Business School studying "technology transfer" and the National Planning Association developing "indicators of trends in social and political change" or "social indicators."[25] Publication, however, lagged.

Four volumes were planned; MIT Press published only three. The fourth, apparently a summary of committee activities, never appeared due to "insurmountable" problems of "choice and format."[26] The three published volumes were: 1) Bruce Mazlish, ed. *The Railroad and the Space Program. An*

---

21. "Introduction," in Raymond A. Bauer with Richard S. Rosenbloom and Laure Sharp and the assistance of others, *Second-Order Consequences. A Methodological Essay on the Impact of Technology* (Cambridge, MA: MIT Press, 1969), p. 9.

22. Earl P. Stevenson, "Report of the Committee on Space," *Records of the Academy (American Academy of Arts and Sciences)*, 1963/1964, 141-142.

23. "The Profession: Reports and Opinion," *American Sociological Review* 27, 4. (August 1962), 595; Committee on Space Efforts and Society, "Space Efforts and Society: A Statement of Mission and Work," (Boston, MA: AAAS, January 1963), reprinted in Bauer et al., op. cit., p. 212).

24. W. Henry Lambright, op. cit., pp. 136-139.

25. Earl P. Stevenson, "Report of the Committee on Space," *Records of the Academy (American Academy of Arts and Sciences)*, 1964/1965, 18.

26. Earl P. Stevenson, "Report of the Committee on Space," *Records of the Academy (American Academy of Arts and Sciences)*, 1965/1966, 22; Raymond A. Bauer, "Preface," in Raymond A. Bauer with Richard S. Rosenbloom and Laure Sharp and the assistance of others, *Second-Order Consequences. A Methodological Essay on the Impact of Technology* (Cambridge, MA: MIT Press, 1969), p. ix. This may explain the change of editor, too.

Exploration in Historical Analogy (1965); 2) Raymond A. Bauer, ed., *Social Indicators* (1966); and 3) Raymond A. Bauer with Richard S. Rosenbloom and Laure Sharp—and the Assistance of Others, *Second-Order Consequences. A Methodological Essay on the Impact of Technology* (1969).[27]

The Committee on Space divided its mandate into three categories: 1) Studies on the Anticipation of the Effects of Space Efforts; 2) Studies on the Detection of Such Effects; and 3) Studies on the Evaluation and Feedback of Information about Effects.[28]

Although its goal "should be to develop devices for anticipating, detecting, evaluating and acting (on)" the inevitable consequences of technical change, the committee claimed limits of time and money necessitated a focus on the first two kinds of devices.[29] Deciding that the development of the railroad provided the most fruitful historical approach to analyzing "what the social consequences of the space program *might* be," the committee commissioned eight papers under the guidance of Bruce Mazlish, a study group member, in anticipation that "In all of these studies an effort will be made to move from the impact of the railroad in the specific area under consideration to an analogy with the possible space impact today in similar areas."[30] In this, the Committee on Space would be disappointed.

STUDIES ON THE ANTICIPATION OF THE EFFECTS OF SPACE EFFORTS

For an effort whose first product was historical analogies, historians were curiously absent. The Committee on Space had two social psychologists and a political scientist, among others, but no historians. According to Bruce Mazlish, no historians attended the 1962 summer workshop, nor did the members of the Committee play an active role in choosing the historians or integrating their efforts into the larger project.[31] History seems to have been an afterthought that was added

---

27. This volume was originally intended to be edited by Robert N. Rapaport, and entitled *Social Change: Space Impact on Communities and Social Groups.* The different title may have been an attempt to appeal a larger audience (Raymond A. Bauer, op. cit., p. 19).

28. Earl P. Stevenson, "Report of the Committee on Space," *Records of the Academy (American Academy of Arts and Sciences),* 1963/1964, 150.

29. Committee on Space Efforts and Society, "Space Efforts and Society: A Statement of Mission and Work," (Boston: AAAS, January 1963), reprinted in Raymond A. Bauer with Richard S. Rosenbloom and Laure Sharp and the assistance of others, *Second-Order Consequences. A Methodological Essay on the Impact of Technology* (Cambridge: MIT Press, 1969), p. 193.

30. Earl P. Steveson, "Report of the Committee on Space," *Records of the Academy (American Academy of Arts and Sciences),* 1963/1964, 150-51. Emphasis in original.

31. Mazlish interview; July 11, 2007. Only recently did the AAAS hire its first archivist; consequently, its records are not accessible.

because of its potential value or possibly because the committee realized that the historians could provide a product for NASA faster than the social scientists.

Mazlish became involved because he was a Fellow at the AAAS, where he would eventually meet Bauer. A professor at MIT since 1955, Mazlish had an interest in methodological problems and helped found the journal *History and Theory* in 1960. Bauer asked Mazlish to develop the theme of historical analogies.[32] Mazlish found the authors, including some of the foremost historians of technology, business, and economics. In an interesting action—or inaction—the writers did not meet to discuss their work due to limited funding. [33]

The papers represented the usual range of collected works: some taken from previous writings, some transitional new work, and some interesting expostulations. However, the AAAS's hope for major analogous comparisons was not realized. The papers instead focused on the railroad with a few paragraphs at most about the space program bolted on the end, which, as reviewer Kenneth Boulding noted, "remind me, I am afraid irresistibly, of the libations to Marxism-Leninism which usually accompany quite sensible Russian works."[34]

The volume's greatest contribution is Mazlish's article on historical analogy, a piece that stands by itself as a major theoretical analysis of that widely used, easily abused, and poorly understood activity. According to Mazlish, analogies often evolve into myths, which not only provide "needed emotional continuity and support, but pass readily into models" that can mislead as easily as lead. A possibly insurmountable problem was "historically conditioned awareness." How could researchers base analogies on events that occurred only once (like the 17th century discovery of microscopic life) and changed perceptions forever (like the railroad altering people's concepts of time and space in a way that reduced the novelty of future advances)?[35]

Faced with these challenges, Mazlish stated, "I am tempted to state categorically that, *for purposes of scientific knowledge*, only a historical analogy that 1) allows for progressive trends, and 2) rises above the comparison or resemblance of two simple elements can be of any real value." More realistically, the best research should treat the space program, like the railroad as "a complex *social invention*" in a specific (and evolving) environment.[36] Any serious historical analogy had to be based on detailed, informed empirical studies; focus on the complex relationships within the larger system, and not simply comparing two

---

32. Ibid. No AAAS records have yet been found of the workshop.

33. Ibid.

34. Kenneth E. Boulding, "Space, Technology, and Society: From Puff-Puff to Whoosh," *Science*, (February 25, 1966): 979.

35. Mazlish, "Historical Analogy," pp. 9–10.

36. Ibid, p. 11. Emphasis in original.

isolated elements; and, use as large a "fair sampling" as possible to ensure study of the right elements.[37]

Mazlish concluded with five generalizations:

1. Beware simplistic conclusions. "All social inventions are part and parcel of a complex—and have complex results. Thus, they must be studied in multivariate fashion."

2. There are usually alternate technological approaches to attain economic goals.

3. "All social inventions will aid some areas and developments, but will blight others."

4. "All social inventions develop in stages and have different effects during different parts of their development."

5. "All social inventions take place within a national 'style,' which strongly affects both their emergence and their impact."[38]

These generalizations have held up well, though often are not heeded. Simplistic comparisons abound today, especially in the political arena. Particularly neglected in both historical and contemporary analyses is "asset and liability bookkeeping," including paths not taken. Economic historians have proved best at constructing such alternative realities.

Building on an excellent overview of the early decades of American railroad technology, Thomas P. Hughes provided more comparative analysis than the other papers, including a compelling definition that encompassed the railroad, space program, and many other areas: "Wherever and whenever nature in her nonanimal manifestations frustrates man in the pursuit of his objectives, there exists a technological frontier."[39]

Space exploration surely satisfies the "most extreme result of technological frontier penetration is the creation of a man-made environment and the rendering of nature imperceptible."[40] But nature in the form of a hostile environment is not imperceptible; rather, it is held at bay to the point that robotic probes can be sent on decades-long missions.

Hughes noted that one challenge of engineers is to compromise *economically* with nature, to solve problems in ways that are technologically but also financially feasible.[41] The importance of economics in shaping the trajectory

---

37. Ibid, pp. 18–20. The preferred word for multi-causal, complex explanations was "multivariate," showing historians can be as trendy as any other group.

38. Ibid, pp. 34–35.

39. Thomas P. Hughes, "A Technological Frontier: The Railway," in Bruce Mazlish, ed. *The Railroad and the Space Program. An Exploration in Historical Analogy* (Cambridge, MA: MIT Press, 1965), p. 53.

40. Ibid.

41. Ibid, p. 55.

of space exploration and exploitation remains a significant, underappreciated topic. The political economy of space remains based on the fact that the high cost of reaching Earth orbit and working in space continues to limit the players in space to those who have deep pockets—primarily national governments and large corporations (themselves often dependent upon government orders).

An aspect of the political economy of railroads Mazlish noted was that over 120 British Members of Parliament served on railroad boards in 1872. Any study of the American political economy of space today would have to include fundraising and other favors for the senators and representatives on the congressional committees overseeing NASA, the military, and, as former Representative Randy Cunningham demonstrated, the intelligence community.[42]

For Hughes, historical awareness can sensitize the observer to future probabilities and suggest questions. Perhaps most importantly, what fields will languish as a result of resources expended on space? Will the institutionalization and reification of this knowledge create a momentum that will transfer into other areas? What style will characterize engineers and scientists who have learned to operate in space?[43] Four decades of experience should enable us to now answer these questions.

Economic historians Robert Fogel and Paul Cootner emphasized the need to compare the costs of alternate approaches to accomplish similar work and the fact that the full impact of the railroad took decades to emerge.[44] Drawing on his 1964 *Railroads and American Economic Growth*, Fogel considered the main question from an investment perspective: "Will the increase in national income made possible by the space program exceed the increase in income that would be obtained if the same resources were invested in other activities?"[45]

Viewing the railroad's main effects as reducing transportation costs of processes and activities already underway, Fogel suggested that the space program would not revolutionize transport, generate transcendent inventions, or expand access to knowledge. Instead, he postulated that the space program's most radical and important contributions may come from the knowledge gained from exploration, exploration impossible without access to space. Unlike the railroad, where transportation alternatives existed, rockets provided entrance

---

42. Mazlish, "Historical Analogy," p. 31.

43. Hughes, op. cit., p. 72.

44. Paul H. Cootner, "The Economic Impact of the Railroad Innovation," in Bruce Mazlish, ed. *The Railroad and the Space Program. An Exploration in Historical Analogy* (Cambridge, MA: MIT Press, 1965), pp. 112, 118.

45. Robert William Fogel, "Railroads as an Analogy to the Space Effort: Some Economic Aspects," in Bruce Mazlish, ed. *The Railroad and the Space Program. An Exploration in Historical Analogy* (Cambridge, MA: MIT Press, 1965), p. 74. Fogel's book was both groundbreaking, leading to a Nobel prize in economics for Fogel in 1993, and a counterargument to Walt W. Rostow's influential concept of stages of economic takeoff, *The Stages of Economic Growth: A Non-Communist Manifesto* (Cambridge, MA: Cambridge University Press, 1960).

to a world hitherto unavailable.[46] This was the most succinct, accurate, and ignored prediction made in the book.

That prediction more than compensated for another prediction. In an excellent example of extrapolating from expectations, Fogel assumed the forth-coming arrival of the supersonic transport would negate any advantage of the rocket for point-to-point transportation on Earth as the maximum time saved by rocket would be five hours.[47]

The Committee on Space had noted "one of the most widely discussed second-order consequences of the space program is the diffusion of space-generated technology into the civilian economy," a consequence space supporters promoted optimistically.[48] Significantly, Fogel decisively dismissed what would be called spinoff in the case of the railroad. NASA ignored Fogel's unwanted conclusion: Tracking and promoting technology transfer absorbed approximately half of the $35 million NASA spent on impact studies in its first decade. Indeed, NASA has long proclaimed and promoted the value of spinoffs.[49]

In a stepping stone to his magisterial *The Visible Hand*, Alfred Chandler, together with Stephen Salsbury, offered very general hypotheses about innovative inventions encouraging new methods of management and administration, needs that emerge as the invention evolves instead of being immediately obvious. Often, operational crises—usually in the form of deadly visible disasters—produce the political attention (including from the press, public, and politicians) needed to introduce large and complex organizations to manage these large and complex technologies.

As the history of space programs amply illustrate, management has been as challenging as the actual technologies with visible disasters often producing major administrative changes. The continuing focus on management indicates the space program is still a major work in progress.

From a NASA perspective, Thomas Cochran wrote the most disappointing paper, not even adding a speculative paragraph at the end. From a railroad perspective, however, Cochran served up a stimulating view of the demographic, institutional, and social-psychological impacts of the railroad.

In an article that local and state governments seeking to attract businesses should ponder, Robert Brandfon examined what happened when a powerful railroad monopoly, the Illinois Central, entered a poor state, Mississippi, with goals quite different than those held by politicians and citizens. For Brandfon, the key analogy was with NASA's then new Mississippi Test Facility (now the

---

46. Fogel, p. 106.

47. Ibid, p. 104.

48. Earl P. Stevenson, "Report of the Committee on Space," *Records of the Academy (American Academy of Arts and Sciences)*, 1963/1964, 153.

49. See for example Marjolijn Biejlefeld and Robert Burke, *It Came From Outer Space. Everyday Products and Ideas from the Space Program* (Westport, CT: Greenwood Press, 2003).

Stennis Space Center). Would NASA act as a colonialist or contributor to the state? How would it handle race relations, an explosive issue in the mid-1960s? Would NASA improve education so locals could be hired, or would NASA import the skilled workforce from outside the state?[50]

Based on his significant 1964 *The Machine in the Garden*, Leo Marx examined why so little was known about the impact of technological progress upon the collective consciousness. Commenting on Marx, Mazlish noted, "in some ways the most difficult to trace and establish, the railroad's impact on the imagination seems almost to be the most fundamental." Just as the iron horse altered conceptions of the pastoral landscape, "one of the most significant impacts" of the space program could be new perceptions of Earth and space.[51] The rise of the environmental movement has affirmed this impact.

Reception to *Railroad* was positive. Academic book reviews admired this "thought-provoking and intriguing book," though some considered the analogy "tremulous."[52] For Bauer and the Committee on Space, the value of *Railroad* was demonstrating that, after a century of writing, the scope and nature of the technology-society relationship had not been fully evaluated and that causation and change were more complex than assumed. By implication if not analogy, the space program would prove equally academically challenging.[53]

Unfortunately, the book's impact was restricted. In a serious blow to its diffusion, MIT Press never issued a paperback version, the fate of many collected works. Consequently, the influence of *Railroad*, especially on graduate students, remained limited.

A 1979 NASA-sponsored study on the space program from the perspectives of the social sciences and humanities placed *Railroad* under the category of "Impact Analysis," which was "an intellectual invention of the late 1960s and early 1970s and evolved as part of the burgeoning academic study of technology in its social context."[54] The main directions of the social study of technology as well as the AAAS project, however, moved away from *Railroad*.

---

50. Robert L. Brandfon, "Political Impact: A Case Study of a Railroad Monopoly in Mississippi," in Bruce Mazlish, ed. *The Railroad and the Space Program. An Exploration in Historical Analogy* (Cambridge, MA: MIT Press, 1965), p. 200.

51. Mazlish, "Historical Analogy," pp. 33, 41.

52. John F. Stover, "The Railroad and the Space Program," *American Historical Review* 72, 1 (October 1966) 280–281; Julius Rubin, "The Railroad and the Space Program," *Business History Review* 41, 3 (Autumn 1967): 334.

53. Raymond A. Bauer, "Detection and Anticipation of Impact: The Nature of the Task," in Raymond A. Bauer, ed., *Social Indicators* (Cambridge, MA: MIT Press, 1966), p. 20.

54. T. Stephen Cheston, "Space Social Science: Suggested Paths to an Emerging Discipline," *Space Humanization Series* 1 (1979): 1.

STUDIES ON THE DETECTION OF SUCH EFFECTS

The second research area focused on three questions:

1. Can effects which have been guessed at or discerned be measured with accuracy?

2. Can procedures be devised for locating effects which have not been thought of?

3. Is it possible to segregate effects of the space program from study of the effects of other factors in our society?[55]

Under the direction of Robert N. Rapoport, an anthropologist and sociologist at Northwestern University, the Committee on Space commissioned papers in 1964 to look at the impact of NASA installations on local communities; of NASA on functional groups like businessmen, students, and engineers; of NASA needs on education and labor; and of the process of technology utilization.

Second-Order Consequences appeared in 1969. Congressman Emilio Q. Daddario (D–CT) introduced the studies as "an important initial contribution to the development of technology assessment" and predicted analyzing secondary consequences would "become an integral part of the research-development-application sequence."[56]

Unlike the other books, *Second-Order Consequences* received poor reviews and vanished into obscurity. The criticism addressed "simply trite and fragmentary" findings, "the unsystematic attack on substantive phenomena, and the lack of a broad theoretical orientation," but also reflected the more skeptical academic and political environment of the late 1960s. Had the researchers been captured by their client, producing supportive reports that did not question NASA goals or costs? Why were the results so passive instead of identifying "the need for action"?[57]

---

55. Earl P. Stevenson, "Report of the Committee on Space," *Records of the Academy (American Academy of Arts and Sciences)*, 1963/1964, 151-152.

56. Emilio Q. Daddario "Foreward," in Raymond A. Bauer with Richard S. Rosenbloom and Laure Sharp and the assistance of others, *Second-Order Consequences. A Methodological Essay on the Impact of Technology* (Cambridge MA: MIT Press, 1969), p. vi. Daddario's interest was more than perfunctory: he later served as director of the congressional Office of Technology Assessment and president of the American Association for the Advancement of Science.

57. Ilkka Heiskanen, "Second Order Consequences," *Administrative Science Quarterly*, 16, 2. (June 1971): 232; see also, William D. Nordhaus, "Economics of Technological Change," *Journal of Economic Literature* 8, 3 (September 1970), 864-867.

STUDIES ON THE EVALUATION AND FEEDBACK OF
INFORMATION ABOUT EFFECTS

The third area, social trends, was the heart of the AAAS project. Social indicators was such a new area of study that it suffered "not only a general lack of consensus as to what should be measured, but also disagreement on goals, purposes, and the nature of our society." The Committee on Space sought "to see if it could raise the quality of such evaluations by examining carefully the bases for making such evaluations—the social indicators used in measuring trends."[58] Bauer and his colleagues were among the leaders in recognizing the importance and potential of, as a 1962 President's Science Advisory Committee stated, systematically collecting behavioral data and providing advice to the government.[59]

Social Indicators was written not just to determine how to measure specific impacts of the space program but to propose a total information system that would provide "*the earliest possible detection or anticipation of impacts that bear on the primary mission*" of "NASA or some similar institution."[60] Contributor Bertram M. Gross claimed the book was "the first occasion on which the entire field has been surveyed and a comprehensive set of proposals, based upon careful analysis, has been developed."[61]

Appearing in 1966 to favorable reviews, *Social Indicators* was the most influential of the three volumes. Political scientist Ithiel de Sola Pool, who worked with Bauer and later organized a retrospective technology analysis of the telephone, stated that the AAAS project pushed the idea of social indicators into "the mainstream of American social thought."[62] Perhaps a more objective indicator of the book's value is the fact that, four decades after its appearance, MIT Press still sells *Social Indicators*. The concept has taken root and flourished: a search of Google Scholar for "social indicators" returns roughly 36,000 hits compared with 47,000 for "economic indicators."[63] Several internationally prominent composite indicators, such as Transparency International, are as much social as economic.

---

58. Earl P. Stevenson, "Report of the Committee on Space," *Records of the Academy (American Academy of Arts and Sciences)*, 1963/1964, 154–155.

59. Life Sciences Panel of the President's Science Advisory Committee, *Strengthening the behavioral sciences; statement by the Behavioral Sciences Subpanel* (Washington, DC: The White House, 1962), pp. 13–19.

60. Raymond A. Bauer, "Detection and Anticipation of Impact: The Nature of the Task," op. cit., pp. 10–11, 63. Emphasis in original.

61. Bertram M. Gross, in Bauer, *Social Indicators*, op. cit., "Preface," p. xv.

62. Ithiel de Sola Pool, "In Memoriam," *PS* 10, 4 (Autumn, 1977): 517. See also Ithiel de Sola Pool, ed., *The Social Impact of the Telephone* (Cambridge, MA: MIT Press, 1977).

63. Accessed at *http://scholar.google.com/scholar?hl=en&lr=&q=%22economic+indicators%22&btnG= Search* (downloaded August 7, 2007).

Less successful were efforts to employ the concept bureaucratically. Bauer and others, including Minnesota Senator Fritz Mondale, employed *Social Indicators* to promote action by the federal government, including the establishment of a Council of Social Advisors similar to the Council of Economic Advisors and more advice to Congress.[64] While such a council did not appear, a bipartisan Congress established the Office of Technology Assessment (OTA) in 1972. OTA lasted until 1995, when terminated by the new Republican Congress.

ANALYSIS

In his introduction, Mazlish outlined five issues to address:

1. What were the theoretical problems of historical analogy?

2. What was the impact of the railroad on 19th century America?

3. Could the railroad's impact be used as a "device of anticipation" to study the impact of the space program?

4. Could this AAAS effort possibly become the prototype of future "impact" studies?

5. Could this volume serve as an example of the difficulties involved in organizing such a project?[65]

As he noted, this effort was an initial exploration, designed to probe possibilities, not prove. The volume indeed provided a much richer appreciation of the theoretical challenges of creating and using historical analogy as well as the many impacts of the railroad on 19th century America. The grander goals and visions, however, remained unfulfilled. The first two issues were the province of the historian and the most successfully developed. The last three fell into the province of NASA and the AAAS as well as the historian, and they must be answered either negatively or, to use the Scottish legal concept, not proven.

What happened to the last three goals of the AAAS and *Railroad*? Or, more accurately, what did not happen? Was the problem a lack or loss of AAAS and NASA support, a lack of effort to link historians with social scientists and NASA policy-makers, or a more fundamental mismatch between historians and policymakers?

That is, were the grand AAAS expectations killed by factors beyond their control, executed poorly and thus unsuccessfully, or doomed from the

---

64. Talcott Parsons, "Report of the President," *Records of the Academy (American Academy of Arts and Sciences)*, (1968 - 1969), 11; Otis Dudley Duncan, "Developing Social Indicators," *Proceedings of the National Academy of Sciences* 12 (December 1974), 5096-5102; Elmer B. Staats, "Social Indicators and Congressional Needs for Information," *Annals of the American Academy of Political and Social Science* 435 (January 1978): 277-285.

65. Mazlish, op.cit., "Preface," pp. vii-xi.

beginning by the inherent inability of historians and social scientists to create information and package knowledge in a form useful to policymakers?

Did the Committee on Space ever ask itself or NASA, "What sort of product would be most useful to NASA policymakers?" *Railroad*, from a practical or theoretical policy perspective, was useless. There were no conclusions, no lessons learned, no set of bulleted issues to serve as guide points, or any other packaging of information in a useful form.

Similarly surprising was what else did not occur. The AAAS did not convene a conference of historians, social scientists, and NASA policymakers to discuss the book. Indeed, the contributors to *Railroad* never met or coordinated their efforts.[66] If initiated today, at a minimum, the contributors would hold a workshop to discuss the topic and their plans. After receiving the papers, the planners would then convene a conference with the intended audience, NASA managers, and policymakers. This process of consultation and feedback would ensure greater focus, feedback, and relevance.

To the Committee on Space, the railroad appeared the logical subject to study. As the Fogel essay suggests, would studying other historical analogies, such as exploring and colonizing hostile environments such as the oceans or Arctic have proved more fruitful?[67] Would studying frontiers—real and imagined—have provided insights valuable to NASA?[68]

Did NASA ask, "What can we learn from history, and what is the best way for historians and social scientists to work together with managers and engineers?" NASA, along with the Department of Defense, is among one of the major government Agencies that uses its history. The History Office at NASA, established in 1959, not only creates and contracts histories but also serves as a source of information for NASA as well as business and the public.[69]

Has NASA learned? In one sense, no. To take a recent example, the 2004 Administrator's Symposium focused on risk and exploration. In addition to administrators, astronauts, and scientists, the speakers included explorers of the earth and sea—but not one historian or social scientist, even though the NASA History Office provided significant support.[70]

---

66. Robert Brandfon, interview, January 25, 2008.

67. I am grateful to Peter Stearns for raising this point. See Albert A. Harrison, Yvonne A. Clearwater, and Christopher P. McKay, eds., *From Antarctica to Outer Space: Life in Isolation and Confinement* (New York, NY: Springer-Verlag, 1991) and Jack Stuster, *Bold Endeavors: Lessons from Polar and Space Exploration* (Annapolis, MD: Naval Institute Press, 1996).

68. David F. Noble, *The Religion of Technology. The Divinity of Man and the Spirit of Invention* (New York, NY: Alfred A. Knopf, 1997), 115-141; Howard E. McCurdy, *Space and the American Imagination* (Washington, DC: Smithsonian Institution Press, 1997); Carl Abbott, *Frontiers Past and Future. Science Fiction and the American West* (Lawrence, KS: University Press of Kansas, 2006)

69. Roger D. Launius, "NASA History and the Challenge of Keeping the Contemporary Past," *Public Historian* 21, 3 (Summer 1999): 63-81.

70. Steven J. Dick and Keith L. Cowing, eds., *Risk and Exploration. Earth, Sea and the Stars.* NASA

In another sense, NASA has learned, but two highly visible disasters were needed in order for the Agency to do so. One result of the 1986 *Challenger* explosion was the superb investigation by sociologist Diane Vaughan, who pinpointed the sociocultural factors that contributed to the Shuttle's loss.[71] Seventeen years later, when *Columbia* disintegrated on reentry, the investigation board had John Logsdon as a member and Dwayne Day as an investigator, while consulting with Henry Lambright, Roger Launius, and Howard McCurdy—all outstanding scholars of the nation's space efforts.

In addition to preserving and studying the past, the NASA History Office has tried to be accessible to policymakers and managers and to produce products aimed at them. While most of these efforts are reactive (e.g., responding to questions and requests), some are proactive or more than the mere delivery of information. The History Office has held annual conferences, of which this is the third, addressing large themes and trying to reach larger audiences beyond academia.

Attention to history has informed some recent and current policy. The developers of President George W. Bush's Vision for Space Exploration sought historical analysis (and analogies) of the ill-fated Space Exploration Initiative on which to base their work.[72] In a 2007 presentation on systems of lunar governance, William S. Marshall of the Ames Research Center suggested including "the use of historical checks to prevent society from repeating its mistakes."[73]

Judged by its ambitious objectives, the AAAS project failed. Institutionally, NASA has no Office of Impact staffed by social scientists and historians earnestly working away to chart and guide the secondary consequences of space exploration and exploitation. Predicting and shaping first-order—let alone second-order—consequences has proven far more challenging than Bauer and his colleagues anticipated, reflecting the problems of applying systems management to that unruly aggregate we call society.

Viewed by discipline, historians and social scientists continue to communicate poorly with policymakers and the public, since most neither know how or care to write or "package" (to use a more jarring but useful word) relevant history for policymakers. Institutional mechanisms for encouraging such efforts are greatly lacking, and I suspect many historians would flee if offered the chance to contribute to the shaping of policy.

Administrator's Symposium. September 26-29, 2004. Naval Postgraduate School. Monterey, California (Washington, DC: NASA, 2005).

71. Diane Vaughan, *The Challenger Launch Decision. Risky Technology, Culture, and Deviance at NASA* (Chicago: University of Chicago Press, 1996).

72. Thor Hogan, "The Space Exploration Initiative: Historical Background and Lessons Learned," Rand PM-1594-0STP (Santa Monica, CA: RAND, September 2003). This was part of Hogan's larger *Mars Wars. The Rise and Fall of the Space Exploration Initiative* (Washington, DC: NASA, August 2007).

73. Padma Tata, "Jury duty on the Moon? *http://www.newscientist.com/blog/space/* October 3, 2007 (downloaded October 8, 2007).

Nor have historians embarked on many future impact studies. Indeed, rarely do historians work in groups or with other disciplines.[74] When they do, which requires finding funding on a much larger scale than they are accustomed, the best analogy may be that of herding cats. A notable exception is the Tensions of Europe network and research collaboration funded by the European Science Foundation to encourage cooperation among European academics.[75]

Yet history and historical analogies are powerful tools, especially when used well.[76] Historical understanding, analogy, and questioning can be employed profitably and wisely.[77] It behooves historians and social scientists to try to accomplish this because we know that if we don't, history will be misused to influence and justify policy. Look at the widely used example of Munich: appeasement is bad, an argument used by supporters of the Vietnam War in the 1960s and the second Iraq war in the 2000s. As Peter Stearns noted in 1981, Munich in 1938 was not Vietnam in 1968. The same is true for Iraq in 2007.[78]

Good history, good analogies, and good guidance are necessary, but they are not enough. What also must be considered is if anyone is listening, not just in NASA but also in the legislative branch and wider public. Organizations like History News Service (*http://www.h-net.org/~hns/index.htm*) and History News Network (*http://hnn.us/*) provide historians with a public forum to address contemporary issues within a historical context. The problem of audience, unfortunately, is not confined to historians and social scientists.[79] We should think more about our responsibilities as public intellectuals and act accordingly.

We expect leaders and administrators to make errors. We want them, however, to make smart rather than dumb ones. Good history—accurate and aimed at policymakers—can and should help them to avoid dumb errors.

Academics tend to end papers with calls for further research. I shall continue this tradition with two recommendations. First, the history profession and NASA should examine the Department of Defense history programs and the field of military history to learn what the military and military historians are doing right

---

74. The situation since 1981 has not changed significantly (Peter N. Stearns, "Applied History and Social History," *Journal of Social History* 14, 4 (Summer, 1981): 533-537.

75. Accessed at *http://www.histech.nl/tensions/* (downloaded August 7, 2007).

76. For a fascinating study of how physicists used analogy, see Daniel Kennefick, *Traveling at the Speed of Thought. Einstein and the Quest for Gravitational Waves* (Princeton, NJ: Princeton University Press, 2007).

77. Richard E. Neustadt and Ernest R. May, *Thinking in Time: The Uses of History for Decision Makers* (New York, NY: Free Press, 1986).

78. Peter N. Stearns, "Applied History and Social History," *Journal of Social History* 14, 4 (Summer, 1981): 533. A less used but equally important lesson of Munich is that all the major players should be at the negotiating table. The inclusion of up-and-coming as well as established spacefaring nations in discussions about coordinating future Moon exploration is a good sign that that lesson has been learned.

79. Barbara Kline Pope, "Because Science Matters," *Science.* (June 1, 2007): 1286.

and wrong.[80] Second, over two decades have passed since Richard E. Neustadt and Ernest R. May published their important *Thinking in Time: The Uses of History for Decision Makers*. It is time to update that classic with lessons for the 21st century.

Let me end by returning to Mazlish's statement that "In short, we are really attempting to set up a new branch of comparative history: the study of comparative or analogous social inventions and their impact on society." Judged by this goal, did he succeed? After all, there is no school of history analogy. But perhaps his words should be thought of as another way of describing the history of technology and of urging historians to expand their theoretical tool chests.

---

80. For a sense of the extensive military programs and their challenges, see Pat Harahan and Jim Davis, "Historians and the American Military: Past Experiences and Future Expectations," *Public Historian* 5, 3 (Summer 1983): 55-64; Richard H. Kohn, "The Practice of Military History in the U.S. Government: The Department of Defense," *Journal of Military History* 61, 1 (January 1997): 121-147. For specific applications of history, see Andrew J. Bacevich, Preserving the well-bred horse," *The National Interest* (September 22, 1994): 43-49; Conrad C. Crane, *Avoiding Vietnam: The U.S. Army's Response to Defeat in Southeast Asia* (Carlisle, PA: Army War College, 2002); Brian McAllister Linn, *The Echo of Battle. The Army's Way of War* (Cambridge, MA: Harvard University Press, 2007).

Part II.

Remembrance and Cultural Representation of the Space Age

~

CHAPTER 10

FAR OUT:
THE SPACE AGE IN AMERICAN CULTURE

Emily S. Rosenberg

Space has long provided a canvas for the imagination. For me, the early
Space Age intertwined with a sense of youth's almost limitless possibilities—
the excitement of discovery, the allure of adventure, the challenge of competition,
the confidence of mastery. As a girl in Montana, I looked up into that Big Sky
hoping to glimpse a future that would, somehow, allow my escape from the
claustrophobia of small towns separated by long distances.

But the Space Age was also bound up with the encroaching cynicism of
my young adulthood: the fear of a future driven by thoughtless fascination with
technique and a Vietnam-era disillusionment with the country's benevolence
and with the credibility of its leaders. The night that the first American landed
on the Moon, I was in the audience at the Newport Folk Festival. Someone from
the audience yelled "What were the first words on the Moon?" The announcer
replied, "They were: 'The simulation was better!'" A cluster of people grumbled
that the Moonwalk was probably faked, a suspicion that my barely literate
immigrant grandmother—and a few others in the country—shared.

The new Space Age could promise giant leaps and also threaten Hal of
2001: A Space Odyssey. Space could be far away or "far out."

Anyone who has been around for the past half century harbors private
memories of the early Space Age. A toy, a TV program, a book, a painting,
a school science fair project can each touch off remembrance of a place, an
emotion, the person we once were. For each individual, the Space Age offered
an array of visual representations and symbolic threads that could, intimately
and personally, weave a unique tapestry.

But the Space Age was not simply an infinitely personalizable canvas for
individual memories. It also offered national and global imaginaries that projected
assumptions about, and debates over, national identities and global futures.

The Space Age, of course, is in one sense as old as historical time—
humans have long looked to the heavens for meaning. And it is also an age
still of the present as the current schemes to militarize space and the renewed
public visibility of public and private missions into space remind us. But this
essay addresses that shorter moment of the Space Age, the couple of decades
beginning in the early 1950s when transcending Earth's atmosphere and

gravitational pull so stirred emotions that space exploration became an intense cultural preoccupation.

Focusing on representations that comprise collective, not individual memory, this essay seeks to suggest some of the diverse symbols and narratives of the Space Age as they circulated in American culture. As a complex of collective signs and symbols, the Space Age intertwined with other rival designations for the postwar era: the Cold War, the Media Age, what Zbigniew Brzezinski called the Technetronic Age, and the Age of a Mid-century Modernist aesthetic. Space exploration augmented the Cold War with the space race, enhanced the Media Age with truly amazing dramas and visual spectacularity, heightened the Technetronic Age's moral and philosophical concerns over the implications of Technocracy and a so-called "Spaceship Earth," and inspired Mid-century Modernist impulses that emerged as Googie and abstract expressionism. Refracting aspirations and fears, the Space Age held multiple meanings for foreign policy, politics, media, engineering, morality, art, and design.[1]

1. THE COLD WAR: SPACE RACE

In October 1957, Sputnik I became a media sensation. Hurled into orbit by a massive rocket, the Soviet-launched space satellite, circling Earth every 95 minutes, appeared to demonstrate urgent strategic dangers. This "Sputnik moment," in which fear mingled with fascination, prompted significant changes in America's Cold War landscape. It by no means, however, began America's fascination with a new Space Age.

A vibrant spaceflight movement comprised largely of science fiction writers and engineers had preceded Sputnik and helped set a tone for the space race that emerged in Sputnik's wake. A team of mostly German rocket-scientists headed by Wernher von Braun had worked for the U.S. Army since the summer of 1950 under order to develop a ballistic missile capable of delivering a nuclear weapon.[2] On the side, von Braun had energetically promoted popular interest in spaceflight, and his efforts during the mid-1950s became part of a boom in both science and science fiction writing about space. A group that the scholar De Witt Douglas Kilgore has called "astrofuturists"—writers who based their tales of an intergallactical future on new scientific breakthroughs in physics—included Isaac Asimov, Robert Heinlein, Arthur C. Clarke, Willy Ley, and others.[3]

---

1. The author wishes to express special thanks to Norman L. Rosenberg for his contributions to this essay.

2. Tom D. Crouch, *Aiming for the Stars: The Dreamers and Doers of the Space Age* (Washington, DC: Smithsonian Institution Press, 1999), p. 118.

3. De Witt Douglas Kilgore, *Astrofuturism: Science, Race, and Visions of Utopia in Space* (Philadelphia, PA: University of Pennsylvania Press, 2003) examines the major scientific and literary productions.

These astrofuturists offered especially powerful images and narratives about a new "age of discovery" in which brave individuals would guide interplanetary explorations. Walt Disney employed von Braun and Ley, both powerful advocates of human piloted spaceflight, as consultants to help design rocket ships and Moon rides for Disneyland's Tomorrowland, which opened in 1955, and a series of TV episodes such as "Man is Space" (March 1955), "Man and the Moon" (December 1955), and "Mars and Beyond" (December 1957). Chesley Bonestell carved out a specialty as a spaceflight artist, illustrating in colored ink during the 1950s much of the equipment and procedure that later NASA scientists would construct for real. Bonestell's collaboration with Ley in *The Conquest of Space*, for example, exuded technological authority in both words and illustration, moving the subject of space travel away from the interwar Flash Gordon style and into scientific respectability.[4] Likewise, comics and popular magazines frequently featured human-piloted space travel, and Hollywood also filled screens with visions of space. *Destination Moon* (1950), a film whose images and messages influenced a generation of movie makers as well as scientists, celebrated the idea of a Moon landing.[5] In the realm of popular music, songwriter Bart Howard's *Fly Me to the Moon* (1951) became such a hit, especially after Peggy Lee sang it on the *Ed Sullivan Show* in the mid-1950s, that Howard was able to live out his life on its royalties.

Fiction writers and rocket scientists such as von Braun, in elaborating their dreams of manned flight and space stations, implied that control of the Moon and of outer space by any other nation would leave the United States abjectly defenseless. Hollywood's *Destination Moon* had especially contributed to this idea. In addition, the well-developed popular fears associated with atomic power led credence to the idea that an enemy's penetration of space might pose an existential threat. Might the rockets that launched Sputnik indicate that the Soviet Union's intercontinental ballistic missiles (ICBM) had the power to send a nuclear weapon to the United States? Might Sputnik signal the enemy's capability of mounting a Pearl Harbor-style attack from the skies, this time with atomic bombs coming from orbiting satellites?

Many scholars have argued that the ideas and literary productions of the astrofuturists "prepared the American public for the conquest of space with elaborate visions of promise and fear" and helped shape the nation's cultural and political responses.[6] As Sputnik orbited overhead, these space-exploration

---

4. Kilgore, *Astrofuturism*, pp. 72-74; Willie Ley, *The Conquest of Space* (New York, NY: Viking, 1951).

5. Kilgore, *Astrofuturism*, pp. 52, 56-58; Howard E. McCurdy, *Space and the American Imagination* (Washington, DC: Smithsonian Institution Press, 1997), pp. 41-43.

6. Crouch, *Aiming for the Stars*, pp. 118-121; Kilgore, *Astrofuturism*, pp. 31-48; McCurdy, *Space*, pp. 54-74 [quote p. 54]. Roger E. Bilstein, *Flight in America: From the Wrights to the Astronauts* (Baltimore, MD: Johns Hopkins University Press, 1984) traces the development of interest in early aerospace flights.

boosters, who had long advocated more energetic efforts, fused their previous visions of human-piloted voyages of discovery together with the heightened Cold War national security concerns to frame the parameters of an urgent new international competition—the space race.

President Dwight David Eisenhower tried to calm the alarm. His scientific experts saw no ICBM gap or even any parity in missile know-how between the United States and the Soviet Union. Had the White House pushed a program similar to that which produced Sputnik, they advised, a U.S. satellite could already have been aloft. While von Braun pressed for a crash program, promised that his team could launch a satellite in 90 days, and called for building a space station, Eisenhower embraced a measured approach with lower costs and greater focus on scientific and military applications. The chair of Eisenhower's science advisory committee, James R. Killian, issued a short *Introduction to Outer Space* that downplayed manned flight and advocated carefully constructed scientific projects that employed automation and robotics. Eisenhower ordered the government printing office to distribute Killian's pamphlet to the public for 15 cents a copy.[7]

As a seasoned military strategist, the President had always been his own most-trusted national security adviser. By 1957, Eisenhower believed he could see Soviet capabilities and likely military intentions more clearly than ever before. The public did not know that he recently had gained access to reconnaissance photographs taken by cameras carried on the newly operational U-2 spy plane. U-2 flights over the Soviet Union, begun during the summer of 1956, secretly confirmed the President's judgment that military necessity required no sudden change in strategic course. The U.S.S.R. had not raced ahead in military might. Moreover, a U.S. satellite-based surveillance system designed to replace the U-2 flights already had Ike's full support. (Satellite-based cameras would take their first pictures of the Soviet Union several months before Ike left office in 1961.) As a general, Eisenhower understood the value of aerial reconnaissance, and his backing of scientific satellites before 1957 had aimed to establish the precedent of free access in space—a principle that could then be adapted to the advantage of military intelligence. Sputnik, ironically, established this precedent, and Eisenhower thus saw advantages to Sputnik that military secrecy kept shrouded from the public.[8]

7. McCurdy, *Space,* pp.56-58; Crouch, *Aiming for the* Stars, pp.143-150. Matthew A. Bille and Erika R. Lishock, *The First Space Race: Launching the World's First Satellites* (College Station, TX: Texas A&M Press, 2004) provides a history of satellite development before 1958.

8. McCurdy, *Space,* pp. 58-59; Robert A. Divine, *The Sputnik Challenge* (New York, NY: Oxford University Press, 1993), pp. 11-12; On the background to and aftermath of Sputnik, see especially Walter A. McDougall, . . . *the Heavens and the Earth: A Political History of the Space Age* (New York, NY: Basic Books, 1985), and Paul Dickson, *Sputnik: The Shock of the Century* (New York, NY: Walker and Company, 2007).

The more Eisenhower tried to reassure the nation about the implications of Sputnik, however, the more his critics could portray him as inept and out of touch with Cold War dangers.[9] Ike's popularity declined as an avalanche of scientific reports, newspaper editorials, and political speeches warned that the United States was losing its military lead because of Moscow's presumed technical superiority. The Democrats especially smelled blood in the water, and most Republican politicians joined in the alarm over Sputnik lest they become its victims.[10]

As the Soviet's 184-pound sphere circled the Earth, Sputnik's beeps, which people could hear on most home radios, appeared to dramatize Soviet technological expertise and military power. Appearances, of course, comprised a significant part of foreign policy calculations during the Cold War era, as capitalist and communist worlds vied for international prestige and waged a global contest over hearts and minds in developing nations.[11]

The war of appearances turned even worse for Americans. On November 3, 1957, the fortieth anniversary of the Bolshevik Revolution, Moscow launched a second Sputnik. Weighing more than 1000 pounds, this satellite carried scientific instruments and temporary life-support equipment for a dog named Laika, the first mammal to orbit Earth. In early December, the U.S. answer to Soviet missilery, a Vanguard TV-3 rocket, lifted a full four feet off its Florida launch pad before toppling back to Earth. In response to the Sputniks, media wags quipped, the U.S. offered "Flopnik" and "Stayputnik." Soviet leader Nikita Khrushchev, recognizing his opportunity, gleefully ridiculed U.S. missile capability.

The *New York Times* saw the United States as entering a "race for survival" against the U.S.S.R., and the Democratic Speaker of the House, John McCormick of Massachusetts, claimed that the country faced "virtual extinction" if it failed to achieve dominance of outer space. Senator John F. Kennedy also endorsed a crash program to advance U.S. capabilities in space. And Lyndon B. Johnson, the Democratic majority leader in the Senate and head of the Defense Preparedness Subcommittee, judged Sputnik to be a disaster comparable to Pearl Harbor. He opened hearings into why the Soviets had beaten the United States into

---

9. David Callahan and Fred I. Greenstein, "The Reluctant Racer: Eisenhower and U.S. Space Policy," in *Spaceflight and the Myth of Presidential Leadership*, eds. Roger D. Launius and Howard E. McCurdy (Urbana, IL: University of Illinois Press, 1997). Divine, *The Sputnik Challenge* also emphasizes Eisenhower's reluctance to join an expensive space race.

10. McCurdy, *Space*, pp. 62-63; Divine, *The Sputnik Challenge*, pp. 74-78.

11. Important works on the space race, in addition to those already cited, include Rip Bulkeley, *The Sputniks Crisis and Early United States Space Policy: A Critique of the Historiography* (Bloomington, IN: Indiana University Press, 1991), Alan J. Levine, *The Missile and Space Race* (Westport, CT: Praeger, 1994), Matthew Brzezinski, *Red Moon Rising: Sputnik and the Hidden Rivalries that Ignited the Space Age* (New York, NY: Times Books, 2007), and Von Hardesty, *Epic Rivalry: The Inside Story of the Soviet and American Space Race* (New York, NY: National Geographic, 2007).

space.[12] *Time* made Soviet leader Nikita Khrushchev its "Man of the Year," and its editors wrote that "the U.S. had been challenged and bested."[13]

As "space race" and "crisis" became the dominant media frames of the Sputnik moment, Eisenhower recognized that his assurances, even if secretly informed by surveillance photographs and knowledge of America's own reconnaissance and military satellite programs, offered insufficient response. The publicity value of U.S. rockets blasting from launch pads, of American satellites circling Earth, and of homegrown adventurers cruising outer space was inescapable. Eisenhower endorsed a speeded-up space program and supported the creation in July 1958 of NASA. The President, in effect, entered a seven-person team, the Mercury astronauts, into the manned-flight event of the space race. NASA and manned spaceflight—featuring astronauts with "the right stuff"—became the public focus of the space race.[14]

The establishment of NASA placed the human piloted space program in the spotlight and under civilian control, but the outcry over Sputnik also strengthened the military's case for stepped up offensive and defensive systems. Less visible to the public than NASA, the Strategic Air Command (SAC) successfully promoted a great acceleration in the ballistic-missile arms race. And deploying military reconnaissance satellites took on greater urgency. Moreover, spending increased for many other unmanned satellites that specialized in weather, communications, and scientific investigations.[15] Strong disagreements over the proper emphasis of space spending (scientific vs. military; manned vs. unmanned) persisted. Still, the Sputnik moment of 1957 intensified both the civilian and military aspects of superpower competition in space.

The responses to the two Sputniks reverberated far beyond bankrolling programs for space exploration. Who could run such programs? Were the American schools failing to produce the scientists and engineers of the future? A great fever of education reform gripped post-Sputnik America. In September 1958, Congress passed the National Defense Education Act, which authorized the allocation of one billion dollars over seven years to develop "those skills essential to the national defense." Eisenhower had earlier opposed the principle

12. Divine, *The Sputnik Challenge*, pp. 62-65. Some prominent scientists broke with the Eisenhower administration by seizing on the Sputnik crisis to argue for increased federal spending on scientific research. See Allan A. Needell, *Science, Cold War, and the American State* (Australia: Harwood Academic Publishers, 2000), p. 148.

13. McCurdy, *Space*, pp. 75-76; *Time*, January 6, 1958.

14. Linda T. Krug, *Presidential Perspectives on Space Exploration: Guiding Metaphors from Eisenhower to Bush* (Westport, CT: Praeger, 1991), pp. 23-42 examines the metaphor of a space race. For a compact overview, annotated bibliography, and set of documents on the U.S. space program generally, see Roger D. Launius, ed., *Frontiers of Space Exploration* (Westport, CT: Greenwood Press, 1998).

15. Crouch, *Aiming for the Stars*, pp. 148-166; Divine, *The Sputnik Challenge*, pp. 34-42, 69, 84-85, 110-127.

of federal aid to education, but he reluctantly bowed to space-race clamor and backed this new extension of governmental funding.[16]

Even so, Eisenhower's sense of caution distanced him from the strident space race rhetoric adopted by future Presidents Kennedy, Johnson, and Nixon—all already maneuvering to succeed him. After hearing the news in early 1958 that the United States had finally orbited its own satellite, Eisenhower characteristically advised his press team not to "make too great a hullabaloo" of the event.

John Kennedy had few reservations about "hullabaloo." He shaped his presidential campaign of 1960 around a critique of national complacency. Eisenhower was by now an aging figure whose stroke that occurred just seven weeks after Sputnik's launch attracted much media attention. By contrast, Kennedy offered youth and vigor (one of his favorite words). He warned against a supposed "missile gap" vis-à-vis the Soviet Union, and he portrayed the pre-sumed gap in space technology as a visible sign of the Cold War challenge facing the United States.

Once in the White House, Kennedy drew effectively on the themes already well established in astrofuturist writings and the pervasive space race rhetoric. On April 12, 1961, Soviet cosmonaut Yuri Gagarin became the first human into space. NASA followed up by rushing Alan Shepard into his five minute ride in space. The popular media went wild over America's achievement and its new astronaut hero. Building on the excitement, Kennedy's famous message to Congress on May 25, 1961, set the goal "before this decade is out, of landing a man on the Moon and returning him safely to the Earth." On September 12, 1962, a presidential address at Rice University, given during a trip to tour NASA facilities, elaborated the rationale for his lunar objective. Space was a "new frontier," a "new sea" in the next great age of discovery. The conquest of space, a historic and strategic imperative, would challenge Americans to show their greatness and would signal national prestige and global leadership. Invoking the competition of the space race, the speech nevertheless transcended the Cold War by emphasizing a romantic and visionary national quest. It stressed how practical and technological greatness could mix with the noblest goals of human aspiration. It provided a chronology of urgency: "We meet in an hour of change and challenge, in a decade of hope and fear, in an age of both knowledge and ignorance."[17]

---

16. Barbara Barksdale Clowes, *Brainpower for the Cold War: The Sputnik Crisis and National Defense Education Act of 1958* (Westport, CT: Greenwood Press, 1981); John A. Douglass, "A Certain Future: Sputnik, American Higher Education, and the Survival of a Nation," in *Reconsidering Sputnik: Forty Years since the Soviet Satellite*, ed. Roger D. Launius, et al., (Amsterdam: Harwood, 2000), pp. 327-362; Juan C. Lucena, *Defending the Nation: U.S. Policymaking to Create Scientists and Engineers from Sputnik to the "War Against Terrorism,"* (Lanham, MD: University Press of America, 2005), pp. 29-53; Divine, *The Sputnik Challenge*, pp. 89-93.

17. John F. Kennedy, "Special Message to the Congress on Urgent National Needs," May 25, 1961, at John F. Kennedy Moon Speech, *http://www1.jsc.nasa.gov/er/seh/ricetalk.htm* (accessed

As in so much of his political rhetoric, Kennedy appealed to (and helped construct) notions of "manly" virtues: risk, adventure, difficulty, competition. He decried opponents as those who wanted "to rest, to wait." He constituted space travel within an inevitable trajectory of America's historic mission to move forward, to rise to challenges, to expand.[18] JFK's exhortations to greatness in individual character and in national purpose appear to have motivated many Americans on a personal level as well as a national one. "A lot of people worked day and night" on NASA projects, observed one aerospace executive. "We were all swept up in it."[19]

John Kennedy's inspirational phrase that Americans would "pay any price" in their struggle against communism applied quite literally to the early space race. From 1961 to 1963, the NASA budget soared from 1.7 billion to 3.8 billion to 5.7 billion; funding for NASA surged to make its budget the fourth largest among all government agencies. At the height of the Apollo Program, NASA and its contractors employed 430,000 people.[20]

When Senator William Proxmire (D-WI), a well-known budget hawk, was asked about the huge expenditures for NASA, he replied that government revenues were increasing because of economic growth and "there was a feeling that we wanted to maintain those revenues and not cut taxes. It was argued what we should do, in order not to slow the economy by running surpluses, was give a substantial amount back through revenue sharing. Therefore, there was funding available."[21] Kennedy, of course, also sponsored a tax cut, pleasing business both by tax-cutting and by offering new contracting opportunities from government-financed projects. In the economic thinking of the postwar years, such governmental expenditures would stimulate greater levels of growth that would, in turn, promote still higher levels of government revenue.

September 28, 2007). For background, see John M. Logsdon, *The Decision to Go to the Moon: Project Apollo and the National Interest* (Cambridge, MA: MIT Press, 1970) and Gretchen J. Van Dyke, "Sputnik: A Political Symbol and Tool in 1960 Campaign Politics," in *Reconsidering Sputnik,* eds. Launius, et al., pp. 363–400.

18. John W. Jordan, "Kennedy's Romantic Moon and Its Rhetorical Legacy for Space Exploration," *Rhetoric and Public Affairs,* 6 no. 2 (2003): 209–231. Krug, *Presidential Perspectives on Space Exploration,* pp. 30–42 and James Lee Kauffman, *Selling Outer Space: Kennedy, the Media, and Funding for Project Apollo, 1961-1963* (Tuscaloosa, AL: University of Alabama Press, 1994) examine Kennedy's metaphors for space exploration.

19. Quoted in Crouch, *Aiming for the Stars,* p. 203.

20. House Committee on Science and Technology, *Toward the Endless Frontier: History of the Committee on Science and Technology* (Washington, DC: Government Printing Office, 1980), pp.171–172 on budget. Crouch, *Aiming for the Stars,* p. 203 on employees. James R. Hansen, *The Spaceflight Revolution: NASA Langley Research Center From Sputnik to Apollo* (Washington, DC: NASA, 1995) provides a rich history of the technological and organizational challenges of spaceflight by focusing on one of NASA's space centers.

21. Quoted in Crouch, *Aiming for the Stars,* p. 203 from Wayne Biddle, "A Great New Enterprise," *Air and Space Smithsonian* 4 no. 7 (June/July, 1989): 32–33.

Representative Olin "Tiger" Teague (D-TX) proclaimed in 1963 that space spending "started the blood coursing a little more fervently through the arteries of our economy." It would, Teague predicted, spark "a new industrial revolution."[22]

Space Keynsianism thus joined military Keynsianism as a justification for pumping governmental spending into the economy and, thereby, besting the Soviets in both economic growth and technological prowess. The space race also introduced a new competitive element into the strategy of containment. As the Soviet leaders placed a high priority on winning the race to the Moon, their underdeveloped and increasingly stressed economy struggled to match America's lavish expenditures. The space race appeared to fulfill the hopes of Democrats that enlarged government spending would simultaneously bring benefits to their party, stimulate prosperity while returning revenue to the Treasury in the form of a growing tax base, and help win the Cold War by weakening the Soviet economy.

The excitement and the rapidly mounting appropriations for the space race, however, did not last. The chastening effect of the Cuban Missile Crisis of October 1961 spawned a series of accommodations in both the U.S. and Soviet governments. The Kennedy presidency had demonstrated that Cold War competition could have its rhetorical thrills, but it also risked unspeakable dangers. With the Soviet pullback in the Missile Crisis, the superpowers' high-pitched competitions abated somewhat. Moreover, after celebrating the flights of Alan Shepard (1961) and John Glenn (1962), and witnessing the other Mercury and Gemini missions of the early and mid-1960s, few Americans continued to maintain that the United States seriously lagged the Soviet Union. The Sputnik moment was quickly passing.

Drawing on the political skills of NASA Administrator James E. Webb, President Lyndon Johnson managed to continue Kennedy's legacy by procuring for NASA a nearly blank check from Congress for awhile longer. Gradually, however, the public and their representatives tired of the costs and grew more confident about America's ultimate successes in space. Moreover, the Great Society and the War in Vietnam vied with space programs over spending priorities, and the country spiraled into a paroxysm of dissent over national direction. As Kennedy's soaring political rhetoric about "paying any price" to best Soviet communism slowly came down to Earth, other concerns challenged the imperatives of the space race.

Republican budget-cutters had sheaved their blades in the shadow of the Sputnik moment, but they gradually grew bolder in attacking governmental spending and taxation. As early as 1962, Representative H. R. Gross (R-IA) voted for Kennedy's request for a greatly enlarged NASA appropriation while

---

22. Quoted in Kauffman, *Selling Outer Space*, pp. 125-126.

also asking pointed questions about why so much money was going to the Southern States and California, and why space contractors were paying their executives such high salaries. "It would be my hope that if and when we do get to the Moon," he remarked, "we will find a gold mine up there, because we will certainly need it."[23] In early 1963, former President Eisenhower sent a letter of protest, printed in the *Congressional Record* in April: "I have never believed that a spectacular dash to the Moon, vastly deepening our debt, is worth the added tax burden it will eventually impose upon our citizens . . . . Having made this into a crash program, we are unavoidably wasting enormous sums."[24] The *Saturday Evening Post* in September 14, 1963, proclaimed that "the space program stands accused today as a monstrous boondoggle."[25] Amitai Etzioni summarized much of the developing critique in a book called *The Moon-Doggle* (1964).[26]

Objections also emerged from those opposed to NASA's emphasis on human piloted spaceflights. Some scientists and their allies advocated less costly and potentially more scientifically valuable robotic exploration. Others, such as Representative Donald Rumsfeld (R–IL), stressed that emphasis should be placed on the military aspects of space—the control of the space closer to Earth—and less on NASA's manned explorations into far space. Such views grew out of, and also fed, the rivalry between the military services and NASA.

In addition, some politicians, scientists, and businesses began to question the regional tilt of NASA installations. In 1959, NASA selected Cape Canaveral, Florida, as the site to train the first group of astronauts. It opened as NASA's Launch Operations Center in 1962 and was renamed for Kennedy just after his death in 1963. Observers of Johnson's legislative career noted that an expanded space effort brought Texas lucrative government contracts. Complementing the center at Cape Canaveral, the Johnson Space Center (JSC), established in Texas in 1961, assumed the lead in human space exploration. The regional tilt of space spending, pouring into the newly expanding "sunbelt," became controversial because of its evident political ramifications.[27]

Influenced by the various doubts and by changing priorities, Congress began trying to reduce NASA budget requests after 1963. The space race remained a useful frame that spaceflight promoters could call on, but its metaphorical power weakened, and it no longer connoted an unchallenged agenda or an open-ended flow of appropriations.

---

23. House Committee on Science and Technology, *Toward the Endless Frontier*, p. 124.

24. Ibid, p. 171.

25. Kauffman, *Selling Outer Space*, pp. 116–125 summarizes the critics. [quote, p. 53].

26. Amitai Etzioni, *The Moon-Doggle* (Garden City, NY: Doubleday, 1964).

27. House Committee on Science and Technology, *Toward the Endless Frontier*, pp.185–190 discusses political maneuvers behind the positioning of NASA sites and some of the controversy.

★★★

The Sputnik moment of 1957 had telescoped fear and mobilized resources in response to a seemingly imminent enemy threat. Space exploration had been underway before Sputnik, of course, and had been driven by many factors: the nationalism inspired by World War II; frontier nostalgia for new lands to discover; public relations campaigns by scientists such as von Braun, entertainment moguls such as Disney, corporations interested in aerospace, and astrofuturist writers. But the Cold War's international rivalry shaped its character and accelerated its tempo into a space race. The space race was exhilarating because it seemed dangerous and character-defining. Boring things such as careful deliberation, cost-consciousness, and safety could be effaced as exciting "new frontiers" of risk and daring beckoned. Advocates of Space Keyesianism saw political and economic advantages—at least until arguments about "big government" and "Moon-doggles" gained traction. A remarkable conjuncture of popular culture, pressure from techno-scientific elites, and political imperatives may have initially produced the space race, but, over time, they also sparked contention over priorities.

After America's lunar landing in 1969, the space race abated and provoked neither the intense fear nor the vaunted inspiration of a decade earlier. But the race had made a lasting imprint. It helped deeply embed a rhetoric of peril into the nation's foreign policy and the practices of large-scale governmental contracting into the nation's political economy.

2. THE MEDIA AGE: SPACE SPECTACULARITY

The postwar Media Age fed the dynamics of the space race. New media forms—visible in photography, film, and television—helped project the beauties, mysteries, and dangers of space. Space was a star of this historical moment in which media spectacularity still seemed really spectacular.

The mass media of the era provided an ideal milieu for coverage of the Space Age, a term that suddenly circulated everywhere. A few weekly magazines and news services dominated the print media, and photography and television images—sometimes live—played a growing role in news delivery. It could be argued that Sputnik prompted little initial popular uproar until techno-scientific elites and politicians teamed with these influential media outlets to frame the event as a Cold War crisis.[28] The space race, after all, provided the attractions of a rich storyline punctuated by stunning images. The initial sensationalized sense of crisis flowed into the breathless score-keeping of

---

28. For example, Amitai Etzioni, "Comments," in *The First 25 Years in Space,* ed. Allan A. Needell (Washington, DC: Smithsonian Institution Press, 1983), pp. 33-36.

a race and finally found triumphant resolution in the nationalistic pageants that celebrated the dangers and successes of America's astronauts.[29]

NASA's public affairs officers provided regular interaction with the media and carefully nurtured certain images and narratives. They controlled the media's access to astronauts and coached its people on making public appearances, regularly drawing up talking points for such occasions. They sponsored high-profile events that would attract media and developed close ties with congressional supporters. One study has concluded that NASA shaped its messages around the themes of nationalism (national pride, prestige, strength, and security), romanticism (heroism, individualism, glamour, frontier heritage), and pragmatism (economic, educational, scientific returns on investment). In its sophisticated public relations techniques and its central messages, NASA both exemplified and helped shape the new media strategies of the Space Age.[30]

NASA crafted an image that united individual heroism with a competence arising from teamwork. Certainly there were plenty of failures, but the successes, especially of Alan Shepard in May 1961 and of John Glenn in February 1962, became spectacular national dramas that celebrated both individual bravery and group accomplishment. Both the intangibles of strong character and the practicalities of seemingly flawless engineering were on display. The media coverage of space in the early 1960s was all in the superlative, and when articles critical of the costs of manned flight began to appear after late 1963, NASA redoubled its efforts to put out positive news.[31]

Astrofuturists had attracted a devoted but limited following in the mid-1950s, but by the early 1960s the popularity of space themes had expanded into a broad-based cultural obsession. Kennedy's telegenic presence, exhorting Americans to reach the Moon, fused together politics and media culture and helped place the Space Age at the center of American life. Reported UFO sightings jumped sharply, and the new awareness of space permeated all kinds of cultural discussions and representational forms.[32]

Life magazine, the famously image-laden staple of American living rooms, lavished attention on space themes and developed an especially close relationship

29. Dickson, *Sputnik,* pp. 22–27, summarizes press reaction based on a collection of press clippings at the NASA History Office in Washington, DC, and also summarizes public opinion polls. Writer for *Newsweek,* Edwin Diamond, *The Rise and Fall of the Space Age* (Garden City, NY: Doubleday, 1964) discusses the media's manipulative coverage. Jay Barbree, *"Live from Cape Canaveral": Covering the Space Race from Sputnik to Today* (New York, NY: Collins, 2007) presents another firsthand account from a reporter.

30. Kauffman, *Selling Outer Space*; Byrnes, *Politics and Space.*

31. McCurdy, *Space,* pp. 89–92; Kauffman, *Selling Outer Space,* pp. 50–66.

32. Carl Sagan and Thornton Page, eds., *UFOs: A Scientific Debate* (Ithaca, NY: Cornell University Press, 1972), and Curtis Peebles, *Watch the Skies! A Chronicle of the Flying Saucer Myth* (Washington, DC: Smithsonian Institution Press, 1994) examine the debate over visits by extraterrestrials. See also McCurdy, *Space,* p. 74, and Dickson, *Sputnik,* pp.164–167.

with NASA. A weekly publication that surveyed worldwide events through glossy pictorial features, *Life* heralded the space race. Normally supportive of Eisenhower, *Life* had greeted Sputnik with a warning that it seemed time for his Administration to get "panicky." It later paid the Mercury-7 astronauts the then-hefty sum of $500,000 for exclusive rights to their life stories. Although Ohio's John Glenn emerged as the star of the astronaut contingent, *Life* highlighted the entire Mercury team's small-town, Protestant backgrounds and photogenic families. One story featured the "Seven Brave Women behind the Astronauts." *Life*'s competitors in middle class living rooms, such as *Collier's*, the *Saturday Evening Post*, and *Look*, followed suit as NASA and the press groomed the image of the astronauts as models of strength, honesty, and strong family values. To help wage an ultimately losing battle against the moving images carried on TV, *Life* and the other magazines faithfully monitored, through still pictures, the operations and personnel of the Moon-landing competition.[33]

Television inexorably became the medium-of-record for the space race. Covering the potentially lethal spectacle of propelling all-American astronaut-heroes into space seemed a sure-fire ratings booster and a money-maker for the television industry. Space travel perfectly suited TV. Heroic dramas of triumph and tragedy could attract and hold viewers, and television generated a voracious demand for ever-more-sensationalized stories. Journalists of the new TV Age who wedded themselves to the space program saw their careers flourish. Still smarting from being overshadowed by NBC's Chet Huntley-David Brinkley duo during the 1956 political conventions, CBS TV's star journalist Walter Cronkite made outer space his personal beat. While seven young pilots, including John Glenn and Neal Armstrong, retrained to be astronauts, this veteran war correspondent retrofitted himself as TV's premier space journalist. Displaying his grasp of the technical details of satellite-rocketry and of NASA's jargon, Cronkite made his mark covering John Glenn's flight in 1962 and continued to become almost the quasi official voice of the Apollo program. Honing his image as the "eighth astronaut," Cronkite reassured TV viewers that the United States would emerge as the ultimate victor in the space race.[34] By the mid-1960s, Cronkite had become known as "the most trusted man in America."

As part of the Cold War's competition of appearances, NASA became adept at promoting the astronauts as international, as well as national, celebrities. The Giantstep-Apollo 11 Presidential Goodwill Tour in 1969, for example, touted the willingness of the United States to share its space knowledge with other nations and carried the Apollo 11 astronauts and their wives to 24 countries and 27 cities in 45 days. Indeed, especially from 1969 on, U.S. accomplishments in the space race often provided a public relations cover or counterweight to

---

33. McCurdy, *Space,* pp. 89-93.

34. CBS News, *10:56:20 PM EDT, 7/20/69: The Historic Conquest of the Moon as Reported to the American People* (New York, NY: CBS, 1970) reproduces reporting on the Apollo 11 mission.

the generally negative news from Vietnam. In April 1969, President Richard Nixon's participation in a celebration for the Apollo 13 astronauts in Hawaii quite literally offered public cover for a secret high-level war meeting about stepping up pressure in Cambodia.[35] Space accomplishments projected the United States as cooperative, technologically superior, and successful in this era when news from Southeast Asia often marked the country as high-handed, technologically threatening, and wedded to a failed policy.

If the highly visual media helped promote Space Age projects, so the new technologies looped back to accelerate transformation in the media environment. In 1962, Congress created the Communications Satellite Corporation (Comsat), a public-private venture to manage an international system, Intelsat. Comsat paid NASA a fee for use of rocket and launch facilities, and within five years communication satellites had become a commercial success. In August 1964, a satellite telecast the opening ceremonies of the Olympic games in Tokyo. By 1969, with sixty countries belonging to Intelsat, geosynchronous satellites served the Pacific, Atlantic, and Indian Oceans.[36] Live space spectaculars, which the United States displayed but the Soviet Union concealed, could now go global—in real time.

NASA worked especially closely with Representative Olin Teague, who became the space program's primary rainmaker and one of its most effective publicists. A Democratic representative from Texas, the chair of the Manned Space Flight Subcommittee, and one of Congress's most decorated combat veterans, Teague was in charge of convincing members of Congress to lavish funding on the space program. He kept them aware of how much space spending was going into their districts; brought models of spacecraft and rocketry to the House floor; and stressed the spinoffs of space spending for medicine, computerization, and fabrication of various kinds. Like other space race supporters, he emphasized that putting a man on the Moon was not an end in itself. The real benefit from the program would be to push the nation forward "in many important fields: science, engineering industrial development, design, mathematics, biology—the whole spectrum of scientific and technological accomplishment."[37] The media enthusiastically embraced this Teague/NASA message, which helped translate space accomplishments into the everyday realm of audience interest.

---

35. Robert Dallek, *Nixon and Kissinger: Partners in Power* (New York, NY: Harper Collins, 2007), pp. 191-192.

36. Wernher von Braun and Frederick I. Ordway, III, *History of Rocketry and Space Travel* (Chicago, IL: J. G. Ferguson Publishing, 1966), p. 186. Hugh R. Slotten, "Satellite Communications, Globalization, and the Cold War," *Technology and Culture* 43 no. 2 (2002): 315-350 provides a basic history and cites the relevant literature on this issue. Heather E. Hudson, *Communications Satellites: Their Development and Impact* (New York, NY: Free Press, 1990) is a thorough history.

37. House Committee on Science and Technology, *Toward the Endless Frontier,* pp. 163-172 [quote p. 172].

In cooperation with NASA, the Science Committee in the House of Representatives, beginning in 1960, published "The Practical Values of Space Exploration," a series of frequently updated studies that detailed productive new spinoffs (a NASA-coined word) from the space program. A few of the most celebrated included miniaturized electronics, spray-on foam insulation, microwaves, freeze-dried dinners, and Teflon. Magazines often featured these down-to-Earth bonuses from the space program, and perhaps left the misleading impression that robust consumer innovation ultimately depended on governmental expenditures in space.[38]

NASA's Office of Public Affairs also used film to publicize NASA activities, taking advantage of NASA's advanced satellite imagery from research facilities and space flight centers around the country. Some of the most widely viewed titles from the first two Space Age decades included *The John Glenn Story* (1963), a film biography; *Assignment Shoot the Moon* (1967); *America in Space—the First Decade* (1968), a history of NASA; *The Eagle Has Landed* (1969), on the manned lunar landing; *Who's Out There?* (1975), on the possibility of extraterrestrial life; and *Planet Mars* (1979). These films (some award-winning for their cinema graphic technique), in addition to rich photographic collections, provided then, and preserve now, a stirring visual record of space program history. The National Archives currently holds 250 "Headquarters Films" made between 1962 and 1981.[39]

Hollywood-produced films also found a congenial partner in NASA. Movies filmed at the space centers included *Apollo 13, Contact, Space Cowboys, Armageddon, The Right Stuff,* the 12-part HBO series *From Earth to the Moon,* and a variety of other TV special productions.

The visitors' centers at the Kennedy and Johnson Space Centers likewise worked with the media. Teague had pushed NASA to construct visitors' centers, providing money for them in the federal budget. He argued that public support was essential to sustaining NASA's appropriations, and he understood the tourist potential of space exploration. The centers quickly proved to be popular destinations, with the one at the Kennedy Center topping one million visitors in 1969.[40] The centers also hosted many foreign visitors and dignitaries, thereby serving the Cold War purpose of exemplifying the United States as a country of great prosperity, amazing technological achievement, and unparalleled power over heavens and Earth. In 1966, Congress authorized construction of the National Air and Space Museum, which became one of the most popular destinations on the National Mall and sponsored programs that attracted media attention.

---

38. House Committee on Science and Technology, *Toward the Endless Frontier,*" p. 173; Mark E. Byrnes, *Politics and Space: Image Making by NASA* (Westport, CT: Praeger, 1994), p. 101.

39. NASA History of Space Flight Motion Pictures, *http://video.google.com/nara.html* (accessed September 15, 2007).

40. House Committee on Science and Technology, *Toward the Endless Frontier,* pp. 177-178.

★★★

A clear synergy developed between the space program and the highly competitive world of image-based media. NASA projected itself to be an Agency involved in science and technology, but it proved also to be skilled at image-making and public relations. Sensational stories generated by human-piloted flights meant publicity for NASA, larger audiences for the media networks, and positive projections of America's power in the Cold War world. Many of the themes that had structured both popular science fiction and popular western tales echoed in the Media Age's presentation of the space race: danger, heroism, competition, suspense, and problems overcome through ingenuity. Yet the dramas that played out at Cape Canaveral and Houston, as exciting as fiction, had the added attraction of being "real." The spectacularity of the space race helped sustain the older print-pictorial media, pioneered a compelling early version of "reality TV," and proved attractive to filmmakers and space center visitors. And this fast-changing and competitive media environment, in turn, boosted the visual spectacularity of the Space Age.

3. THE TECHNETRONIC AGE: TECHNOCRACY AND SPACESHIP EARTH

The complexity of research and development in the Space Age raised moral and practical questions. How might new technologies change life and politics? How might people manage the interrelated systems that comprised the planet Earth within its solar system? Issues about technology and global management were not new to the Space Age, but the rapidity of scientific and technological change made them seem more urgent. Moreover, the penetration of space, by helping to focus attention on Earth's future, provided new terrain for reimagining age-old concerns about the ultimate fate of humans and their planet. "Technocracy" and "Spaceship Earth" became key words in Space Age-era discussions.

Although a Technocracy Movement, which envisioned greater prosperity and social progress through the systematic application of technical expertise, had flourished during the 1930s, the word "technocracy" became a much-discussed concept of the Space Age.[41] Techno-scientific and governmental elites seemed fused together as never before, as NASA's budgets soared and

---

41. On the pre-World War II Technocracy movement, see Henry Elsner, Jr., *The Technocrats: Prophets of Automation* (Syracuse, NY: Syracuse University Press, 1967); William E. Akin, *Technocracy and the American Dream: The Technocrat Movement, 1900-1941* (Berkeley, CA: University of California Press, 1977); and Howard P. Segal, "The Technological Utopians," in *Imagining Tomorrow: History, Technology, and the American Future*, ed. Joseph J. Corn (Cambridge, MA: MIT Press, 1986), pp. 119-136.

government embraced the funding for research and development (R&D). Panels of experts, paid through government grants, became a regular feature of defense and space planning.[42] Steering the enormous space bureaucracy and its complex contracting processes even spawned a new management style called "systems engineering." Just after Sputnik's launch, *Newsweek* pointed out a "central fact" that had to be faced: "As a scientific and engineering power, the Soviet Union has shown its mastery. The U.S. may have more cars and washing machines and toasters, but in terms of the stuff with which wars are won and ideologies imposed, the nation" now had a frightful opponent.[43] But what might be the impact of the fusion between government and technical/scientific expertise in creating this stuff? Could technocracy, which the Soviet system seemed able simply to impose, be reconciled with democracy?

In films, comics, and literature of the pre-Sputnik 1950s, space travel had provided an ideal venue for elaborating various utopian and dystopian visions of a technological future directed by techno-scientific and political elites. Films such as *Destination Moon* presented a positive view, but others, such as *Rocketship XM,* predicted that technology (and the life in space that it sustained) would ultimately fail, bringing death and destruction as the primary outcome.[44]

The same year that Sputnik prompted calls for building new cadres of space scientists and technicians the film *The Incredible Shrinking Man* (1957) presented a dark fantasy about a man who, after exposure to radioactivity, became gradually smaller and more insignificant until he disappeared entirely. Drawing on fears of atomic power, the film advanced a thoroughly alarming vision of the inexorable prospects of man's "shrinkage" in an expanding universe, a victim of his own technology. (A few years later the *Jetsons* brought this theme to TV in "The Little Man," an episode in which a faulty compression technique reduces George Jetson to six inches tall.) The theme of human insignificance resulting from an almost God-like technology and an awareness of Earth's smallness in a vast cosmos ran through Space Age culture.

It was within this broad debate over technocracy, of course, that NASA's own public affairs offices weighed in. By emphasizing group competence and the good individual character of those in the space program, NASA depicted science and technology as being under control and debunked popular worries of shrinking men and overbearing machines. Moreover, NASA's stress on the innovative products and better living arising from space research aimed to diffuse the darker fears of technology's impact.

---

42. See, for example, Ann Finkbeiner, *The Jasons: The Secret History of Science's Postwar Elite* (New York, NY: Penguin Books, 2006).

43. Quoted in Lucena, *Defending the Nation,* p. 29.

44. Frederick I. Ordway, III, and Randy Leiberman, eds., *Blueprint for Space: From Science Fiction to Science Fact* (Washington DC: Smithsonian Institution Press, 1992) deals with the popular culture of spaceflight.

Popular culture's consideration of technological themes had counterparts in political philosophy and religion, in literature and history. Major works contributed thoughtful, yet highly diverse, elaborations of the cautions and promises of the Space Age. Lewis Mumford's book *The Pentagon of Power*, for example, disparaged colonization of space as a waste of resources and, like the building atomic weapons, a pathological use of technology. Zbigniew Bzrezinski's book *Between Two Ages* examined the dawn of the "technetronic age," a new era that would reorient the customary relationships of the industrial age and bring inevitable dislocations and challenges. One of the most popular science writers of the Space Age, Carl Sagan, extolled space exploration but at the same time warned that the siren song of "sweet" science and engineering projects could also turn sinister if pursued with single-mindedness. These works, and so many others, prompted broad consideration of the new role that science and technology assumed in the Space Age.[45]

In religious thought, the "Is God Dead?" controversy contained subtexts about the spiritual meanings of the Space Age. Was the total secularization of the modern world bringing about the death of God "in our time, in our history, in our existence?" The exaltation of science and rationality, many theologians agreed, was helping to fuel a reexamination of the doctrine of God, which in such a secular world stood as an almost empty and irrelevant idol. Still, might the mysteries and infinity of the cosmos provide proof of a divine being with creative powers of unfathomable magnitude and splendor? Appearing in theological treatises, in pulpits of every faith, and even on a highly controversial cover of *Time*, the "death of God" controversy laced the Space Age with momentous philosophical questions about faith and its connection to social action.[46]

Many of the most memorable portrayals of the Space Age similarly centered on the consequences of technology and technocracy. Stanley Kubrick's film *2001: A Space Odyssey* (1968)—developed along with a novel by Arthur C. Clarke that was based on some of Clarke's earlier stories—presented the wonders of space, the potential hazards of technology, and the inevitability of humans' pursuit of new techniques and new modes of being. Norman Mailer in *Of a Fire on the Moon* (1970), an account of the Apollo 11 flight, stated "that he hardly knew whether the Space Program was the noblest expression of the Twentieth Century or the quintessential statement of our fundamental insanity."[47] In 1985,

45. Kilgore, *Astrofuturism*, pp. 54-56; Lewis Mumford, *The Myth of the Machine: The Pentagon of Power* (New York, NY: Harcourt, Brace, Jovanovich, 1964); Zbigniew Brzezinski, *Between Two Ages: America's Role in the Technetronic Era* (New York, NY: Viking Press, 1970); Carl Sagan, *The Cosmic Connection: An Extraterrestrial Perspective* (Garden City, NY: Anchor Press, 1973), and Ray Bradbury, Arthur C. Clarke, Bruce C. Murray, and Carl Sagan, *Mars and the Mind of Man* (New York, NY: Harper and Row, 1973).

46. "The 'God Is Dead' Movement," *Time*, (October 22, 1965); cover photo, *Time*, (April 8, 1966).

47. Norman Mailer, *Of A Fire on the Moon* (New York, NY: Little Brown, 1969), p. 15.

Walter A. McDougall's prize-winning history of the Space Age examined the dilemmas raised by the nation's expensive and expansive networks of scientific and technological expertise. The space program, he argued, led Americans to accept a greater concentration of governmental power and the enlistment of technological change for state purposes. McDougall ended with a plea to neither worship nor hate technology; to neither expect utopia nor fear distopia. [48] In Kubrik's, Mailer's, and McDougall's very different kinds of representations that occurred years apart, humans had no choice but to continue to embrace technology and confront its challenges.

Like "technocracy," the phrase "Spaceship Earth" echoed in a broad range of cultural products during the Space Age. In 1963, Buckminster Fuller published *Operating Manual for Spaceship Earth*; in 1966 Kenneth Boulding wrote *Human Values on the Spaceship Earth*. Both aimed to map a new consciousness for a sustainable environment that would abandon reliance on fossil fuels and develop sources of renewable energy. Along with so many other works of the era, they sought to unite science, engineering, humanities, and art in an integrated effort to focus upon ameliorating human problems. Some connected the current fears of overpopulation and the "population bomb" with the prospects of space colonization.[49]

The Apollo crews in 1968 and 1969 captured from outer space the now famous images of a Spaceship Earth. Perhaps the best known photo, called "Earthrise," showed Earth ascending over the Moon. Such visions of a whole Earth, drifting in space, became among the age's most meaningful icons. On the front page of the *New York Times*, poet Archibald MacLeish wrote that these images might transform human consciousness. "To see the Earth as it truly is, small and blue and beautiful in that eternal silence where it floats, is to see ourselves as riders on the Earth together."[50] To many people, especially the young who were beginning to call for a counterculture—a new way of living and relating— such images signified a global consciousness that might spur transnational and global networks of non-governmental organizations (NGO) to work beyond

48. Walter A. McDougall, . . . *the Heavens and the Earth,* and Joseph N. Tatarewicz, *Space Technology and Planetary Astronomy* (Bloomington, IN: Indiana University Press, 1990) examine the interaction between government and "big science."

49. R. Buckminster Fuller, *Operating Manual for Spaceship Earth* (New York, NY: Simon and Schuster, 1969); Kenneth Boulding, *Human Values on the Spaceship Earth* (New York, NY: National Council of Churches, 1966). The idea that technocratic skills needed to be wedded to more humanist values was a common theme of Space Age writers. See, for example, R. Buckminster Fuller, Eric A. Walker, and James R. Killian, Jr., *Approaching the Benign Environment* (Auburn, AL: University of Alabama Press, 1970). Norman Mailer, *Of a Fire on the Moon* explored the tension between NASA's appeal to nationalism and the countercultural humanism of the 1960s.

50. Archibald MacLeish, "A Reflection: Riders on Earth Together, Brothers in Eternal Cold," *New York Times*, (December 25, 1968); discussed in Finis Dunaway, *Natural Visions: The Power of Images in American Environmental Reform* (Chicago, IL: University of Chicago Press, 2005), pp. 207-208.

nation-states. They also, to some, signified a new ecological awareness about the interrelatedness of planetary systems and called for greater stewardship of the Spaceship Earth on which humans live. Steward Brand's *Whole Earth Catalog* and its famous cover, which came out in 1968, powerfully expressed the goal of linking the stewardship of Earth to individual empowerment. It opened with the words "We are as gods and might as well get good at it." The catalog, a kind of Bible for the counterculture including many of those innovators who would later magnify Brand's ideas in creating the Internet, promised a broad access to whatever tools might save Earth and foster self-improvement. If technology was to be the future, it should be a technology that empowered everyone, not just technological elites and their political masters.[51]

In the early 1980s, the Disney Corporation opened an exhibit called "Spaceship Earth" as the center of its new Epcot exhibit in Florida. Advised by Ray Bradbury and presumably inspired by Fuller's ideas about the advantages of geodesic dome architecture, the Disney Spaceship reached 18 stories tall. Its intricate system of some 11,000 triangles formed cladding that absorbed rainwater and channeled it into a lagoon. Upon its opening, "Spaceship Earth" presented a story of human enlightenment beginning with early cave dwellers and ending with a spacecraft launch. Disney's rendition of civilization as a linear arc of progress flattened the complexities of many of the era's other representations of Earth as a spaceship, but it surely attracted the largest crowds.

Images of Earth in space raised complex questions about the future role of nations and nationalism on a Spaceship Earth. Such tensions between nation and planet, of course, preceded Sputnik and recalled the astrofuturist visions of the pre-Sputnik years. In the 1951 movie *The Day the Earth Stood Still*, for example, the dangerous combination of atomic power, rocketry, and nationalistic competition prompted a visit from a superior civilization from outer space. The emissary, Klaatu, issued a warning that unless nations of Earth began to live peacefully, superior beings would blow up their planet. Nationalism and international conflict, this early Space Age movie suggested, were obsolete and threatened the extraterrestrial order.

Other science fiction scenarios, especially those from *Star Trek*, which debuted in 1966, played imaginatively with the idea that space exploration might provide new configurations of power and authority. The 23rd century "starship," the *Enterprise*, cruised space to explore rather than to dominate other worlds through violence. The creation of Gene Roddenberry, *Star Trek* aired for three years, after which it went into syndication, developed a global following of loyal fans, and ultimately spun off five television series and nearly a dozen movies. In its much-quoted introduction, Captain James T. Kirk (William Shatner) presented the *Enterprise*'s purpose in traditional astrofuturist and Kennedyesque terms: "to

---

51. Andrew G. Kirk, *Counterculture Green: The Whole Earth Catalog and American Environmentalism* (Lawrence, KS: University Press of Kansas, 2007).

explore strange new worlds, to seek out new life and new civilizations, to boldly go where no man has gone before." The show revolved around thinly veiled Cold War themes, as the Federation's *Enterprise* dealt with rivalries with the Klingon and Romulan civilizations. Would new kinds of policing be able to enforce rules within a new kind of intergalactic, or internationalist, order?

The more practical minded turned to forging international space law in the real world. Development of international norms might create precedents for turning space-race competition into Spaceship Earth cooperation and reconfigure the landscape of the Cold War. The United Nations Committee on the Peaceful Uses of Outer Space (COPUOS) worked to develop international space law, and the U.S. Congress undertook various cooperative initiatives.[52] The Outer Space Treaty of 1967, for example, banned weapons of mass destruction from space, demilitarized the Moon and other non-terrestrial bodies, and promised peaceful international cooperation in space. In 1975, Apollo astronauts and Soyuz cosmonauts orchestrated a symbolic handshake in space.[53]

In 1969, U.S. astronauts posed for a much-debated iconic image in which they planted an American flag on the Moon. They also left behind a gold olive leaf and a plaque that stated "We came in peace for all mankind." Throughout the Space Age, a multitude of such representations persistently and unproblematically mixed rhetoric of a national "conquest" of space with invocations of peace and cooperation; they embedded calls for national greatness within universalistic justifications. The tensions between serving the nation and humanity as a whole may have seemed insignificant, indeed even invisible, to most Americans because such juxtapositions sounded so familiar. A long rhetorical tradition avowing America's unique national mission to and for the world, after all, stretched from the Puritans through America's long experience of frontier expansionism to Woodrow Wilson and Franklin Roosevelt and into Kennedy's New Frontier. In classic American tradition, Space Age representations both raised and quieted or masked the tensions between serving the nation and representing all of humanity.

Another question implied in the concept of Spaceship Earth concerned the social make-up of its denizens. How, for example, might Earth-bound racial and gender differences appear when rendered in outer space? Some historians have seen science fiction (like early space travel itself) as a rather exclusionary

52. Eilene Galloway, "Organizing the United States Government for Outer Space, 1957-1958," in *Reconsidering Sputnik*, ed. Launius, et al., pp. 309-325; Joan Johnson-Freese, *Changing Patterns of International Cooperation in Space* (Malabar, FL: Orbit Books, 1990) examines cooperation in space. House Committee on Science and Technology, *Toward the Endless Frontier*, pp. 367-450 details congressional efforts.

53. The text of the Outer Space Treaty of 1967 may be found at the NASA History Division, *http://history.nasa.gov/1967treaty.html* (accessed September 15, 2007).

preserve of white males, but imagined forays into space often provided a forum for envisioning and confronting assorted futures.

Stories about interplanetary space travel often featured encounters with alien "others." As U.S. leaders and experts projected their nation's power into areas of the world that required dealing with the dilemmas of cultural and racial differences, imagined encounters with aliens living outside of the planet Earth could mirror the complexities of addressing the problem of "otherness." *The Thing, The Blob, Invaders from Mars,* and *War of the Worlds,* among many others, presented aliens as monsters. But *Forbidden Planet,* one of the most acclaimed films in the space genre, went beyond the simplistic formula, probing the monstrous ineffectiveness of good intensions and of presumably benevolent interventions. The film features a protagonist (Walter Pidgeon) who tries to understand the alien Krell but, despite his high-minded motives, ultimately fails.

Moreover, the interrelationships within groups of people on small crafts hurtling through space raised issues of gender, race, and class, allowing discussions related to the contemporaneous civil rights and feminist movements. In the much analyzed *Star Trek,* for example, the *Enterprise* has a multiracial crew of women and men and aliens (the half-human First Officer Mr. Spock, Leonard Nimoy) who live and work in a spirit of (mostly) cooperation. The program literally took its crew into new territory when it offered audiences a highly controversial, if compelled, interracial kiss. *Star Trek,* of course, has attracted an enormous amount of analysis and commentary, and some commentators have seen the centrality of white men and the marginality of others as a reinforcement of existing hierarchies. DeWitt Douglas Kilgore, however, makes a compelling argument that *Star Trek,* like other astrofuturist imaginings of life in space, invites "speculation about alternatives" and can operate as a "liberatory resource" for those who wish to stake a claim in a more egalitarian future. Astrofuturist narratives, he argues, are multivalent and "unusually porous, with consumers regularly seizing the reigns of production." *Star Trek* both reflected and also scrutinized contemporary issues of gender, race, and class by safely projecting them into an imagined future.[54]

★★★

In America's Space Age imaginings, the present seemed poised to make an unprecedented leap, and the future came in many styles. The terms *Technocracy* and *Spaceship Earth,* appearing and reappearing in diverse contexts, raised seemingly urgent moral and practical questions that revolved around three interrelated concerns: the political and moral impact of technocracy; the

---

54. Kilgore, *Astrofuturism,* pp. 28-29. In making this argument he draws effectively on the racial politics that surrounded George Takei's and Nichelle Nichols's participation in the show and an analysis of class in Homer H. Hickam, Jr.'s *Rocket Boys* and the film *October Sky.*

problems of managing the health of the planet; and the national or international (or intergalactic?) government of and social relationships in space. Would extraterrestrial travel become a terrain for renewal and betterment, or for hubris and subsequent failure? Would spacecrafts and space colonies transcend or simply transplant the divisions that beset humans on Earth? Space exploration brought no answers to dilemmas over technology, planetary consciousness, and nationalism—versions of which had long preceded spaceflight—but it did refresh imaginations and reignite philosophical, religious, and practical controversies.

4. MID-CENTURY MODERNISM: GOOGIE DESIGNS AND "FAR OUT" ART

The term Space Age, in addition to signifying a Cold War competition, a media sensation, and a debate over future political and social structures, signified the look called Mid-Century Modernism. Space Age design—in architecture, signage, decorative arts, and painting—elaborated an aesthetic of risk, individualism, and confidence. It emphasized eclecticism, mixing retro primitivism with futuristic styles. It juxtaposed calm, planet-shaped curvatures with abrupt, spaceship-style thrust. If the science of space penetration suggested sleek exactitude, the wonder of the cosmos encouraged unpredictable pastiche.[55]

Space Age modernism brought spherical and angular motifs into a cacophony of the unexpected. American automobiles sprouted their storied tailfins, suggesting rocket propulsion. Manufacturers redesigned children's playground equipment: Space Age kids ascended ladders into rockets, played house in spaceships, and scrambled around on faux Moon surfaces. Toys, coloring books, wallpaper, and storybooks adapted space themes. Space mania revolutionized the design of household items. Chandeliers resembled space platforms; dinnerware assumed the elliptical shape of a satellite orbit; vases, ashtrays, and appliances disguised their functions within new forms and facades.

French designer André Courreges launched his Moon Girl Collection in 1964. It featured angular, geometric shapes; space-style hats; short skirts in white and silver; and high, shiny-white plastic "go-go" boots. His 1968 Space Age Collection continued to display simple, stark lines with metallic silver as the design color of the age. The look of these "moon fashions" swept through the worlds of famous designers and of street faddists. They adorned covers of *Vogue* and percolated into the sew-it-yourself pattern catalogues that set the styles on Main Street.

---

55. Thomas Hine, *Populuxe* (New York, NY: Alfred A. Knopf, 1986); George H. Marcus, *Design In The Fifties: When Everyone Went Modern,* (New York, NY: Prestel, 1998). For background, see John H. Lienhard, *Inventing Modern: Growing Up with X-rays, Skyscrapers, and Tailfins* (New York, NY: Oxford University Press, 2003).

New theme parks, especially Disneyland, popularized the look of Space Age modernism as belonging to the future. Popular entertainment (and education) in America had long been structured around a trajectory from past to future: the Buffalo Bill Wild West Shows replayed the popular clichés about the transition from barbarism to civilization; the 20th century World's Fairs displayed visions of progress extending from the drudgery of the unenlightened past into the pleasures of the technologically driven future. Reprising, but always improving upon, these culturally embedded narratives, Disneyland offered two of its Kingdoms as "Frontierland" and "Tomorrowland." True to cultural archetypes, an imagined future derived definition from an imaged past: Tomorrowland enacted the new comforts and ease offered in the Space Age, yet it also extended the individual heroism of the legendary frontiersman into the new era. At Disneyland, as in so much of the era's political rhetoric, the "new frontier" of space and the "endless frontier" of science could confirm the national and personal virtues that popular culture of the 1950s and early 1960s still associated with the winning of the West.

The two international exhibitions held in America during the Space Age—in Seattle and New York—also emphasized space themes expressed in the look of Mid-century Modernism. The Seattle Century 21 Exposition's Space Needle set the tone for an exhibition that claimed to represent the summit of human (well, really *American*) accomplishment. The Seattle World's Fair Commission sought some kind of restaurant in space as a central symbol and engaged an architect, John Graham, Jr., who had created a revolving restaurant in Honolulu. Graham joined with other partners to design a slim steel tower anchored to Earth by 74 32-foot-long bolts topped by a large rounded structure containing a revolving restaurant, which stood eight hundred thirty-two steps away from the base. Built in a year and opening slightly before the Fair began in April of 1962, the Space Needle's color scheme included "Orbital Olive," "Reentry Red," and "Galaxy Gold." NASA had its own exhibit at the exposition, including John Glenn's space capsule that was then touring the world rather than soaring above it.[56]

The 1964 New York World's Fair centered around a Unisphere, a large sculpture of Earth circled by orbiting bands. The Space Park displayed America's aerospace superiority, and corporate pavilions had futuristic themes. General Motors's "Futurama" featured an extraterrestrial-looking building holding models of futuristic cities built on land, under the sea, and in space. Monsanto showcased a Space Age home. At the end of the World's Fair, some of the exhibits migrated to become features at Disney's Tomorrowland.

These international exhibitions helped promote an architectural style that has become popularly known as "Googie." Space evocations predominated in

---

56. "History of the Space Needle," *http://www.spaceneedle.com/about/history.asp* (accessed September 15, 2007); McCurdy, *Space*, p. 93.

Googie styles, which reached their apogee in the slice of southern California that stretched between Hollywood and Disneyland. In 1949, the architect John Lautner had designed a building for Googie's Coffee Shop on Sunset Boulevard in Los Angeles. Architecture critic Douglas Haskell wrote an article in *House and Home Magazine* in 1952 in the form of a playful interview with a ficticious expert on "Googie" architecture. "It seems to symbolize life today," his imaginary expert explained, "skyward aspiration blocked by Schwab's Pharmacy." The article closed with the rumination that seemed to be at the heart of Cold War Googie: "It's too bad our taste is so horrible; but it's pretty good to have men free."[57]

Googie quickly moved beyond coffee shops. Architect John Lautner's own 1960 home, the Cemosphere (recently saved and rehabilitated by the German book publisher Benedikt Taschen) shimmered above the horizon like some extraterrestrial hovercraft. Built in the Hollywood Hills off Mulholland Drive, Lautner's design responded to the challenge of building on a 45-degree slope. Erected on top of a 30-foot concrete pole, it appeared to defy not just conventional forms but gravity itself.[58] Googie, Douglas Haskell wrote, "was an architecture up in the air."

Googie's influence flowed out into highways and towns throughout the nation. "Serious" architects picked up the Googie designation as a slur, but it became the roadside look of a Space Age nation-on-the-go. Just as the space race defied gravitational laws, so representations of space offered suggestive mixtures of lines and curves that flaunted the conventions of Earthbound realities. Travelers in the late 1960s could stay in the Space Age Inn or the Cosmic Age Lodge. After taking in Disneyland (a Googie paradise), they might shop in Satellite Shopland and cruise by the fabulous Anaheim Convention Center. Gas stations might rest in the shade of characteristically upswept roofs and aerospace-inspired flying buttresses. In Googie, domes hugged Earth as spires and starbursts (see Las Vegas and Holiday Inn) transcended terra firma. The original McDonald's golden arches projected a Space Age ellipse. If Googie had any rules of form, they were the embrace of abstraction and surprise.[59]

As in other Space Age representations of the future, Googie's futuristic elements often mixed anachronistically with primitivist motifs: tiki-hut roofs, South sea island-style lava rock walls, frontier themes. Long before the postmodern architecture of the end of the twentieth century self-consciously (and

---

57. Douglas Haskell, *House and Home Magazine,* (February 1952) *http://www.spaceagecity.com/googie/index.htm* (accessed September 15, 2007).

58. Alan Hess, *The Architecture of John Lautner* (New York, NY: Rizzoli International Publications, 1999).

59. Alan Hess, *Googie: Fifties Coffeehouse Architecture,* (San Francisco, CA: Chronicle Books, 1985), and Philip Langdon, *Orange Roofs, Golden Arches: The Architecture of American Chain Restaurants* (New York, NY: Alfred A. Knopf, 1986).

controversially) conglomerated elements of style, Space Age avatars were already jumbling together time and geographic space. They were stage-setting the future within the past and, with a wink, presenting fantasies of both to the present.[60]

Googie not only brought the excitement of a jet-propelled look into the everyday activities of American life, but it also colonized the new "fast" medium of television. True to the age's most popular entertainment formula—the toying with both past and future—the creators of *The Flintstones*, Hanna-Barbara Productions, introduced *The Jetsons*. This animated series of 24 episodes played in the prime Sunday night spot from September 1962 to March 1963. After that, it became a staple of Saturday morning cartoon reruns for decades. (Additional episodes were made between 1985 and 1987, followed by movies and television specials.) Over the years, merchandise spinoffs from *The Jetsons* continued to attract a market.

The life of the Jetsons fairly bristled with Googie style. Their neighborhood boasted houses raised high above the ground on poles—suggesting the Cemosphere and anticipating the Seattle Space Needle. The family flew around the air in its private rocket-ship and traversed the ground in individual people movers that look just like today's Segways. Sets were spare, brightly colored, modernistic.

The Jetson's Googie world of spheres and angles and turquoise and pink projected an automated future. George Jetson worked three hours a day, three days a week for Mr. Spacely of Spacely Space Sprockets. He mostly pressed buttons. Although the Jetsons enjoyed the standard fare of family sitcom mixups, frustrations, and travails, labor-saving devices of all kinds provided abundant leisure. Space references abounded in this future: a Moon Side Country Club, a space-club trip to the Moon, an auto shop called Molecular Motors. Football was played by robots. The family dog, Astro, was acquired after a comparison with an electronic, nuclear-powered dog. A used robot maid, Rosie, made a couple of appearances, but the future, after all, was fairly work-free.

Googie was one highly popular part of a broader aesthetic that had emerged along with abstract expressionism in high art. The abstract expressionists during the 1940s and 1950s, too, explored the modern as a statement of freedom, an acceptance of risk, and a willingness to shock. The connections between the mid-century visions of space and postwar art seem almost too obvious, as so many artists of the age employed lines, spheres, and vast canvasses to project enigmatic representations of unknowability. Both artists and astronauts drifted beyond the rules that governed their atmospheres; both projected a kind of outlaw masculinity that combined an extraordinary endurance for the regularity of hard work with a confident ability to improvise and transcend boundaries.

---

60. Stephen Lynch, "Excursion Roadside Retro Be It Space Age, Cocktail or Tiki: Orange County Has Gobs of Googie," *Orange County Register*, (June 27, 1998) ID# 1998178044, *http://www. ocregister.com/ocregister/archives/* (accessed September 15, 2007).

In some cases the trajectory of vision for abstract expressionist painters seemed quite directly in synch with space exploration, even if the connections often stood largely unnoticed or unarticulated at the time. Richard Pousette-Dart, for example, brought to canvass Space Age motifs in *Night Landscape* (1969-71), a blue, black, white, and yellow sky dense with layered planets and rotations; *Starry Space* (1961); and *Earth Shadow in Time* (1969). Other artists also explored Space Age concerns. Robert Rauschenberg, for example, produced his "Stoned Moon" cycle of paintings after being invited by NASA to witness the launch of Apollo 11.[61] There were many, many more artists who drew from new understandings in physics and astronomy to fashion commentaries on perspective, on the fungibility of matter and energy, and on the universe's enigmatic proportions. Fascination with the "far out" provided the ethos of the era, in art as well as in science and politics.

★★★

For a couple of decades, Googie design and its many offshoots shined as brightly as the Moon and stars. Googie was a style of optimism, an exemplar of free and unregimented spirits who broke the rules, an effervescence of populist self-confidence. If the Space Age coincided with an increasingly powerful American imperium, then Googie represented the imperial signature of what one historian has termed America's "empire of fun." Its bold and shiny surfaces revealed few dark sides.

5. CONCLUSION

From the 1950s to the 1970s, space held many meanings: it was a symbol-laden arena in which people and nations staged Cold War competitions, a "star" in the media firmament, an ultimate challenge for scientists and engineers, and an inspiration for artists and designers.

In 1966, Wernher von Braun ended his book on the history of space travel with a vision of what steps would follow after the projected Apollo Moon landing. He asserted that there would soon be semi-permanent bases on the Moon, growing vegetables and chickens. He then predicted a flyby to Mars or Venus by the late 1970s, landings on Mars in the 1980s, and the exploration of other planets and moons until the process of discovery became routine.[62]

---

61. Robert Rauschenberg; *Stoned Moon, http://www.orbit.zkm.de/?q=node/277* (accessed September 15, 2007).

62. Von Braun and Ordway, *History of Rocketry*, p. 222. For a well-illustrated, recent attempt to reimagine the past and future of space exploration, see Roger D. Launius and Howard E. McCurdy, *Imagining Space: Achievements, Predictions, Possibilities, 1950-2050* (San Francisco, CA: Chronicle Books, 2001).

Even by the time of these breathless predictions, however, the most storied days of the Space Age were already coming to a close. An image-saturated public seemed to be tiring of spectacles in space. Only a major triumph, such as the walk on the Moon in 1969, or a major disaster—such as the death of three Apollo astronauts in January 1967, the Apollo 13 travails in April 1970, or the *Challenger* explosion in January 1986—could reclaim a large viewing public. Between 1969 and 1972, the United States landed six sets of astronauts on the Moon. The successes of the Moon program, first amazing and then routine, became its greatest burden. Media attention ebbed along with the public's investment of emotional and monetary assets. In 1979, Tom Wolfe's *The Right Stuff* seemed the stuff of nostalgia. Wolfe recalled the early 1960s fascination with the buccaneer days of space and concluded with the epitaph "the era of America's first single-combat warriors had come, and it had gone, perhaps never to be relived."[63]

Even as the exuberant high of the Space Age slipped away, it nonetheless left an enduring array of creative and rhetorical resources in American culture. Like any star celebrity, the legacy of meanings for national identities and global futures were complex and multiple. Space exploration in this era—entangled in Cold War rivalries, magnified by the explosion of new image-based media, intertwined with discussions over the role of technology and planetary stewardship, and expressed through innovative artistic products and designs— anchored diverse images and representations.

The Space Age boosted national pride—and placed it under threat. It forged pipelines to pump money into fantastic new projects—and prompted warnings about the size of a "Moon-doggle" and an enervating dependence on government largesse. It promoted techno-science—and stimulated new fears about "technocracy." It encouraged the triumph of rational endeavor— and a mystical faith about the meanings of the heavens. It promised peace and social justice—and more frightening forms of hierarchy and war. It offered the excitement of new modes of living—and apprehensions about the unknown. It inspired creativity—and created bureaucracies that could stifle it.

In its intersections with the Cold War, the Media Age, the Technetronic Age, and Mid-century Modernism, the Space Age provided a canvas for many visions, a setting for multiple narratives about who "we" were and could be. Loaded with so many meanings, space indeed seemed infinite. And in its undefinability and semiotic expansiveness, space was—and still is—*far out.*

---

63. Tom Wolfe, *The Right Stuff* (New York, NY: Farrar, Straus, Giroux, 1979), p. 436.

CHAPTER 11

A SECOND NATURE RISING: SPACEFLIGHT IN AN ERA OF REPRESENTATION

Martin Collins

INTRODUCTION

R ecently novelist and essayist Barbara Kingsolver began a reflection on the virtues of the local food movement with the following sentence: "In my neighborhood of Southwest Virginia, backyard gardens are as common as satellite dishes."[1] She casually invokes then subverts the cultural notion— vestigial and romantic—that the garden, backyard or otherwise, stands as the "natural" against which ubiquitous communications and its machines might be defined and measured. She makes clear our contemporary tendency to grant priority to the human made in creating our sense of what the world is.[2] Even in rural southwest Virginia, it is the garden that is the surprising presence, one that needs to be placed in relation to an alternate ontology represented by the satellite dish. Media and machines, she implies, have become the embodiment of a *new* natural, the tip of a vast, globe-connecting system of technology, of capital, of first world and other world cultural transactions, of a condition of semiotic super-abundance as "in your face" and compelling existentially as beans, corn, and garden dirt. Indeed, the new natural is more so. In this cultural condition, the semiotic realm enabled by globally connected satellite dishes frames the very way in which we think about intimate rituals of local food cultivation and consumption.

Her matter-of-factness in this regard offers incidental proof of a thesis initiated in the humanities in the late 1960s and regnant in sociology, anthropology, geography, and literary theory since the late 1970s: That representation—the signs of things, rather than things themselves—had over-spilled its pre–World War II channels of circulation, spread luxuriantly, and established a new order of experience. To put this in the passive voice, of course, is deceptive. Kingsolver's satellite dish, as thing, media conduit, and symbol situated in one locale and

---

1. Barbara Kingsolver, "The Blessings of Dirty Work," *Washington Post*, September 30, 2007.

2. Of course, to be accurate the "backyard garden" also is human made, a particular construct of what counts as nature, a symbol of a romantic notion of nature that is apart from the human.

standing in relation to other very different locales elsewhere in the world, arrived through a specific process of historical agency. Distilled, this humanities literature makes two deep claims. First, that our regime of representation is ontological, that semiotics perform and act—as discourse and signs, especially as instantiated in commodities and the ever expanding presence of electronic media. They touch nearly all geographic nooks of the globe and order our experience: There is, if you will, a there there. Second, this literature claims that this semiotic-ness coincides with a historic transformation of capitalism in the post-World War II period.[3]

Thus, it is an argument about historical basics: about the way the world is structured, operates, and feels. It is, too, if semiotics may be taken to perform and act, about fundamental sociological categories: of how individuals constitute themselves and are constituted by ambient cultures, about identity and politics. And it presents a rousing challenge: it places at the center of the historical playing field two conjoined and reinforcing vectors of agency—representation and capitalism—that many historians of the Cold War and spaceflight sort might see as inferior or ancillary to two other organizing concepts, state action and elite politics. The title of this paper comes from this literature— Frederic Jameson, preeminent literary theorist and exponent of historicizing the relations between capital and culture coined the phrase "second nature" to describe this remapping of the human experience in the postwar years.[4]

---

3. A range of authors have advanced these points, in varying combination and degree of emphasis. Most important are the works of Jean Baudrillard, Francois Lyotard, and Frederic Jameson, referenced in succeeding notes. Also crucial is David Harvey, *The Condition of Postmodernity: An Enquiry into the Origins of Cultural Change* (Oxford, UK: Blackwell, 1989), as well as various works by Zygmunt Baumann, e.g., *Culture As Praxis* (London: Sage Publications, 1999). A sampling of additional works includes Arjun Appadurai, *Modernity at Large: Cultural Dimensions of Globalization* (Minneapolis, MN: University of Minnesota Press, 1996); Marc Augé, *Non-Places: Introduction to an Anthropology of Supermodernity* (London: Verso, 1995); Ulrich Beck, *Risk Society: Towards a New Modernity* (London: Sage Publications, 1992); Pierre Bourdieu, *The Logic of Practice* (Stanford, CA: Stanford University Press, 1990); Judith Butler, Ernesto Laclau, and Slavoj Žižek, *Contingency, Hegemony, Universality: Contemporary Dialogues on the Left* (London: Verso, 2000); Frederick Cooper, *Colonialism in Question: Theory, Knowledge, History* (Berkeley, CA: University of California Press, 2005); Guy Debord, *Society of the Spectacle* (Detroit, MI: Black & Red, 1983); Michael Denning, *Culture in the Age of Three Worlds* (London: Verso, 2004); Terry Eagleton, *The Idea of Culture* (Oxford, UK: Blackwell, 2000); Paul N. Edwards, *The Closed World: Computers and the Politics of Discourse in Cold War America* (Cambridge, MA: MIT Press, 1996); Mike Featherstone, *Consumer Culture and Postmodernism* (London: Sage, 1990); Scott Lash, *The End of Organized Capitalism* (Cambridge, UK: Polity, 1987); Edward W. Said, *Culture and Imperialism* (New York, NY: Vintage Books, 1994); Graham Thompson, *The Business of America: The Cultural Production of a Post-War Nation* (London: Pluto Press, 2004); Anna Lowenhaupt Tsing, *Friction: An Ethnography of Global Connection* (Princeton, NJ: Princeton University Press, 2005); and Reinhold Wagnleitner and Elaine Tyler May, eds., *Here, There, and Everywhere: The Foreign Politics of American Popular Culture* (Hanover, NH: University Press of New England, 2000).

4. Jameson is the focal point for the literature (loosely grouped under the rubric of critical theory) making this claim. See variously: Fredric Jameson, *Postmodernism, Or, the Cultural Logic of Late*

The subtitle raises a question: What are the implications of this literature and its associated historical claims for our understanding of the development of spaceflight, its cultural meanings, and its integration into the broader field of history? Even so, why should this question merit our curiosity? Because the changes mapped and claimed by this swath of literature are coeval *and* intimately bound to the development of spaceflight in its many dimensions—as significant genre of postwar technology; as site of knowledge creation; as state activity, business undertaking, military venture, and global utility; as a cultural zone for contesting the era's values and beliefs; and as national and international trope extraordinaire.

First, an admission: What I am offering is an analytic sketch—of period history, of a diverse group of theoretical literatures and positions—and proceeds primarily by feeding off of theory rather than empirical data. Despite such simplifications, I think one can argue for an alternative perspective on the field's historiography and research problematic. Let me first historically situate the question of representation in slightly more detail, and then offer a couple of thumbnail case studies to suggest the historical stakes when we juxtapose capitalism, semiotics, and spaceflight.[5]

THE CHALLENGE OF REPRESENTATION

Based on the theoretical literature already cited, almost all a product of the 1970s and after, one might offer a periodization of the postwar years that traces the trajectory of representation and its cultural importance. In the 1950s and 1960s, the image and semiotics take on a stronger, more pervasive cultural function, with emphasis on their phenomenological everywhere-ness and density—especially in the media-rich West—resulting in an incipient problem on a transnational scale. Think of McLuhan's global village and his Western-centered geopolitical perspective reflected in his 1964 thought

---

Capitalism (Durham, NC: Duke University Press, 1991); *The Geopolitical Aesthetic: Cinema and Space in the World System* (Bloomington, IN: Indiana University Press, 1992); *The Cultural Turn: Selected Writings on the Postmodern, 1983-1998* (London: Verso, 1998); *A Singular Modernity: Essay on the Ontology of the Present* (London: Verso, 2002); and Jameson and Masao Miyoshi, eds., *The Cultures of Globalization* (Durham, NC: Duke University Press, 1998). For a discussion of Jameson's importance to this discussion and his centrality to the related issue of postmodernity as a descriptor of the postwar condition see Perry Anderson, *The Origins of Postmodernity* (London: Verso, 1998); as regards the argument for a "second nature," see especially p. 53.

5. This essay is a companion to two earlier explorations by the author on space history and its historiography, see: Martin Collins, "Community and Explanation in Space History (?)," in *Critical Issues in the History of Spaceflight*, edited by Steven J. Dick and Roger Launius (Washington, DC: NASA, 2006); and "Production and Culture Together: Or, Space History and the Problem of Periodization in the Postwar Era," in *Societal Impact of Spaceflight*, by Steven J. Dick and Roger Launius (Washington, DC: NASA, 2007).

that "in the electric age, we wear all mankind as our skin."[6] By the 1970s, changes in capitalism and technology intensified these developments and made representation a (perhaps *the*) central problem of the human condition—an analytic perspective one might trace through seminal authors Jean Baudrillard, Francois Lyotard, and Jameson.[7]

Let's venture into this postwar circumstance, though, by considering a scholar firmly in the center of the historical profession, Daniel Boorstin.[8] In 1961, Boorstin published *The Image, or What Happened to the American Dream*, a book-length disquisition on the ascendance of the image and its consequences for the American experience. After presenting the reader with a broad inventory of the image's ubiquity and modes of use in contemporary life, he offered a first-pass assessment:

> In nineteenth-century America the most extreme modernism held that man was made by his environment. In twentieth-century America, without abandoning the belief that we are made by our environment, we also believe our environment be made almost wholly by us. This is the appealing contradiction at the heart of our passion for pseudo events: for made news, synthetic heroes, prefabricated tourist attractions, homogenized forms of art and literature (there are no "originals," but only the shadows we make of other shadows). We believe we can fill our experience with new-fangled content. Everything we see and hear and do persuades us that this power is ours.[9]

---

6. Marshall McLuhan, *Understanding Media; the Extensions of Man* (New York, NY: McGraw-Hill, 1964), p. 56.

7. On Jameson's work, see prior note. On Baudrillard, see Jean Baudrillard, *Selected Writings* (Stanford, CA: Stanford University Press, 2001). The introduction by Mark Poster provides useful insight on the arc of Baudrillard's thinking. He began publishing on these issues in 1968 and continued through his death in 2007. Lyotard's writings have been equally seminal; see, as his best known example, Jean François Lyotard, *The Postmodern Condition: A Report on Knowledge* (Minneapolis, MN: University of Minnesota Press, 1984).

8. On Boorstin and the U.S. historical profession in the first decades after World War II, see Peter Novick, *That Noble Dream: The "Objectivity Question" and the American Historical Profession* (Cambridge, UK: Cambridge University Press, 1988).

9. Daniel J Boorstin, *The Image: A Guide to Pseudo-events in America* (New York, NY: Atheneum, 1971), pp. 182-183. It must be noted that Boorstin changed the subtitle of the book within the first years after publication. Originally published in Great Britain as *The Image, or, What Happened to the American Dream* (London: Weidenfeld and Nicolson, 1961), the book was reissued with the revised title *The Image: A Guide to Pseudo-Events in America*, Harper Colophon Books (New York, NY: Harper & Row, 1964). The change is indicative of the tensions in Boorstin's thought on how to integrate the problem of the image into his notions of political economy. Also, note that Boorstin's analysis was roughly contemporaneous with Marshall McLuhan's first articulations of the notions of the global village and the medium as message in the late 1950s and early 1960s.

The cultural condition that Boorstin described as different-than-modern quickly became identified with a name—postmodernity—that grew in usage and application in the years to follow. In writing this passage he carefully sidestepped a narrative common in American history, the people versus the interests. This decades-old motif dominated immediate postwar critiques of advertising and consumer culture, and included prominent instances such as Theodor Adorno's "The Culture Industry" and defined the early work of McLuhan.[10] Yet Boorstin made clear that the condition he found so unsettling was a consensual creation: of the masses and elites, of consumers and producers, a field of experience that all inhabited and in which all participated, a basic reorganization of the perceptual and social order.

The stakes were high. The image or pseudo-event gave a new cast to a problem as old as philosophy: How do we know what we know? The social practices Boorstin detailed through rich example had a powerful consequence: they undermined the idea of the real as an independent referent for human thought and action *and* as a fundamental motivation for human engagement with the world.[11] The former had a long contested history in epistemology; in the context of post-Enlightenment thought, the latter seemed a newly emerged view and the heart of Boorstin's concern: it was not merely the coexistence of the real and pseudo, it was our avid preference for the pseudo. The image was a challenge in collective ethics. He did not belabor these implications in his main text, tucking his strongest concern in a concluding bibliographic note. Here he neatly combined the ethical and ontological implications: "The trivia of our daily experience are evidence of the most important question in our lives: namely, what we believe to be real."[12] The rise of the image was not just a lament but a foundational shock.

Why did Boorstin put the most concise, potent statement of his thesis in the back-matter of his book? The simplest answer is that he was uncomfortable with two broad issues raised by the real-to-image turn. One concerned politics. For him, the question of the real was not a mere philosophical problem à la

---

10. Adorno's seminal essay "The Culture Industry" was published in 1947; around the same time, McLuhan began a long run of media and advertising critiques. See: Theodor W. Adorno, *The Culture Industry: Selected Essays on Mass Culture* (New York, NY: Routledge, 2001) and, as one example, Marshall McLuhan, *The Mechanical Bride: Folklore of Industrial Man* (New York, NY: Vanguard Press, 1951).

11. The absence of any foundation (e.g., "reality") or of access to a priori truths became a leitmotif of the postmodern. The nearly contemporaneous work of Thomas Kuhn on the role of non-science in establishing scientific knowledge became an intellectual touchstone of this position. See Thomas S. Kuhn, *The Structure of Scientific Revolutions* (Chicago, IL: University of Chicago Press, 1962). The most relevant discussion of issues of epistemology in relation to the literatures covered here is Bruno Latour, *We Have Never Modern* (Cambridge, MA: Harvard University Press, 1993).

12. Boorstin, *The Image*, p. 265.

Plato's allegory of the cave (an allusion he invokes with his "shadows of other shadows"). It was a genuine problem of the everyday, a particular historical condition of American life. If the real and unreal, the actual and simulations, coursed through the polity without distinction and with equal status, then how could citizens be rational actors, sorters, and evaluators of the world around them and serve as the bedrock of political life? But Boorstin's use of the word "trivia" signaled the problem was not about politics in isolation but in its American-style bred-in-the-bone connection with market capitalism. Image-ness posed a conundrum for the culture in full. His American "dream" assumed individual rational actors as an essential foundation, yet the robust pursuit of this ideal over the 20th century created a condition—the turn to the pseudo-event—that threatened the possibility of making and nurturing such actors, and thus the dream itself. The ethics of the image mirrored the ethics of the system of which it was a part. Still, Boorstin could not bring himself to a vigorous analysis of a main engine of this change—market capitalism—and made the emergence of image-ness seem only a causeless development or a collective shift in taste.[13]

This set of issues led to a second—the intellectual basis of history and the organization of knowledge in the academy. Boorstin held to a view of history compatible with his notion of the ideal citizen—an instrumental view in which nations, institutions, and individuals acting as purposive, rational agents provided the best means for describing and accounting for historical change. But the image and its semiotic kin, Boorstin concluded, stood as a new form of agency, structural and diffuse rather than localized, a tide of the trivial and the serious that only loosely and imperfectly fit with an instrumental view of the world. Among his reflections in the book's back matter, Boorstin confided it had been his personal, in-the-moment experience with everywhere-ness that stimulated this insight. He sketched a day-in-the-life, from waking to sleep, in which he found the semiotic ever-present—billboards, radio and television programs, newspapers, magazines, movies, advertisements, commodities in stores, sales pitches, street conversation and the "desires I sense all around me"— and entering into the very constitution of the world. He saw the limitations of his intellectual framework *and* the disciplinary organization of the academy, which, he averred, when confronted with new phenomena "continues to pour almost exclusively into old molds." His epiphany-stimulated study, he concluded, "might offer a rough map of some too little known territories in the

---

13. This concern received fuller expression just over a decade later in Daniel Bell, *The Cultural Contradictions of Capitalism*, 20th anniversary ed./with a new afterword by the author (New York, NY: Basic Books, 1996) [originally published 1976]. Boorstin's notion of rationality should be situated in the postwar environment; see S. M. Amadae, *Rationalizing Capitalist Democracy: The Cold War Origins of Rational Choice Liberalism* (Chicago, IL: University of Chicago Press, 2003).

new American wilderness. It might suggest how little we still know, and how slowly we are learning of the inward cataclysms of our age."[14]

Yet Boorstin's reflections on the "inward cataclysms of our age" were not original, and perhaps provided an unwitting commentary on the limitations of the consensus school of history then dominant in the academy. The "too little known territories" long had been part of Marxist critique (centered on Marx's notion of the fetish of the commodity). Boorstin's emphasis on the everyday phenomenology of the semiotic was a decades-later echo of German critical theorist Walter Benjamin's Arcades Project, begun in the 1920s, which examined the effects of early 20th century urban commercial-media environments on the Western experience.[15] Benjamin argued that this new condition, distinguished by a kinetic and overlapping environment of signs, reoriented vision and perception, diminishing the ability of individuals to separate out and contemplate the constituents of experience—whether, to use Boorstin's terminology, such constituents were real or pseudo events or objects.[16] Benjamin's analytic made clearer the intellectual task of analyzing underlying cultural structures as embodied in and expressed through day-to-day immersion in things and images associated with industrialization, new forms of consumption, and modernity.[17] Benjamin and later critical theorists argued that these fields of experience, new and fundamental, shaped the behavior of individuals and groups in ways that were bound yet distinct from the purposive ideologies and actions connected to elite politics or the institutions of capitalism.

Boorstin's hybrid fascination-worry about an image-culture was a descriptive statement overlaid with misgivings—not a theory. The Marxist tradition, which Benjamin exemplified, advanced in the 1960s new ways of looking at this post-World War II phenomenon—by linking it to (not surprisingly) a transformation in the basis of capitalism. Daniel Bell's 1973 *The Coming of Post-Industrial Society: A Venture in Social Forecasting* helped make the case that, as Cold War modes of

---

14. Ibid., p. 264.

15. On the history of the semiotic and the importance of Benjamin, see two valuable works by Jonathan Crary: Jonathan Crary, *Techniques of the Observer: On Vision and Modernity in the Nineteenth Century* (Cambridge, Mass: MIT Press, 1990) and Jonathan Crary, *Suspensions of Perception: Attention, Spectacle, and Modern Culture* (Cambridge, MA: MIT Press, 1999).

16. See Crary, *Techniques of the Observer*, pp. 19-20.

17. This analytic line became an academic growth industry in the 1960s, leading to new fields of study such as material culture. As a partial measure of these developments, see Pierre Bourdieu, *The Logic of Practice* (Stanford, CA: Stanford University Press, 1990); Daniel Miller, ed., *Materiality* (Durham, NC: Duke University Press, 2005); and Mark Poster, *Information Please: Culture and Politics in the Age of Digital Machines* (Durham, NC: Duke University Press, 2006), especially chapter 10. For the post-World War II period, a particularly useful treatment of the theoretical issues is Patricia Ticineto Clough, *Autoaffection: Unconscious Thought in the Age of Teletechnology* (Minneapolis, MN: University of Minnesota Press, 2000).

knowledge production moved from the state to the market and became more closely connected to the world of commodities, representation and practices of representation became a more integral part of capitalism.[18] By the early 1980s, a range of authors came to see this complex of changes as the seedbed of globalism (conceived as the relative enhancement of the power of markets in relation to states across the transnational landscape) and postmodernism (conceived as a new cultural condition associated with this mode of production).[19] With this turn, Boorstin's argument was reframed: Yes, images and semiotics broadly conceived existed as a quasi-ontological fixture of life, but they operated in and through the power-relations of this emerging thing called globalization.[20]

---

18. To be clear: Bell did not advance this argument himself; others, such as the authors cited here, did. The best delineation of the shift from "organized" to "disorganized" capitalism (and in the Marxian tradition correlating this change to distinctive cultural orders) is Scott Lash, *The End of Organized Capitalism*; also see Nick Heffernan, *Capital, Class and Technology in Contemporary American Culture: Projecting Post-Fordism* (London; Sterling, VA: Pluto Press, 2000). For a dense, contemporaneous account, less attuned to the enhanced status of knowledge seen by Galbraith and Bell, see Ernest Mandel, *Late Capitalism* (London: NLB, 1975) [First published as *Der Spätkapitalismus*, Suhrkamp Verlag, 1972]. For an overview of changes from the 1960s in corporate structure and strategy as firms moved from primarily national to broadly transnational modes of operation, see Naomi R. Lamoreaux, Daniel M. G. Raff, and Peter Temin, "Beyond Markets and Hierarchies: Towards a New Synthesis of American Business History," *American Historical Review*, 108 (April 2003): 404–433. On the rise of private market ideology and accompanying policy reorientation in this period see Daniel Yergin and Joseph Stanislaw, *The Commanding Heights: The Battle Between Government and the Marketplace That Is Remaking the Modern World* (New York, NY: Simon & Schuster: 1998). On changes in modes of production from the perspective of history of science and technology, see Paul Forman, "The Primacy of Science in Modernity, of Technology in Postmodernity, and of Ideology in the History of Technology," *History and Technology* 23 (2007): 1–152; and Philip Mirowski, *The Effortless Economy of Science?* (Durham, NC: Duke University Press, 2004).

19. See, as a range of examples: Arjun Appadurai, ed., *Globalization* (Durham, NC: Duke University Press, 2001); Francis Fukuyama, *The End of History and the Last Man* (New York, NY: Perennial, 2002); Anthony Giddens, *The Consequences of Modernity* (Stanford, CA: Stanford University Press, 1990); Harvey, *The Condition of Postmodernity*; Jameson, *Postmodernism*; Jonathan Xavier Inda and Renato Rosaldo, eds., *The Anthropology of Globalization: A Reader* (Malden, MA: Blackwell Publishers, 2002); and Frank Webster, *Theories of the Information Society* (London: Routledge, 2002). Several case studies in space history engage these changes. See Martin Collins, "One World One Telephone: Iridium, One Look at the Making of a Global Age." *History and Technology* 21(2005): 301–324; Lisa Parks, *Cultures in Orbit: Satellites and the Televisual* (Durham, NC: Duke University Press, 2005); and Peter Redfield, "The Half Life of Empire in Outer Space," *Social Studies of Science* 32 (2002): 791–825. More broadly, in the historical profession these changes have given new life to the transnational (in contrast to the long-standing preference for the "national") as a key unit of analysis. See Thomas Bender, ed., *Rethinking American History in a Global Age* (Berkeley, CA: University of California Press, 2002).

20. In this analytic sketch one needs to be careful not to assume a unitary global capitalism. For deeper discussion of this point see *Varieties of Capitalism: The Institutional Foundations of Comparative Advantage*, eds. Peter A. Hall and David W. Soskice (Oxford, UK: Oxford University Press, 2001), and *Varieties of Capitalism, Varieties of Approaches*, ed. David Coates (New York, NY: Palgrave Macmillan, 2005).

Spaceflight mapped onto this tangle of production and culture in particular and seemingly contradictory ways. Earth-serving satellites (communications, meteorological, remote sensing, science—civilian, commercial, military) served as a prominent institutional and material element of this regime of production and consumption, exemplifying in the postwar period the robust capabilities of state-market collaborations. These arrangements enlarged and amplified the very condition of image-ness examined by Boorstin. Benjamin's work focused on the early 20th century urban experience; Boorstin's on the U.S.; and subsequent critical theory (using, say, Baudrillard's 1968 *The System of Objects* as a starting point) began to see semiotic immersion as a transnational condition.[21] As such, spaceflight, mostly in tacit and un-remarked ways, became bound to the politics of globalization and the myriad points of contestation that gave new weight to terms such as local and identity. However, in U.S. culture in particular, spaceflight also operated as an explicit, widely circulating symbol bearing important, transcendent connotations. As exploration, as frontier, as a place apart from the corruptions of Earth, spaceflight suggested the possibility of individuals and humanity collectively achieving Enlightenment ideals of universal values fulfilled, either via travel beyond Earth or drawing space experience and knowledge back into worldly experience.

As a window onto this set of issues, consider any of the late 1960s or early 1970s Earth-as-seen-from-space images, such as Apollo 8's Earthrise or selected covers from the Stewart Brand-created Whole Earth catalog. Such images invigorated a romantic discourse in which humanity via spaceflight perspectives and machines might find harmonious balance with nature. This discourse served as a powerful counter narrative to a century-plus series of writings in European and American critical thought about the machine in the garden—a critique that reached a crescendo in 1950s and 1960s in the work of authors such as Hannah Arendt, Herbert Marcuse, and Lewis Mumford.[22]

But such space-based Earth images, too, in other contexts, could be consistent with notions of the machine run rampant, of bureaucratic or corporate control on a planetary scale—a level of technological hubris that Lewis Mumford railed against in his nearly contemporaneous *The Myth of the Machine*. And such iconography helped make concrete a uniquely important practical Cold War ambition: that the totality of Earth could and should serve as a stage of action. This ambition, made possible by numerous, discrete civilian

---

21. Jean Baudrillard, *Le Système Des Objets* (Paris: Gallimard, 1968).

22. See, for example, Hannah Arendt, *The Human Condition*, 2nd ed. (Chicago, IL: University of Chicago Press, 1998) [originally published 1958]; Herbert Marcuse, *One-dimensional Man: Studies in the Ideology of Advanced Industrial Society* (Boston, MA: Beacon Press, 1964); and Lewis Mumford, *The Myth of the Machine*, 1st ed. (New York, NY: Harcourt, Brace & World, 1967) and *The Pentagon of Power* (New York, NY: Harcourt Brace Jovanovich, 1974) [Volume 2 of *Myth of the Machine*].

and military accomplishments over the 1960s, followed the main lines of force in the Cold War assemblage of politics, knowledge, and institutions. Total war, in its U.S. Cold War incarnation, assumed the notion of complete action across the globe, even if it had not yet been realized. With the shift to market-driven economic policies in the 1980s, multinational corporations stood ready if the military (as a source of scientific and technical tools and as a guarantor of security) and international regulatory regimes made planetary-scale markets possible. While one might squirm at a hackneyed invoking of classic explanatory "go-to" guys, the story of the last 30 years—the drama of globalization—is intimately bound to the elaboration and working out of national, military, and business interests in making the planet an honest-to-God, no-messing-around stage of action. And a new tighter integration between military and business was a significant part of that undertaking.[23] As a matter of historical framing, the emphasis is properly U.S.-centric—for it is U.S. actors who are most motivated and most able to effect this planetary ambition, to pursue, as a growing body of literature argues, a distinctly U.S. form of empire, hegemony, dominance—choose your word.[24]

But let us return to those Apollo-era Earth-from-space images and their connotations of universal ideals and an abstract, collective humanity. In a recent study, historian Frederic Turner argues that Whole Earth creator Stewart

---

23. This was an unstated thesis of Bell's, *The Post-Industrial Society*; it was the institutional basis for observations on the emergence of scientific and technical knowledge communities as a distinct and important sociological formation. A number of works in 1950s and 1960s pointed to the close collaboration between the military and business; the most sustained argument regarding this collaboration as an economic system is John Kenneth Galbraith, *The New Industrial State* (Boston, MA: Houghton Mifflin, 1967).

24. The argument for U.S. centrism as a valid historiographic angle is present in several literatures— and, of course, not without contestation. It is a prominent thread in the critical theory literature, reflecting Jameson's influence. On this, see Anderson, *Origins of Postmodernity*, chapter 3. From the perspective of "empire," see Charles S Maier, *Among Empires: American Ascendancy and Its Predecessors* (Cambridge, MA: Harvard University Press, 2006). In the globalization literature, the assumption is rife, but with attention to the countervailing or resistive effects of the local. A sample of this literature is in note 3. As an indicator, consider how Reinhold Wagnleitner and Elaine Tyler May, eds., *Here, There, and Everywhere*, frame their study of the global reach of U.S. popular culture: "It could easily be argued that the products, icons, and myths of American popular culture represent the single most unifying and centripetal cultural force for the global triumph of the American century. On the other hand, in many areas of the world, American cultural products are potentially among the most disruptive and centrifugal cultural forces of the twentieth century." (p. 1) Either way it is the "American" that is at the center of their analytic. To apply this argument to the 1970s and 1980s seems more problematic given the antagonistic positioning of the U.S. and U.S.S.R. But several authors point to relevant similarities between the two as inheritors of a modernist ideology that linked politics, technology, and notions of global control. See, for example, David C. Engerman, et al., ed., *Staging Growth: Modernization, Development, and the Global Cold War* (Amherst, MA: University of Massachusetts Press, 2003); and Odd Arne Westad, *The Global Cold War: Third World Interventions and the Making of Our Times* (Cambridge, UK: Cambridge University Press, 2005).

Brand saw post-industrial capitalism, rooted in Cold War knowledge and technological practices, as the means to transport counterculture values into the large society and to invigorate Enlightenment beliefs in the individual as the measure of all things—one instance of how the global construct emerging in the 1970s provided a home for seemingly contrary belief systems, for the mega-machine, *and* for its fundamental transformation.[25] The distinctive feature of the U.S. style of empire was to perform that very conflation—to conjoin the Enlightenment heritage of abstract universals with an actual, rubber-meets-the-road total global everywhere-ness of state and market actors. Courtesy of space-based capabilities, idealism and practice now confronted each other literally everywhere. This twist on old colonial and imperial modalities gave rise to different and nuanced intersections of the local and global—as suggested by Kingsolver's view from southwestern Virginia—whether as sites of contrast and difference, of acceptance or rejection, or of absorption and transformation—whether of music, film, television, hamburgers, or IMAX space adventures.

In the Era of Representation, Two Examples: GPS and Iridium

Let's shake this mix of semiotics, capitalism, spaceflight, the global and the local, and consider a couple of examples. In public discourse on globalization, capitalism—restless, U.S. and European-centric, with Asia on the rise—has drawn the most attention. But the U.S. military, as already inferred, has played an essential part, creating seemingly strange linkages between national security and a rampant transnational consumer culture. Consider the example of the Global Positioning System (GPS), a network of satellites designed and operated by the U.S. Air Force (USAF). Conceived in the early 1970s and only becoming fully operational in 1995, GPS's history straddled the Cold War and its market-oriented aftermath. The system provided a soldier, ship, airplane, or missile with information on their exact position anyplace on the planet via signals encoded with highly accurate time data, its profound effects symbolized by its use in guiding "smart" bombs and missiles with deadly, precise accuracy in the post-September 11 conflicts in Afghanistan and Iraq.[26] In recent years, satellite photographic images of these "smart" actions have been a staple of television news and are widely available on the Internet.[27]

---

25. Fred Turner, *From Counterculture to Cyberculture: Stewart Brand, the Whole Earth Network, and the Rise of Digital Utopianism* (Chicago, IL: University of Chicago Press, 2006).

26. A useful account of GPS to 1995 is Scott Pace, et al., *The Global Positioning System: Assessing National Policies*, MR–614–OSTP (Santa Monica, CA: RAND Corporation, 1995).

27. As one example, see this "before and after" account of a strike in Afghanistan from the Aerospace Corporation: *http://www.aero.org/publications/crosslink/summer2002/05.html* (accessed on February 10, 2008).

In the 1990s, though, GPS became not just a tool for the U.S. military, but for anyone, anywhere in the world—for a hiker in the Rocky Mountains, for mom and dad driving the family car, as well as for past and present adversaries such as a Chinese soldier or a terrorist. In its posture toward users, the system became egalitarian in the extreme. Not surprisingly, in the go-go market-driven post-Cold War world, business and consumer use of GPS vastly outstripped that of the military. Imagine a use for location, tracking, or accurate time information and one found GPS there, often in the intimate contours of daily life— say, tracking a spouse suspected of having an affair. Indeed, one can "google" a phrase such as "cheater tracker" and find GPS products marketed for that purpose. Combined with a geographic information system (such as satellite-based Google maps), one can track and visualize the itinerary of an errant mate.[28] Such uses are not just confined to the U.S.: as the *New York Times* observed, "the world has incorporated our GPS into its daily life as rapidly as Americans took up the ATM banking network."[29] Nothing, perhaps, speaks more to the distinctive conjunction between production and semiotics in the global era, to the total, actual, not merely metaphorical, planetary scale of this conjunction, than GPS, its radio signals equally available to friends and foes, to weapons in flight, and to off-the-shelf products offering to meet every consumer need.

The GPS story is not the same as that by-gosh, by-golly story of the origins of the Internet—of "isn't it strange that a research project into maintaining command and control during a nuclear holocaust gave us this wildly diverse, unpredictable, electronic social universe." It is different and more revealing of the transformation in the world order since the early 1970s. It still is, in essence, controlled by the U.S. military, and in a way unprecedented in U.S. history unites a classic function of empire—controlling and maintaining its perimeters—with the churning demands of capital and consumer appetites. And not just those based in United States, but everywhere. GPS has become a military-consumerist hybrid, in which each political-cultural domain has continually redefined the other. We—a transnational we—know the precision bomb blast from Afghanistan or Iraq and the "cheater tracker" originate in and depend on the same system of production, yet in our everyday cultural frame of semiotics, we allow them to maintain their separateness. It is tensions such as these that continually redefine the global and the local, geographically and in time, and keep the United States—and its preeminence and exploitation of spaceflight— in the center of this dialectic.

Consider another example, drawn from my current research that tracks Boorstin's concern about semiotics and the structuring of our sense of the real. Like GPS, Iridium was and is a satellite constellation that completely embraces

---

28. A fun example may be seen here: *http://www.brickhousesecurity.com/catch-a-cheater.html* (accessed on February 10, 2008).

29. James Hitt, "Battlefield: Space," *New York Times Sunday Magazine*, (August 5, 2001): 63.

the planet—but with a different purpose, to provide telephony and data services, and with a different institutional actor in the lead, a multinational corporation. Conceived in 1987 at Motorola, a Fortune 500 company and a leading firm in cellular phone equipment and systems business and in semiconductors, the Iridium satellite project seemed to epitomize the historical moment: as the Cold War waned and collapsed, markets rather than government would lead into a techno-democratic future, and corporations rather than nations would articulate the pathways through which the local and the global took shape. The largest privately-financed technology project ever undertaken, and with an array of international investors, including the newly constituted Russian Federation and the People's Republic of China, Iridium stood as symbol of this fusion of technology, corporations, markets, and international politics. In 1998, as the system neared completion, *Wired* magazine proclaimed, "It's a bird, it's a phone, it's the world's first pan-national corporation able to leap geo-political barriers in a single bound."[30]

Part of my challenge in untangling this story has been to understand the varied ways in which semiotics functioned in a multinational corporation (MNC). You might expect that a MNC, deep-pocketed, well-connected politically, at home and internationally, with tens of factory and sales sites around the world would be an instrumental historical actor extraordinaire, a big "them" guy able to exert power in ways unavailable to all the little "us" guys. And, of course, that crude truth is there. But so is another one, one in which Motorola regarded the semiotic realm as real, a reality that required substantive corporate responses that intermingled culture, politics, and identity. As a literally planetary project, incorporating flesh-and-blood actors from around the world, Iridium dramatically highlighted the problem of semiotics—local, global, multiple, contesting, and not readily controlled—and the need for solutions.

The Motorola's response to this condition can be glimpsed in a 1998 book entitled *Uncompromising Integrity: Motorola's Global Challenge*.[31] The concept of culture stood as organizing precept. The narrative provided definitions of culture and related concepts that showed it as a structure, but varied in place and time, and as a process—national culture, subculture, host culture, enculturation, and transcultural. Two key additional notions situated the discussion in the corporate context: "Motorola culture" and "home culture." The first made clear that the organization had a semiotic sphere, derived from its own history and as a U.S.-

---

30. Keith Bradshear, "Science Fiction Nears Reality: Pocket Phone for Global Calls," *New York Times*, (June 26, 1990): pp. A1 and D7; David S. Bennahum, "The United Nations of Iridium," *Wired* 6.10 (October 1998): pp. 134-138, 194-201.

31. R. S. Moorthy and Robert Galvin, *Uncompromising Integrity: Motorola's Global Challenge* (Schaumberg IL: Motorola University Press, 1998). As a measure of the importance Motorola attached to this issue, note that Robert Galvin was the son of Motorola founder Paul Galvin and CEO of the company at the time Iridium was initiated.

centered capitalist institution. The second that that sphere was permeable and in flux because employees hailed from many localities around the world and because, as a multinational, the corporation always was operating in someone else's backyard. Culture was something around which a company had to define itself (Motorola culture). Yet, in the global age, "home" was complex and mobile, reflective of the world's many diasporas—of people, individually and en masse, following the flow lines of capital. Home inhered in individuals even as they moved (with Motorola employees themselves an example) and in those places from which they came. Motorola and home cultures were oppositional and profoundly interpenetrating.[32]

Uncompromising Integrity's preoccupation with culture—perceived as variegated and everywhere, in specific geographical places, in institutions (including Motorola), in individuals, and pulsing through the many channels of the media—had a corporate history. It encapsulated more than 15 years of high-level managerial attention to the global. It led executives in the late 1980s to create a hybrid academic-corporate institution—Motorola University—to engage and comprehend the fauna and flora of culture-world. This book was a product of that—a Motorola University Press publication! Lest this example seem quirky and isolated, note that it exemplified a larger trend: Over a decade, from the mid 1980s to mid 1990s, more than a thousand corporate universities were created in the United States—all of which were a response, in one fashion or another, to the perceived challenge of culture and semiotics to transnational business practice.[33]

The biography of the lead author—R. S. Moorthy—makes concrete some of the issues of identity and politics embedded in these developments. Born into an Indian family and raised in poverty in Malay, as a young man he found work in a Motorola facility in that country. His professional life at Motorola became one of reconciling his origins in a place with a specific history, one tied to colonialism and the new globalism, with the purposes and outlooks of a multinational firm. He found a way to marry his interests with Motorola's culture preoccupation and he came to play a major role in establishing Motorola University, creating a subunit of that enterprise, the Center for Culture and Technology.

This vignette only is meant to suggest the complicated and non-obvious ways in which social boundaries got created and negotiated and how semiotics constrained and enabled this process at different levels of corporate activity. As one instance, consider this graphic (see figure) outlining the manufacturing flow for the Iridium project, one that required a transnational "virtual factory"—

---

32. A particularly cogent analysis of culture in its post-1970 global dimensions is Zygmunt Baumann, *Culture As Praxis,* pp. vii-lv.

33. For an overview of this trend from a policy perspective, see Stuart, Cunningham, et al., *The Business of Borderless Education* (Canberra: Commonwealth of Australia, Department of Education, Youth, and Training, 2000).

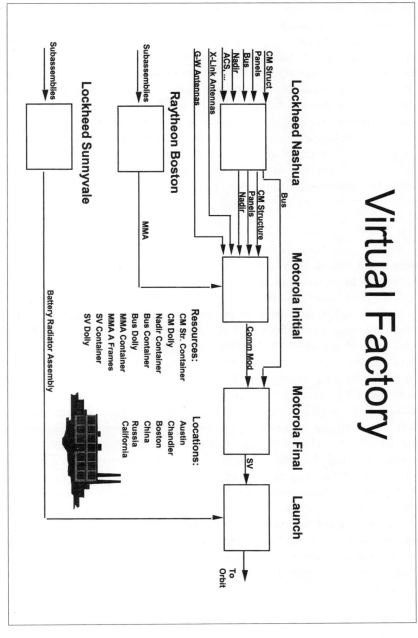

Virtual Factory

Lockheed Nashua
- CM Struct.
- Panels
- Bus
- Nadir
- ACS,...
- X-Link Antennas
- G-W Antennas

Raytheon Boston
Subassemblies

Lockheed Sunnyvale
Subassemblies

Motorola Initial
- Bus
- CM Structure
- Panels
- Nadir
- MMA

Comm Mod.

Battery Radiator Assembly

Resources:
- CM Str. Container
- CM Dolly
- Nadir Container
- Bus Container
- Bus Dolly
- MMA Container
- MMA A Frames
- SV Container
- SV Dolly

Motorola Final

SV

Launch

To Orbit

Locations:
- Austin
- Chandler
- Boston
- China
- Russia
- California

Diagram prepared by Motorola engineers in the mid 1990s that shows how the varied institutional actors and geographic sites involved in the Iridium project were envisioned as an unified construct—a factory (Courtesy of Andrew Feller)

Motorola's own phrase. In the lower right corner, we find included in the virtual factory Baikonaur, Kazakhstan and Taiyuan, China. That Kazakhstan and China could be integral, functioning elements of a U.S.-based business project made sense only in the context of this belief in the everywhere-ness of culture and the reality of semiotic structures—whether in dealing with transnational elites or with questions of politics and identity in specific geographic locales. But as the project moved from planning and manufacturing execution to marketing, the instrumental view—think the way I, Iridium, want you to think, buy my phone—comes to the fore and you find a different way of presenting geographical specificity and cultural accommodation, bleached and abstracted back to the aims of the corporation and neo-liberal capitalism, and using Enlightenment universals to facilitate those aims.[34]

Conclusion

Where do these examples leave us? What might we draw from this mix of state action and capitalism, of semiotics and on-the-ground practice, of geographical specificity, local meanings, and the meta-narratives of the West? And where does spaceflight fit into this contested field of action? I would point to two things.

First, my recitation here advances a particularly modest claim: Deep and important issues become visible when we recalibrate our interpretive lens and see spaceflight *in* history, rather than expecting it to be sui generis. Spaceflight—especially those near-Earth applications cited here—has been a major element in creating the incarnation of the global we have experienced over the last 40 years. It has provided images and practices that have made the category of the global, natural and insistent, even when different actors give it different meanings. It has been a primary site in which prior categories of the modern—the nation state, the military, civil society, capitalism—have been refashioned and given new meanings. And in helping to elevate the importance of the global—to distribute its effects across regions and places, to recalibrate our sense of distance and time, of identity as a creation of community or the flows of transnational semiotics—it has helped to invigorate the meaning of the local. This has led to an intensified scrutiny of globalism's origins in the West—emanating from its military, economic, political and cultural institutions—and in its ideological underpinnings, the legacy of the Enlightenment, of universal values inhering in universal humans. Spaceflight thus has enabled the dominant vectors of the global and its critique.[35]

---

34. Advertising images and language conveyed corporate aims through use of universals. Language, in particular, served this function. Two of the most widely used taglines, at different times, were "One world, one telephone" and "Freedom to communicate."

35. I don't think it is a coincidence that in the post-1970s literatures of critical theory, post-colonialism, and globalization some of the most influential work is coming from scholars in India

Second, this essay grapples with the problem of the image and semiotics identified by Boorstin. Is the question "what is real" as transformative as he suggests? Does it stand as a fundamental change in the relation of citizens to politics? Of what politics means and makes possible when interest-driven actors *and* cultural structures are taken as historical agents? These musings are not mere abstractions; they filter down to the everyday—in what we take to be credible, in what we trust, and in what we question rather than give assent. If they have historical weight, they represent a reorientation of existential and value structures in the postwar period. Is that not a historical problem of the first rank, a deep argument for placing spaceflight in the broadest frames of analysis?

For spaceflight itself, there is a perhaps surprising blowback in this cultural condition. In exposing the contingency of the global, as a product with a geography and history in the West, of universal values as a specific cultural creation not as given absolutes, one can find some empirical oomph behind two of those tenets of the postmodern that some love to hate. One is Lyotard's claim about the "death of meta-narratives," principally those Enlightenment universals. What Lyotard meant is not the *disappearance* of such narratives but skepticism as to their truth and general applicability. They still run wild in the semiotic transnational landscape. And with this skepticism, comes the second tenet, the one from which Boorstin recoiled: the end of that abstract Enlightenment individual enacting Enlightenment values.

And the blowback is this: spaceflight as *application* helped make this world. But spaceflight as *exploration*, particularly human exploration, encountered and encounters this condition somewhat differently. Granting human spaceflight's grounding in Cold War real politik, that experience gained credibility because exploration as a culture trope drew strength from Western meta-narratives and explorers as universal human subjects. If human exploration is only a narrative and not a meta-narrative in competition with other narratives, then space exploration as an exemplification of Enlightenment values fades. The exploration narrative still resonates, but in a much diminished way. And this ties back to Boorstin's concern that image-ness threatened to change the very nature of politics—from a field of experience built on elite actions, meta-narratives, and Enlightenment rationality to one in which these elements are transformed and conjoined with the ontology of everywhere semiotics. This insight offers an analytic hint: traditional explanatory modes that rely on interest groups and elite power to account for the history of human spaceflight in the last 40 years miss the changed foundations (as presented in critical theory and other literatures) of politics and culture. The Moon journeys, it may be argued, exemplify the modern temperament; the subsequent years of human spaceflight illustrates the intrusion of the postmodern into the modern, a marker

---

or the Indian diaspora. See as one important example Dipesh Chakrabarty, *Provincializing Europe: Postcolonial Thought and Historical Difference* (Princeton, NJ: Princeton University Press, 2000).

of "second nature," of the complex ways in which spaceflight and culture have been mutually reconfigured.

At the beginning of the Space Age, applications stood as the stepchild of the more glamorous meta-narrative of human exploration. Over the 1970s and 1980s she came to supplant her parent and helped make a new realm of meaning and experience—her own world, a second nature, a new chapter in the saga of the machine and the garden.

CHAPTER 12

CREATING MEMORIES: MYTH, IDENTITY, AND CULTURE IN THE RUSSIAN SPACE AGE[1]

Slava Gerovitch

The Nobel Prize laureate Orhan Pamuk's novel, *The White Castle*, is a subtle reflection on the power of memory. Living in 17th-century Istanbul, two main protagonists, an Italian scholar and a Turkish noble, share their most intimate memories and gradually adopt each other's memories as their own. Their distinct identities begin to blur until they (and the reader) can no longer recognize who is who. Eventually they switch their original identities as the power of memory overwhelms them. The Turk becomes a scholar and leaves for Italy, while the Italian abandons science to enjoy luxurious life at the sultan's court.[2] Our memories determine who we are, and manipulating these memories affects the very core of our identity.

Key events in the Space Age are especially memorable—this is why it is called "the Space Age" in the first place. The triumphs of Gagarin's first flight and Armstrong's first step, and the tragedies of Apollo 1, Gagarin's death, *Challenger*, and *Columbia* are among recent generations' most vivid and emotional memories. But what do we really remember when we remember the Space Age? In 1986-1988, the cognitive psychologist Ulric Neisser conducted a study of 44 student subjects, who were asked how they first heard the news of the *Challenger* disaster. The first round of questioning took place the next morning after the event, the second round—with the same participants—two and a half years later. It turned out none of the later accounts fully coincided with the original report, and over a third were, as Neisser put it, "wildly inaccurate." Moreover, even when confronted with their own earlier written reports, the subjects were convinced that the later memory was true. The original memories quite simply disappeared from their minds.[3]

---

1. The author wishes to thank Asif Siddiqi for his insightful comments on an early draft of this article. Research for this article has been supported by Fellowship in Aerospace History from the American Historical Association.

2. Orhan Pamuk, *The White Castle*, trans. Victoria Holbrook (New York, NY: Braziller, 1991).

3. Ulric Neisser and Nicole Harsh, "Phantom Flashbulbs: False Recollections of Hearing the News about *Challenger*," in *Affect and Accuracy in Recall: Studies of "Flashbulb" Memories*, ed. Eugene Winograd and Ulric Neisser (New York, NY: Cambridge University Press, 1992), pp. 9-31.

Recent research in cognitive, social, and clinical psychology and in cognitive neuroscience indicates that our memory is a much more dynamic and malleable process than previously thought. Our memories are not stored in a fixed form; we do not pull them out of a permanent storage and then put them back intact. According to the constructivist approach to memory, every act of recollection is re-creation, reconstruction of a memory.[4] Every time we "recall" a memory, we relive the event that caused it, we emotionally relate to it, we remake that memory, and we store a new version, totally overwriting the old one. At the moment of recollection, memory becomes unstable, and it can be modified and even "erased," or a false memory can be planted.[5] Recalling something is essentially similar to making a new, original memory. In the language of neuroscience, "reconsolidation . . . is qualitatively strikingly similar to consolidation";[6] in the psychology parlance, "recollection is a kind of perception, . . . *and every context will alter the nature of what is recalled.*"[7] As a result, we do not really remember the original event; we remember only our last recollection of that event. The more we remember and the more often we recall something, the more we rebuild and change that memory and the farther we get from the original event.

According to the school of "narrative psychology," linking individual memories into a coherent narrative, which supplies meaning to past events, plays an essential role in the formation of one's self.[8] As the neurologist Oliver Sacks has put it, "We have, each of us, a life story, an inner narrative—whose continuity, whose sense, *is* our lives. It might be said that each of us constructs and lives a 'narrative,' and that this narrative *is* us, our identities."[9] When our

---

4. The idea of memory as a dynamic and constructive process goes back to Frederic C. Bartlett's book *Remembering* (Cambridge, UK: Cambridge University Press, 1932). For overviews of recent studies, see Daniel L. Schacter et al., "The Cognitive Neuroscience of Constructive Memory," *Annual Review of Psychology* 49 (1998): 289-318; Daniel L. Schacter, "Memory Distortion: History and Current Status," in *Memory Distortion: How Minds, Brains, and Societies Reconstruct the Past,* ed. Daniel L. Schacter (Cambridge, MA: Harvard University Press, 1995), pp. 1-43; and Daniel Schacter, *Searching for Memory: The Brain, the Mind, and the Past* (New York, NY: Basic Books, 1996).

5. On experiments with "erasing" fear conditioning in rats, see Karim Nader, Glenn E. Schafe, and Joseph E. Le Doux, "Fear memories require protein synthesis in the amygdala for reconsolidation after retrieval," *Nature* 406 (August 17, 2000): 722-726. On experiments showing the possibility of implanting false memories in humans, see Elizabeth F. Loftus and Katherine Ketcham, *The Myth of Repressed Memory* (New York, NY: St. Martin's Press, 1994).

6. Karim Nader, "Memory Traces Unbound," *Trends in Neurosciences* 26:2 (February 2003): 70.

7. Israel Rosenfeld, *The Invention of Memory: A New View of the Brain* (New York, NY: Basic Books, 1988), p. 89 (emphasis added).

8. See Jerome S. Bruner, *Acts of Meaning* (Cambridge, MA: Harvard University Press, 1990), chap. 4, "Autobiography and Self"; and Ulric Neisser and Robyn Fivush, eds., *The Remembering Self: Construction and Accuracy in the Self-Narrative* (Cambridge, UK: Cambridge University Press, 1994).

9. Oliver Sacks, *The Man Who Mistook His Wife For a Hat and Other Clinical Tales* (New York, NY: Summit Books, 1985), p. 110.

present self constructs and distorts our memories of the past, the very fallibility of these memories serves a purpose—to establish a continuity between our present and past selves. The literary scholar Paul Eakin has argued that memory is "not only literally essential to the constitution of identity, but also crucial in the sense that it is constantly revising and editing the remembered past to square with the needs and requirements of the self we have become in any present."[10]

We are what we remember, and this is equally true for individuals and societies.[11] By focusing on the notions of "collective memory" and "social memory," cultural history draws on the metaphor of society as a remembering subject, which constructs its identity based on collective remembrance and can go through a psychological "trauma" profoundly distorting collective memories.[12] Collective memory—culturally sanctioned and publicly shared representations of the past—shapes social identities and provides narratives through which individuals publicly describe their selves, remember the past, and interpret the present.[13]

---

10. Paul John Eakin, "Autobiography, Identity, and the Fictions of Memory," in *Memory, Brain, and Belief*, ed. Daniel L. Schacter and Elaine Scarry (Cambridge, MA: Harvard University Press, 2000), pp. 293-294. On the "false memory syndrome" as an adaptive mechanism, see Daniel L. Schacter, *The Seven Sins of Memory: How the Mind Forgets and Remembers* (New York, NY: Houghton Mifflin, 2001).

11. For recent attempts to bring together specialists from cognitive psychology, psychopathology, psychiatry, neurobiology, social psychology, sociology, and history to discuss the phenomenon of memory from different disciplinary perspectives, see Thomas Butler, ed., *Memory: History, Culture and the Mind* (Oxford, UK: Blackwell, 1989); Schacter, ed., *Memory Distortion*; Schacter and Scarry, eds., *Memory, Brain, and Belief*; and the newly established academic journal *Memory Studies*.

12. For recent general works on collective memory in social and cultural history, see Alon Confino and Peter Fritzsche, eds., *The Work of Memory: New Directions in the Study of German Society and Culture* (Urbana, IL: University of Illinois Press, 2002); Paul Connerton, *How Societies Remember* (Cambridge, UK: Cambridge University Press, 1989), John R. Gillis, ed., *Commemorations: The Politics of National Identity* (Princeton, NJ: Princeton University Press, 1994); Pierre Nora, ed., *Realms of Memory: Rethinking the French Past*, trans. from the French, gen. ed. Lawrence D. Kritzman, 3 vols. (New York, NY: Columbia University Press, 1996-1998); Pierre Nora, ed., *Rethinking France: Les Lieux de mémoire*, trans. from the French, gen. ed. David P. Jordan, 2 vols. (Chicago, IL: University of Chicago Press, 2001-2006); Jeffrey Olick, *The Politics of Regret: On Collective Memory and Historical Responsibility* (New York, NY: Routledge, 2007); Jeffrey Olick, ed., *States of Memory: Continuities, Conflicts, and Transformations in National Retrospection* (Durham, NC: Duke University Press, 2003); and Eviatar Zerubavel, *Time Maps: Collective Memory and the Social Shape of the Past* (Chicago, IL: University of Chicago Press, 2003). Among the recent works that examine "traumatic" events in American historical memory are Edward Tabor Linenthal and Tom Engelhardt, eds., *History Wars: The Enola Gay and Other Battles for the American Past* (New York, NY: Metropolitan Books, 1996); Edward Tabor Linenthal, *The Unfinished Bombing: Oklahoma City in American Memory* (Oxford, UK: Oxford University Press, 2001); and Emily S. Rosenberg, *A Date Which Will Live: Pearl Harbor in American Memory* (Durham, NC: Duke University Press, 2003).

13. More precisely, "collective memory" in this article is understood as both a set of cultural norms that regulates practices of remembrance and a body of texts and other types of symbolic

When the constructivist model of individual memory is applied to cultural history, the implications are profound. Like individual memory, collective memory is continuously recreated, supplanting original memories with later versions. Cultural memory thus becomes self-referential: it feeds on itself and recollects its own recollections. The more a particular society or group remembers an event, the more intensely collective memory is at work, the more we mythologize and the more we forget. Remembering and mythologizing are the same thing. Just like false private memories reinforce the continuity of the individual self, cultural myths shore up national or group identity.

Taking seriously the view that culture is the myths we live by, historians have focused on the cultural functions of collective myths—to structure and pass on historical memory, to create the basis for a dominant "master narrative," and to shape social identities. In this context, whether the myth is literally true or not is not particularly significant. What is important is the political and cultural force of collective myths—ethnic, religious, ideological—their ability to act, to create a public appeal, to tell a story to identify with and an ideal to imitate. Most recent studies have shifted the focus toward the historical origins—the genealogy—of myths, their deliberate construction by political elites, and their repressive power to marginalize alternative stories and identities.[14]

The institutionalization of memory by nation states—the establishment of national archives, the public celebrations of various anniversaries, the dissemination of favorable historical narratives—often serves the political purpose of reinforcing national identity and marginalizes individual memories and other social identities. Private memories become "contaminated by national projects of remembrance," writes the historian Peter Fritzsche.[15] The French cultural historian Pierre Nora even argues that the old age of memory and tradition gave way to the new age of history and conscious narrative-construction. "Memory is constantly on our lips," he writes, "because it no longer exists."[16]

Space history has its own recurrent myths. The historian Asif Siddiqi has identified four cultural archetypes, or "tropes," structuring master narratives of space exploration in different countries: the myth of the founding father,

---

representations that a particular culture produces based on these norms. The most authoritative texts function as instantiations of the "master narrative," setting an effective norm for a wider discourse of remembrance. The term "collective" here does not imply uniformity of individual memories or a monolithic character of culture. Different groups within a larger society may have distinct collective memories that reinforce their group identities; narratives produced by these groups may come into conflict with the "master narrative" prevalent in larger culture.

14. Peter Fritzsche, "The Case of Modern Memory," *The Journal of Modern History* 73 (March 2001): 87–117.

15. Ibid., 107.

16. Pierre Nora, "General Introduction: Between Memory and History," in *Realms of Memory*, vol. 1, p. 1.

the myth of exclusively domestic space technology, the myth of spaceflight as expression of national identity, and various stereotypical justifications for spaceflight—the destiny of humanity, the glory for the nation, national security, economic development, scientific exploration, and benefits to the ordinary people.[17] Every nation develops its own variations, such as the "myth of presidential leadership"[18] and the "astronaut myth" in the United States. The historian Roger Launius has identified several key elements of the popular image of Apollo astronauts as a "cultural icon": the astronaut represented "everyman" and yet personified the American ideal, embodying the image of a masculine hero, a young, fun-loving, vigorous warrior, guided by an older, wiser leader, and showing the nation the path of progress toward utopian future.[19]

Like the Turk and the Italian in Pamuk's novel, who change their identities by listening to each other's stories, the astronauts could hardly remain unaffected by their image in popular culture. A recent documentary, *In the Shadow of the Moon*, is made entirely of interviews with Apollo astronauts illustrated with fragments of archival footage.[20] The film is not organized as a collection of separate stories of individual missions; instead, it weaves together bits and pieces of astronauts' stories to create a meta-story that blurs distinctions among different missions and even among different astronauts. It is as if a composite image of the astronauts is telling a composite story of lunar landings. Another recent documentary, *The Wonder of It All*, uses a similar technique, interleaving commentaries from seven astronauts who walked on the Moon.[21] As one reviewer has noted, "the editing has been done so skillfully that instead of seven individuals talking, it seems more like one—each of them often continues a sentence that the other started."[22] Individual stories—and individual astronauts' identities—blend together seamlessly. How did this blending occur? Is this a trick of the filmmakers or a fundamental cultural mechanism at work in real life, squeezing individual identities to conform to the dominant cultural stereotype of an astronaut? What happens to alternative memories? What are the mechanisms by which a culture decides which memories to erase and which to write over them?

---

17. See Siddiqi's article in this volume.

18. See Roger D. Launius and Howard E. McCurdy, eds., *Spaceflight and the Myth of Presidential Leadership* (Urbana, IL: University of Illinois Press, 1997).

19. See Roger D. Launius, "Heroes in a Vacuum: The Apollo Astronaut as a Cultural Icon," 43rd AIAA Aerospace Sciences Meeting and Exhibit, January 10-13, 2005, Reno, Nevada. AIAA Paper No. 2005-702 (available at *http://klabs.org/history/roger/launius_2005.pdf*).

20. *In the Shadow of the Moon*, directed by David Sington (Discovery Films, 2007).

21. *The Wonder of It All*, directed by Jeffrey Roth (Jeffrey Roth Productions, 2007).

22. Ronald A. Wells, "Review: *The Wonder of It All*," *The Space Review*, (accessed November 12, 2007). (available at *http://www.thespacereview.com/article/996/1*).

THE SPACE AGE IN AMERICAN CULTURE

The cultural historian Emily Rosenberg has set up an appropriate system of coordinates to analyze the role of the Space Age in American culture: a four-dimensional space of politics, the media, philosophy, and the arts. The Sputnik shock and the perceived "missile gap" boosted Cold War anxieties, and these anxieties, in turn, gave a spur to the space race. The media were enrolled in the ideological "battle of appearances," turning astronauts into international celebrities and making spacecraft launches and television broadcasts from space into spectacular public events. The idea of technocracy gained support, and technological elites gained economic and political power, while "counterculture" chose the Spaceship Earth image to promote environmental consciousness and a new global identity, which transcended the political boundaries of a nation state. In architecture, product design, and abstract expressionist painting, new space-inspired shapes and color palettes captured the spirit of a "new frontier" of space in the aesthetic of self-confident progress, futuristic automation, and individual adventure.[23]

The dynamics of the relationship between spaceflight and the media, outlined by Rosenberg, stresses the active, instrumental role of culture in shaping the Space Age. NASA skillfully used the media to create and disseminate a favorable public image of the U.S. space program, and at the same time space technologies engendered a technological revolution in visual media, making electronic communications truly real time and global. Emerged what Rosenberg has called a "synergy" between the Space Age and the Media Age: spaceflight acquired its spectacular character while the media thrived on new popular subjects of interest and on the advanced technologies. Wider culture did not simply reflect developments in the space program; it became a vehicle for specific agendas within the space program.

Rosenberg's analysis highlights tensions and contradictory trends in different aspects of the Space Age culture. The Space Age both threatened the sense of national pride and was enrolled to boost it. It gave birth to mammoth technological projects and raised concerns about uncontrollable government spending. It created a cult of technology and awoke suspicions about the attempts to find technological solutions to political problems. It trumpeted rationality and gave rise to various forms of spirituality. It was wrapped in the rhetoric of global unity and peaceful cooperation, and it led to the militarization of the heavens. It unleashed fantasy in the arts and regularized engineering creativity with systems engineering management techniques. It gave rise to both exciting and frightening visions of the future.

What are the cultural mechanisms that select specific iconic images, prominent figures, and big ideas that end up occupying a central place in the public

---

23. See Rosenberg's article in this volume.

memory of the Space Age? Recent literature begins to tackle the question of how, of all the variety of different visions of the Space Age, only a few survive as the dominant symbols of the era, while others are marginalized and forgotten.[24] As Roger Launius has argued, the American "master narrative" of spaceflight incorporates the mythology of "limitless frontier," the popular image of the "heroic explorer," and futurist visions to tell the story of American triumph in the space race, exceptionalism, and success. Three counter narratives have also emerged: the left-wing criticism of spending funds on space instead of social programs, the right-wing criticism of the space program as an excessive government expense, and various conspiracy theories of secretive space militarization schemes, alien abductions, and alike.[25] The competition among the master narrative and the three counter narratives might provide a template for analyzing the clash of diverse cultural representations of the Space Age outlined by Rosenberg. Each narrative plays out in public discourse through literature, imagery, film, and other media. The competition among Space Age symbols serves as a proxy for the battle of the narratives.

A number of seminal works have explored the relationship between NASA and popular culture. The political scientist Howard McCurdy has examined the links between popular conceptions of space exploration and national space policy, focusing on NASA's deliberate exploitation of the frontier myth and the utopian visions of social progress through technological means, and its encouragement of the Cold War fears of Soviet domination. As the space program after Apollo changed its character, it no longer matched the popular expectations inherited from the previous era. The gradual disillusionment with the NASA space program since the 1970s could be traced to a widening gap between popular sentiment and the reality of spaceflight.[26] The cultural theorists Marina Benjamin, Constance Penley, and others have studied how popular culture responded to the Space Age by reinterpreting NASA's symbolic imagery and generating competing discourses.[27] Broader culture turns space images, artifacts, names, events into

---

24. See, for example, Roger D. Launius, "Perceptions of Apollo: Myth, Nostalgia, Memory, or All of the Above?" *Space Policy* 21 (May 2005): 129–139; William D. Atwill, *Fire and Power: The American Space Program as Postmodern Narrative* (Athens, GA: University of Georgia Press, 1994); Andrew Smith, *Moondust: In Search of the Men Who Fell to Earth* (New York, NY: Fourth Estate, 2005). For a historiographic review of the cultural history of the Space Age, see Asif A. Siddiqi, "American Space History: Legacies, Questions, and Opportunities for Future Research," in *Critical Issues in the History of Spaceflight*, eds. Steven J. Dick and Roger D. Launius (Washington, DC: NASA SP-4702, 2006), esp. pp. 472-477.

25. See Launius's article in this volume.

26. Howard E. McCurdy, *Space and the American Imagination* (Washington, DC: Smithsonian Institution Press, 1997).

27. See Marina Benjamin, *Rocket Dreams: How the Space Age Shaped Our Vision of a World Beyond* (New York, NY: Free Press, 2003), Constance Penley, *NASA/Trek: Popular Science and Sex in America* (New York, NY: Verso, 1997), and Debra Benita Shaw, "Bodies Out of this World: The Space Suit as Cultural Icon," *Science as Culture* 13 (March 2004): 123–144.

"floating signifiers"—symbols without fixed meaning—that are reinterpreted again and again as they pass through different contexts. No single group or agency—even a government agency—can fully control them.

From a cultural anthropologist's perspective, the interaction between NASA and broader culture could be recast as a dialogue of different cultures: NASA's own culture(s) and the diverse subcultures of space fans, activists, educators, and artists. A study of this interaction might finally bring together two disparate research areas—the analyses of the Space Age in popular culture and the studies of NASA's own institutional culture(s).[28] The anthropological models of cultural contact, conflict, translation, mediation, and the "trading zone" may prove useful here.[29]

Combining the notion of historical memory with the model of cultural exchange leads to an investigation of the dynamics of memory in different cultures. Within larger American culture, every distinct group—space engineers, astronauts, and space fans, for example—nurtures its own memories, its own folklore, and its own historical visions of the Space Age. What happens when different groups interact and exchange their memories? What new mythologies and hybrid identities emerge?

Although different groups and different nations may have different memories of the Space Age, the cultural mechanisms by which these memories are exchanged and altered over time prove remarkably similar. If we look beyond American culture and examine the convolutions of the historical memory of the Space Age in Russian and Soviet culture, we will find a similar struggle between a master narrative and an array of counter-stories, even though the dynamics of this struggle will follow a specific Russian political and cultural trajectory.[30]

---

28. On NASA culture(s), see Alexander Brown, "Accidents, Engineering, and History at NASA, 1967–2003," in *Critical Issues in the History of Spaceflight*, pp. 377–402; Yasushi Sato, "Local Engineering and Systems Engineering: Cultural Conflict at NASA's Marshall Space Flight Center, 1960-1966," *Technology and Culture* 46:3 (July 2005): 561-583; Diane Vaughan, *The Challenger Launch Decision: Risky Technology, Culture, and Deviance at NASA* (Chicago: University of Chicago Press, 1996); Vaughan, "Changing NASA: The Challenges of Organizational System Failures," in *Critical Issues in the History of Spaceflight*, pp. 349-376.

29. See Peter Galison, "Trading Zone: Coordinating Action and Belief," in *The Science Studies Reader*, ed. Mario Biagioli (New York, NY: Routledge, 1999), pp. 137–160.

30. On memorialization practices in Soviet and post-Soviet contexts, see Svetlana Boym, *The Future of Nostalgia* (New York, NY: Basic Books, 2001); Frederick C. Corney, "Rethinking a Great Event: The October Revolution as Memory Project," *Social Science History* 22:4 (Winter 1998): 389–414; Geoffrey A. Hosking, "Memory in a Totalitarian Society: The Case of the Soviet Union," in *Memory*, ed. Butler, pp. 97–114; and James V. Wertsch, *Voices of Collective Remembering* (Cambridge, UK: Cambridge University Press, 2002).

RUSSIAN SPACE MEMORIALIZATION

Memories of the Space Age occupy a prominent place in contemporary Russian culture. This year alone, the Russians have celebrated the centennial of the legendary Chief Designer Sergei Korolev, the 150th anniversary of the space visionary Konstantin Tsiolkovskii, the 120th anniversary of the Soviet rocketry pioneer Fridrikh Tsander, the 50th anniversary of the R–7 intercontinental ballistic missile designed by Korolev, and, finally, the 50th anniversary of Sputnik and of Laika's flight on Sputnik II. One anniversary, however, was barely noticed: the ill-fated Soyuz 1 mission, which ended 40 years ago in a crash and the tragic death of the Soviet cosmonaut Vladimir Komarov. That year, 1967, was a significant turning point in Soviet cultural attitudes toward spaceflight: from admiration and pride to grief, cynicism, and, ultimately, indifference. Yet this memory is overwritten by a different, pride-boosting version of history.

The cultural trope of the founding father, as Asif Siddiqi has pointed out, still dominates the Russian cultural perceptions of the Space Age. In January–February 2007, a large conference was held in Moscow to commemorate Korolev's centennial. The conference had 1,650 participants; over 1,000 papers were submitted, and 420 were selected for oral presentation at the conference in 20 sections running in parallel over four days.[31] Although not all the papers were historical (many were devoted to current issues in astronautics), several sections were devoted to history. Such Korolev conferences are organized every year; this year's was the 31st. Also, every April, Gagarin conferences are held at his birthplace, the town of Gagarin (this year, it was the 33rd conference), and every September the town of Kaluga organizes Tsiolkovskii conferences (this year's was the 42nd). The general mood at such conferences is celebratory: veteran cosmonauts wear their ceremonial uniform, dancers in ethnic Russian costumes provide a suitable patriotic background, and Korolev's (or Gagarin's, or Tsiolkovskii's) portrait dominates the stage. During the Korolev conference, a new monument to Korolev was dedicated at the conference site, the Bauman State Engineering University in Moscow. Giant portraits and dominating, larger-than-life monuments serve as symbolic beacons for historical discourse. These conferences provide a suitable setting for hero-worshipping, rather than critical analysis. A chosen set of historical figures—Korolev, Tsiolkovskii, and Gagarin—serve as sources of light rather than objects of study at which light should be directed.

---

31. Analytical report on the XXXI Academic Conference on Cosmonautics, dedicated to the 100th anniversary of academician Sergei Korolev. Moscow, Russia, January 30–February 1, 2007 (available at *http://www.ihst.ru/~akm/ao31.htm*). See also Asif Siddiqi, "From Russia with History," *NASA History Division News and Notes* 24:2 (May 2007): 1-2, 4-5 (available at *http://history.nasa.gov/nltr24-2.pdf*).

This weaving of space history around a handful of key personalities was characteristic of Soviet space history from its early days. If Korolev has traditionally been portrayed as the "founding father" of Soviet cosmonautics, Tsiolkovskii might be christened its "founding grandfather." A deaf schoolteacher in the provincial town of Kaluga, Tsiolkovskii was a self-taught theorist and visionary of space travel. In the 1910s–30s, his writings widely circulated in the growing Russian community of space travel enthusiasts. In the 1930s, the Stalin propaganda machine made him into a national hero, a "poster grandpa" for national technological superiority. This ascribed identity was quite different from his own cultivated image of a humble provincial inventor, science popularizer, and public educator who built rocket models in his home workshop.[32]

In the postwar period, Soviet rocket engineers and the space enthusiasts' community put the government-constructed myth to their own use. In the late 1940s, the name of late Tsiolkovskii was regularly evoked amidst a Party-sponsored nationalist campaign asserting the priority of Russian-born scientists and engineers. Journalists claimed that Tsiolkovskii had invented the airplane and the dirigible.[33] On September 17, 1947, on the 90th anniversary of Tsiolkovskii's birth, Sergei Korolev gave a speech at the commemoration meeting at the Central Hall of the Soviet Army. As Asif Siddiqi has noted, "significantly, Korolev drew attention to Tsiolkovskii's ideas about space travel rather than rocketry or airships, thus beginning the process of relocating Tsiolkovskii within space research rather than aeronautics."[34] Suddenly, Korolev and other rocket engineers interested in space exploration began to recall their prewar meetings with Tsiolkovskii and to present their space projects as "inspired" by Tsiolkovskii. Pilgrimages to Tsiolkovskii's home in Kaluga to meet with the great man came to be seen retrospectively as a "rite of passage" for any major figure among the rocket engineers. A symbolic link with Tsiolkovskii, canonized by the Soviet state, played an important role in legitimizing their proposals in the eyes of government officials. In 1952-1953, in autobiographical materials, accompanying his applications for membership in the Communist Party and in the Academy of Sciences, Korolev wrote about his personal meeting with the late visionary as a starting point for his interest in rocketry. Even though he had met Tsiolkovskii only once in 1932, during

---

32. See James T. Andrews, "K. E. Tsiolkovskii, Ascribed Identity, and the Politics of Constructing Soviet Space Mythology, 1917-1957," paper presented at the 2006 annual conference of the American Association for the Advancement of Slavic Studies in Washington, DC; Andrews, "In Search of a Red Cosmos: Space Exploration, Public Culture, and Soviet Society," *Societal Impact of Spaceflight,* eds . Stephen Dick and Roger Launius (NASA, forthcoming); and Andrews, *Visions of Space Flight: K. E. Tsiolkovskii, Russian Popular Culture, and the Birth of Soviet Cosmonautics, 1857-1957* (Texas A&M University Press, forthcoming).

33. "My – nasledniki Tsiolkovskogo," *Komsomol'skaia pravda* (September 17, 1947).

34. Asif A. Siddiqi, "The Rockets' Red Glare: Spaceflight and the Russian Imagination, 1857-1957," Ph.D. dissertation, Carnegie Mellon University, 2004, p. 293.

Soviet poster commemorating the centennial of Tsiolkovskii's birth, 1957. (Courtesy of the Russian Academy of Sciences Archives)

Tsiolkovskii's visit to Moscow, the story later became embellished to the point of Korolev's vivid recollection of a visit to Tsiolkovskii's house in Kaluga—a visit that evidently never happened.[35] Privately, Korolev admitted that he barely remembered Tsiolkovskii and that the main source of his recollections was his own "fantasy."[36] Yet the official canonization of Tsiolkovskii and the resurrection of his legacy played a crucial role in legitimizing the idea of space exploration in the postwar Soviet Union. By turning a government-sponsored myth into a personal memory, Korolev managed to present his space projects

---

35. See Iaroslav Golovanov, "Korolev i Tsiolkovskii," unpublished manuscript; RGANTD, f. 211, op. 4, d. 150 (available at *http://rgantd.ru/vzal/korolev/pics/006_008.pdf*); Georgii Vetrov, *S.P. Korolev i kosmonavtika: Pervye shagi* (Moscow: Nauka, 1994), chaps. 20, 21.

36. Iaroslav Golovanov, *Korolev: Fakty i mify* (Moscow: Nauka, 1994), p. 110.

as a matter of national prestige and eventually to secure permission to launch Sputnik shortly after the centennial of Tsiolkovskii's birth.[37]

THE MYTH OF THE COSMONAUT

As the Soviet government kept the identity of the true leaders of the space program secret (Sergei Korolev remained an anonymous "chief designer" until his death in 1966), a handful of flown cosmonauts literally had to stand on top of Lenin's mausoleum next to Nikita Khrushchev for the entire space program. State-sponsored memorialization of Soviet space achievements turned such staged events as mausoleum appearances into iconic images of the space era widely disseminated through television, newspapers, posters, and postcards.

The space historian Cathleen Lewis has examined the Soviet "myth of the cosmonaut," which in some aspects mirrors the astronaut myth even though the two were supposed to stand for two ideologically opposite political regimes and systems of values. During the Soviet era, ghost writers produced numerous cosmonauts' biographies that followed a familiar pattern of heroic narrative: humble beginnings, childhood burdened by wartime hardship, encouragement by the family and teachers, good education paid for by the Soviet state, a wise mentor who teaches the core communist values, loyal military service, building up character and physical strength through a "trial of fire," achieving the lifetime dream by carrying out an important mission trusted to the cosmonaut by the Communist Party, and finally coming back with an important message reaffirming the communist values.[38] As the cultural historian Svetlana Boym has noted, "Soviet space exploration inherited the rhetoric of war; it was about the 'storming of space,' and the cosmonaut was the peacetime hero who was ready to dedicate himself to the motherland and, if necessary, sacrifice his life for her sake."[39]

The cosmonaut myth played a major role in Khrushchev's attempts to de-Stalinize Soviet society—to break with the Stalinist past and to reconnect with the original revolutionary aspirations for a communist utopia.[40] In 1961, soon after Gagarin's flight, Khrushchev ordered to remove Stalin's remains

37. Siddiqi, "The Rockets' Red Glare." See also Asif A. Siddiqi, *The Red Rockets' Glare: Soviet Imaginations and the Birth of Sputnik* (Cambridge University Press, forthcoming).

38. Cathleen Lewis, Curator of Russian spacecraft at the Smithsonian National Air and Space Museum, has been working on a book on the social and cultural history of "hero-cosmonauts" in the Soviet Union. She has presented various aspects of her research at numerous scholarly conferences.

39. Svetlana Boym, "Kosmos: Remembrances of the Future," in *Kosmos: A Portrait of the Russian Space Age*, photographs by A. Bartos, text by S. Boym (Princeton, NJ: Princeton Architectural Press, 2001), p. 91.

40. On the Khrushchev period, see Polly Jones, ed., *The Dilemmas of De-Stalinization: Negotiating Cultural and Social Change in the Khrushchev Era* (London and New York: Routledge, 2006), and William Taubman, *Khrushchev: The Man and His Era* (New York, NY: W.W. Norton, 2003).

from Lenin's mausoleum in Red Square and to change the name of the city of Stalingrad, the site of a major battle that turned the tide of World War II and a potent symbol of the Soviet victory over Nazism. As monuments of the Stalin era were being dismantled, new memorials to the Space Age were erected, supplanting the collective memory of Stalinist terror and devastating war with futurist visions of space conquests.

The cosmonaut myth was mostly about the future, not the past. In 1961, on the heels of Gagarin's triumph, Khrushchev proclaimed a new Communist Party Program to build a communist society in the Soviet Union within the lifetime of the current generation. The creation of the New Soviet Man—an honest, sincere, modest, morally pure person and a conscientious worker— was an essential part of the program, and the cosmonauts were hailed as a living embodiment of this human ideal. Cosmonauts themselves often felt uncomfortable playing a public role that had little to do with their own professional identity.[41]

In the Brezhnev period, as conservative ideologues attempted to whitewash the image of Stalin as a political and military leader, memories of World War II again took up a prominent place in public discourse. The conquest of space became symbolically associated with the Soviet victory over Nazi Germany. A typical Brezhnev-era biography pictured Gagarin in his capsule, preparing for his flight and listening to music, which evoked memories of his childhood: life under Nazi occupation, war privations, and the joy of liberation by Soviet soldiers.[42] This ideological appropriation of private memories quite creatively reinterpreted Gagarin's actual experiences. As a boy, Gagarin indeed survived the occupation, but he reportedly had to hide this fact while applying to a flight school; this "dark spot" in his biography could have prevented his admission.[43] He later wondered how the authorities still allowed him to become a cosmonaut after learning about the fact.[44] And the music he listened to during the preparations for his flight could hardly evoke elevated patriotic feelings: he actually listened to *Lilies of the Valley*, a popular love song whose lyrics cosmonauts parodied, turning it into a drinking song.[45]

---

41. See Slava Gerovitch, "'New Soviet Man' Inside Machine: Human Engineering, Spacecraft Design, and the Construction of Communism," *OSIRIS* 22 (2007): 135-157.

42. *Yuri Gagarin: The First Cosmonaut* (Moscow: Novosti Press Agency Publishing House, 1977).

43. Interview with Marina Popovich, *Iakutsk vechernii* (March 18, 2005) (available at http://epizodsspace.testpilot.ru/bibl/intervy/popovich-m1.html).

44. Interview with Pavel Popovich, *Fakty* (Kiev) (July 18, 2003) (available at http://epizodsspace.testpilot.ru/bibl/intervy/popovich.html).

45. Interview with Pavel Popovich, *Meditsinskaia gazeta* (April 13, 2007) (available at http://www.mgzt.ru/article/310). For a transcript of Gagarin's onboard communications, see "Zvezdnyi reis Iuriia Gagarina," *Izvestiia TsK KPSS*, no. 5 (1991): 101-129.

Gagarin monument in Moscow, dedicated in 1980. (Courtesy of Wikipedia)

Like any irrational construction that was to be believed rather than critically examined, the myth of the cosmonaut was full of internal contradictions. The cosmonauts were portrayed as both ordinary people and exceptional heroes. All the first cosmonauts had military ranks but their missions were presented as entirely peaceful. Their flights were praised as daring feats, while official reports of perfectly functioning onboard automatics did not seem to leave much room for human action.[46]

In July 1980, shortly before the opening of the Moscow Olympics, a monument to Gagarin was dedicated in Moscow. Gagarin's giant statue soars 40 meters above the crowd on top of a colossal pillar, evoking the image of a rocket plume. The cosmonaut and his rocket are symbolically fused, presenting Gagarin as a superhuman blend of man and machine. The insurmountable distance between the statue and the viewer emphasizes the mythological proportions of Gagarin's figure, which rises in its futuristic perfection far above today's all-too-human world.

CONSTRUCTING THE MASTER NARRATIVE

Just like the cosmonaut myth in many respects resembled the astronaut myth, the Soviet master narrative of space exploration mirrored essential features of the American story of national exceptionalism, technological progress, and continuous success. Pervasive secrecy and centralized control over the media further streamlined public discourse about space. Bound by secrecy on one side and by propaganda demands on the other, Soviet-era space history was reduced to a set of clichés: flawless cosmonauts flew perfect missions, supported by unfailing technology. All contingencies, failures, and alternative paths were thoroughly purged from history books. Entire programs, such as the manned lunar program, were passed over in silence. The space industry itself, namely its leading think tank, the Scientific Research Institute No. 88 (since 1966, the Central Scientific Research Institute of Machine Building), was charged with the task of clearing all space-related materials for publication in the open press.[47] While Soviet propaganda cultivated an idealized image of the Soviet space program for ideological purposes, space industry officials had their own reasons for deemphasizing failures and contingencies before decision-makers in the high echelons of Soviet power.

The cosmonauts resented the restrictions on information about their flights, having to repeat the same platitudes if not outright lies over and over again. In his private diary, Lieutenant General Nikolai Kamanin, the Deputy

---

46. See Slava Gerovitch, "Human-Machine Issues in the Soviet Space Program," in *Critical Issues in the History of Spaceflight*, pp. 107–140.

47. See Yurii A. Mozzhorin, *Tak eto bylo: Memuary Iu.A. Mozzhorina. Mozzhorin v vospominaniiakh sovremennikov* (Moscow: Mezhdunarodnaia programma obrazovaniia, 2000), p. 298.

The Chief Designer Sergei Korolev reenacting his actions during Yuri Gagarin's flight on April 12, 1961. (Photo from the author's collection)

Chief of the Air Force's General Staff in charge of cosmonaut selection and training, complained about the official ban on reports about equipment failures and cosmonaut errors: "Because of these restrictions, we are actually robbing ourselves by creating an impression of 'extraordinary ease' and almost complete safety of prolonged space flights. In fact, such flights are very difficult and dangerous for the cosmonauts, not only physically, but also psychologically."[48] "The most interesting things in our cosmonautics are classified," he lamented.[49] These sentiments, however, did not translate into an active opposition to the master narrative. When asked to serve as a consultant for Andrei Tarkovsky's feature movie based on Stanislaw Lem's novel *Solaris*, in which space travel turned into an exploration of the human soul, Kamanin blatantly refused.

48. Nikolai Kamanin, *Skrytyi kosmos*, vol. 4, *1969-1978* (Moscow: Novosti kosmonavtiki, 2001), p. 182 (diary entry of June 6, 1970).

49. Nikolai Kamanin, *Skrytyi kosmos*, vol. 1, *1960-1963* (Moscow: Infortekst, 1995), p. 176 (diary entry of October 31, 1962).

Such science fiction "degrades human dignity and denigrates the prospects of humanity," he wrote in the same diary.[50]

An "inner censor" reinforced the master narrative more efficiently than any outside censoring agency. Early Soviet discourse constantly oscillated between "what is" and "what ought to be"—the quality literary scholar Katerina Clark has labeled a "modal schizophrenia."[51] The blurring of this boundary and the desire to replace "what is" with "what ought to be" was characteristic of the later space-related discourse as well. Sergei Korolev was acutely aware of the historical significance of his space projects, but his vision of history reflected a desire to improve on reality to meet an ideal. "What is" was just a messy, error-prone draft, while the history's hall of fame deserved a clean, showcase version of "what ought to be." Korolev did not admit any journalists to the launch site on the day of Yuri Gagarin's pioneering flight, April 12, 1961.[52] Later, however, he sat down for a photo session, pretending to communicate with the cosmonaut in orbit. As Korolev's identity was still a state secret, the photo was not, of course, publicly released at the time. This fake was made for internal consumption—for those who knew about Korolev and his role in the space program—and for future generations as a "clean" version of historical events.

For Korolev, space artifacts were first and foremost symbols, not merely technological objects. Before the launch of Sputnik, two copies of the satellite were made: one for the flight and one for ground tests and simulations. Korolev ordered the satellite surface to be polished in order to maximize reflection of solar light to avoid possible overheating. He was outraged, however, when he learned that his subordinates neglected to polish the test copy: "It will be displayed in museums!" He stressed the aesthetic appeal of the ball-shaped Sputnik, arguing that, as a symbol of human entry into space, it must look "properly."[53]

Korolev's notion of looking "properly" apparently did not include looking authentic. Soon after Gagarin's flight, Korolev suggested to display a make-up of Gagarin's space capsule at an aviation show at the Tushino airfield in Moscow. Since Gagarin's Vostok spacecraft was still classified, Korolev let his subordinates "unleash their fantasy."[54] The result looked impressive but had nothing to do with the actual spacecraft.[55]

50. Kamanin, *Skrytyi kosmos*, vol. 4, p. 152 (diary entry of April 18, 1970).

51. Katerina Clark, *The Soviet Novel: History as Ritual*, 3rd ed. (Chicago, IL: The University of Chicago Press, 2000) pp. 36-38.

52. Iaroslav Golovanov, *Zametki vashego sovremennika*, vol. 1, 1953-1970 (Moscow: Dobroe slovo, 2001), p. 399 (diary entries of January-March 1970).

53. Memoirs by Mark Gallai, in *Akademik S.P. Korolev: uchenyi, inzhener, chelovek. Tvorcheskii portret po vospominaniiam sovremennikov*, ed. Aleksandr Ishlinskii (Moscow: Nauka, 1986), p. 63.

54. Memoirs by Stal' Denisov, in ibid., p. 218.

55. Anton Pervushin, "Glavnaia taina 'Vostoka,'" *Sekretnye materialy XX veka*, no. 8 (April 2004) (available at *http://epizodsspace.testpilot.ru/bibl/pervushin/vostok.html*).

Soviet media skillfully "enhanced" iconic images to stress their ideological message and to eliminate any undesired connotations. For example, the May 1961 issue of the Soviet illustrated magazine *Science and Life* featured a drawing of Gagarin's launch on its cover. The drawing faithfully depicted the actual scene of Gagarin's bidding farewell to a group of administrators, officers, engineers, and technicians, with one exception: all the military personnel at the launch pad were magically transformed into civilians, their military uniforms replaced with colorful cloaks. Recent research has uncovered many instances of retouching or cropping cosmonaut photos to erase "undesirable" individuals (who died in an accident or left the cosmonaut corps) from group shots—a venerable Soviet tradition going back to the Stalin-era iconographic erasure of high-placed "enemies of the people."[56]

To create a "clean" version of space history, both visuals and audio records were edited. On August 8, 1962, at a meeting of the State Commission that confirmed crew selections for the Vostok 3 and Vostok 4 flight, Deputy Chief of the Air Force Marshal Sergei Rudenko mistakenly pronounced the cosmonaut Pavel Popovich's last name as Popov. "This gross error created discomfort for everybody present," wrote Kamanin in his diary. "Too bad, but we'll have to cut 'Popov' out of Marshal's speech."[57] Again, the editing was made not for an immediate public release (the State Commission meeting, attended by Korolev and other "secret" designers, went on behind closed doors), but for a "clean" historical record.

Artifacts and records deposited in museums and state archives were carefully selected to reinforce the master narrative. For example, when a document outlining the instructions for a cosmonaut who accidentally landed on foreign soil came up for declassification, this sparked a internal debate. The instructions explained in detail that the cosmonaut should not disclose any information about the launch site, the booster, the spacecraft, and the leadership of the Soviet space program, and only the last—seventh—item on the list permitted the cosmonaut to ask for contact with a Soviet consul. "How can we give this document to a museum? How will we look like after that?" asked the person responsible for declassification and ordered the document to be destroyed. Valentina Ponomareva, a former cosmonaut candidate and a space

---

56. See James Oberg, "Cosmonauts and Cosmo-NOTS: Image Falsification in the Soviet Manned Space Program," Remembering the Space Age: 50th Anniversary Conference, NASA History Division and National Air and Space Museum Division of Space History, October 22-23, 2008, Washington, DC On the Stalin-era political manipulation of iconography, see David King, *The Commissar Vanishes: The Falsification of Photographs and Art in Stalin's Russia* (New York, NY: Metropolitan Books, 1997).

57. Kamanin, *Skrytyi kosmos*, vol. 1, p. 137 (diary entry of August 8, 1962).

historian, salvaged the document from destruction, but it still was not made available to the public.[58]

The master narrative was literally written in stone—in massive monuments that placed the cosmonauts, the leading engineers, and Soviet political leaders on a pedestal of historical myth. In a revealing symbolic gesture, space industry leadership actually placed space documents and artifacts in the foundation of one such monument in Moscow. A recently declassified petition from a group of industry leaders to the Soviet political leadership read:

> For the memorialization of the outstanding historical achievements of the Soviet people in the conquest of space and for the eternal preservation of documentation and other materials about the flights of Soviet spacecraft, it would be advisable to place in special sealed containers documents, films, and make-ups of Soviet artificial satellites of Earth, of space stations, of space ships, and of the most important research equipment used in flight, and to brick up these containers into the foundation of a monument commemorating the outstanding achievement of the Soviet people in the conquest of space to be erected in Moscow.[59]

An identical set of carefully selected documents and artifacts was put on display at a museum open under the monument. Space history was written once and for all. The master narrative was literally protected from challenge by a stone wall.

SOVIET COUNTER-NARRATIVES

Individual memories that could not fit into the master narrative did not disappear. Beneath the glossy surface of official history, a myriad of private stories circulated informally, and they formed an oral tradition totally separate from written accounts. Historians have traditionally associated such "counter memories in the very shadow of the official history" with groups which are "excluded or overlooked."[60] In the Soviet space program, by contrast, the groups that secretly cultivated such "counter memories" were front and center in official history: the space engineers and the cosmonauts. They were privy to information carefully concealed from an average Soviet citizen, and they

---

58. Valentina Ponomareva, *Zhenskoe litso kosmosa* (Moscow: Gelios, 2002), pp. 118-119.

59. Leonid Smirnov et al. to the Party Central Committee, February 2, 1966; Russian State Archive of the Economy (RGAE), Moscow, f. 4372, op. 81, d. 1944, l. 50.

60. Catherine Merridale, "War, Death, and Remembrance in Soviet Russia," in *War and Remembrance in the Twentieth Century*, eds. Jay Winter and Emmanuel Sivan (Cambridge, U.K.: Cambridge University Press, 1999), quoted in Fritzsche, "The Case of Modern Memory," p. 107.

preserved and passed on their memories as part of professional folklore. Telling and listening to the "true stories" of events hashed up or distorted in official accounts became an essential part of their group culture, a part of being a space engineer or a cosmonaut. Counter memory defined their private identity as much as the master narrative shaped their public persona.[61]

The engineers and the cosmonauts resented the obvious gap between their private memories and the official story. Forced to toe the official line in public, they let off their frustration in diaries and private conversations. "Why are we telling lies?" Korolev's deputy Boris Chertok jotted in his notebook, reflecting on multiple launch failures concealed from the public.[62] "All our reports are half-truths, which is worse than a lie," Iaroslav Golovanov, a leading space journalist, wrote in his notes.[63] While the rest of the world was watching a live report of the Apollo 8 mission, Soviet television broadcasted a children's movie. Golovanov remarked on that occasion, "Are Central Committee officials so thick that they don't understand how foolish and shameful this is?"[64] When his newspaper put off the publication of his article on Apollo 11 indefinitely, he let off steam in his private notebook: "I am tormented with shame. Will they allow such a disgrace again?"[65]

The same people—journalists, cosmonauts, and leading engineers—wrote both official accounts and private counter memories. A discursive split went right through their souls. Lieutenant General Nikolai Kamanin was one of the leading spokespersons for the Soviet space program. He appeared on the radio and television, published popular books and articles, arranged cosmonauts' public appearances, and wrote and rehearsed their public speeches. In December 1968, he wrote an article for *The Red Star*, the Soviet Armed Forces newspaper, about the forthcoming launch of Apollo 8. He entitled his article "Unjustified Risk" and said all the right things that Soviet propaganda norms prescribed in that case. Naturally, he did not even mention that the Soviet Union had its own secret human lunar program. But in his private diary, he frankly admitted that the Americans were getting ahead in the lunar race and railed against those whom he saw as the true culprits: party leadership, military brass, and top administrators of the space program who neglected or misdirected the program

61. On the tension between the professional identity and the public image of Soviet cosmonauts, see Gerovitch, "'New Soviet Man' Inside Machine," pp. 149–152. On how secrecy shaped the identity of space engineers, see Gerovitch, "Stalin's Rocket Designers' Leap into Space: The Technical Intelligentsia Faces the Thaw," *OSIRIS* 23 (2008): 189–209.

62. Boris Chertok, Notebook #16, September–November 1964; Chertok papers, Smithsonian National Air and Space Museum, Washington, DC.

63. Golovanov, *Zametki vashego sovremennika*, vol. 1, p. 383 (diary entries of September 1969– January 1970).

64. Ibid., p. 343 (diary entries of September–December 1968).

65. Ibid., p. 372 (diary entries of June–September 1969).

for far too long. "We have fallen behind the United States by two or three years," he wrote in the diary. "We could have been first on the Moon."[66]

The master narrative dominated Soviet media, but during the relatively liberal "thaw" of the Khrushchev era, newspapers occasionally gave voice to ordinary citizens who did not join in the public expression of enthusiasm for space. For example, in June 1960, a youth newspaper published a letter from one Alexei N., who bluntly asked about the space program, "What's in it for me?" "I, for example, on the eve of the launch of a rocket, received 300 rubles salary, and this is what I still receive, in spite of the successful launch. Doesn't is seem to you that the enthusiasm for these sputniks and the cosmos in general is inopportune and, more precisely, premature?" he asked. "Rocket, rocket, rocket—what's it needed for now? To hell with it now, and with the moon, but give me something better for my table. After that, then it will really be possible to flirt with the moon."[67] Most likely, the newspaper published this critical letter not to generate a genuine debate but simply to provoke an indignant reaction from space enthusiasts and thus further shore up the master narrative. An occasional display of dissenting opinion only stressed the need for the further strengthening of the space propaganda effort. Even such carefully controlled expressions of criticism, however, totally disappeared from public discourse during the Brezhnev period.

The first visible cracks in the master narrative came from those inside the space program who wanted to reassign credit among the major protagonists, while preserving the overall structure of the narrative. In 1974, the chief designer of rocket engines Valentin Glushko, Korolev's longtime opponent, was appointed head of Korolev's former design bureau. For 15 years, as Glushko ruled this central asset of the Soviet space program, he made a determined effort to rewrite Soviet space history by emphasizing his own contributions and downplaying Korolev's. He even ordered to remove spacecraft designed by Korolev from the bureau's internal museum and to replace them with rocket engines of his own design.[68]

The tensions that brewed under the lid of the master narrative over decades eventually came to surface as the policy of glasnost during Gorbachev's perestroika gave voice to the suppressed counter memories.

---

66. Nikolai Kamanin, *Skrytyi kosmos*, vol. 3, *1967-1968* (Moscow: Novosti kosmonavtiki, 1999), p. 335 (diary entry of December 12, 1968).

67. Quoted in Paul Josephson, "Rockets, Reactors and Soviet Culture," in *Science and the Soviet Social Order*, ed. Loren R. Graham (Cambridge, MA: Harvard University Press, 1990), p. 185.

68. Asif A. Siddiqi, "Privatising Memory: The Soviet Space Programme Through Museums and Memoirs," in *Showcasing Space*, eds. Martin Collins and Douglas Millard (London: Science Museum, 2005), p. 107.

THE END OF THE SOVIET UNION AND THE COLLAPSE OF THE MASTER NARRATIVE

In the late 1980s, public revelations of the full scale of Stalin's crimes led to a swift deterioration of the official historical discourse. Space history was also profoundly affected. Some archival documents came to light, private diaries became available, participants began to speak out, and a totally new picture of the Soviet space program emerged like a giant iceberg suddenly lifted out of the water. As Asif Siddiqi has written, "the single narrative of Soviet space history—teleological and Whiggish—fractured into multiple and parallel narratives full of doubt (for the claimed successes of the program), drama (for the episodes we never knew about) and debate (over contesting narratives of history)."[69] Veteran engineers, cosmonauts, and politicians began to tell stories of multiple failures during Soviet space missions, fatal errors and true heroism, favoritism in project funding, and hidden pressures to launch by a politically motivated date.

The collapse of the Soviet Union, as the Russian state largely withdrew both its economic support for the space industry and its ideological oversight over historical discourse, became a truly traumatic event for historical memory of the Space Age. This trauma resulted in a systematic transformation of memory of all previous Soviet space history. Soviet-era political leadership, often depicted as inept and short-sighted in the perestroika-period memoirs, suddenly acquired a better image. Stalin, Khrushchev, and Brezhnev were now portrayed as wise leaders, who appreciated the importance of the rocket and space industry and lent it much-needed political and economic support.

The memory of the Space Age became atomized and decentralized, or, in Asif Siddiqi's expression, "privatized" along with Russian industry itself. Trying to attract Western investors and clients, Russian space companies began advertising their history, opened exhibit halls for the public, and put on display rare space artifacts, including many original spacecraft. Owned and operated by space companies themselves, these "corporate" museums produced versions of space history that placed these companies in the best possible light. A competition in today's marketplace naturally led to competing versions of history, each shored up with its own set of artifacts and corporate collections of memoirs. To this day design bureaus and other Russian space institutions often physically hold or control access to most historical documents related to the Soviet space program, and the insiders have complete control over which, when, and in what form documents are released.

The old mode of hero-worshipping history did not change; only now we witness clashes between followers of different space hero cults. Soviet space history itself is full of acrimonious disputes, including the famous fallout

---

69. Ibid., p. 99.

The unveiling of a monument to the chief designer of rocket engines Valentin
Glushko at the Alley of Space Heroes in Moscow, October 4, 2001. (Photo from the
author's collection)

between Korolev and the chief rocket engine designer Valentin Glushko, or
the equally famous and equally bitter rivalry between Korolev and his main
domestic competitor in the space race, the chief designer of cruise missiles
Vladimir Chelomei. A loyal team of followers gathers around each of these
historical figures, and they construct their own versions of history, trying
to invalidate their opponents' accounts. Korolev's defenders accuse Glushko
of refusing to build rocket engines for Korolev's lunar rockets, and blame
Chelomei for siphoning off a large part of resources of the lunar program,
all this resulting in the Soviet loss in the lunar race. But the rivals have their
own stories to tell. From their perspective, Korolev is often portrayed as a
ruthless competitor and a clever political operator. For example, Khrushchev's
son Sergei, who had worked for Chelomei, has suggested that Korolev had
"focused his energy on what he did best—the elimination of his rivals."[70] A
group of Russian space industry dignitaries are posing in front of Glushko's

---

70. Sergei Khrushchev, "How Rockets Learned to Fly: Foreword," in Von Hardesty and Gene
   Eisman, *Epic Rivalry: The Inside Story of the Soviet and American Space Race* (Washington, DC:
   National Geographic, 2007), p. xviii.

monument, using the monument as a backdrop for a photo opportunity. At the same time, symbolically, they are standing guard to this monument and to a specific version of history that sanctifies this particular hero.

Monuments are not just silent memorials commemorating the past. Monuments do speak. Valentin Glushko reportedly bequeathed to inter his remains on the surface of the Moon. This bequest is cited nowadays as an inspiration for the Russians to go the Moon.[71] An aura of national pride is projected from the glorious past into the promising future. A heroic image of the past is enrolled to promote a specific policy agenda today. "Memorialization has become an essential function of the *current* Russian space program," Asif Siddiqi has noted. For Russians, "truly, their future (e.g., bases on the Moon) exists in simultaneity with their past (e.g., Sputnik, Gagarin). It has become almost impossible to separate them."[72]

The dominant medium for reassessing the past and translating this reassessment into lessons for today and tomorrow has been a steady stream of memoirs written by veterans of the Soviet space program: cosmonauts, engineers, physicians, military officers, and administrators. By revealing hitherto unknown historic details and placing space artifacts into context, these memoirs serve as a major vehicle for exploring Soviet space history. Since archival records are largely unavailable to researchers, new revelations come mostly through such memoirs. Nowhere is the "privatization" of memory as evident as in these highly personal, often emotional and partisan, accounts. Memoirists often try to write not merely an account of their own activities within the space program, but the whole history of specific periods or projects as seen from their partial perspective. In other words, they present coherent alternative versions of space history, not simply collections of bits and pieces of their individual experiences. Thus, even though these memoirs purport to articulate "counter-memory"—an alternative to the official story line—in fact they show a craving and a nostalgia for a Soviet-style single master narrative that would elevate their own patron—be it Korolev, Glushko, or Chelomei— above others.[73] "Counter-memory" ends up reproducing the same stereotypes of the master narrative, for it still serves a propaganda purpose—if not for the central government, then for a particular group within the space industry.

The changes in the way memoirs were written from the Soviet era to the perestroika to the post-Soviet period reflect an adaptation of individual memory to a specific historical context.[74] An oft-cited memoir by Oleg Ivanovskii went

---

71. Aleksandr Zhelezniakov, "V Moskve otkryt pamiatnik akademiku Glushko," *Poslednie kosmicheskie novosti*, no. 206 (October 4, 2001) (available at *http://www.cosmoworld.ru/spaceencyclopedia/hotnews/ index.shtml?04.10.01.html*).

72. Siddiqi, "From Russia with History," p. 5.

73. Siddiqi, "Privatising Memory," p. 108.

74. On memoirs of the Soviet era, see *The Russian Memoir: History and Literature*, ed. Beth Holmgren (Evanston, IL: Northwestern University Press, 2003); Irina Paperno, "Personal Accounts of the

through multiple editions from 1970 to 2005.[75] Ivanovskii was the lead designer on the Vostok mission; he coordinated interaction among multiple participants in the production, testing, and launch of Gagarin's spacecraft. He later headed the space industry department of the Military Industrial Commission, the top government body overseeing the space program. The early editions of his memoirs were published under the pseudonym Ivanov; he wrote about many leading space engineers but could not reveal their names. In the 1980s, he added their real names but still followed the Korolev-centered master narrative. Even in the post-Soviet period, he was not ready to reveal anything about his activity inside the government bureaucracy. In the latest edition, a three-page section on this period of his life is filled entirely with quotations from other people's memoirs.[76] Without access to many original documents, the world of personal memory becomes self-referential. Ivanovskii did openly what others do implicitly or even unconsciously—he presented other people's memories as his own.

In the absence of crucial archival sources, memoirs are becoming a major source for historical scholarship. Among all the memoirs of the post-Soviet era, the most ambitious and the most influential has been the four-volume set of books by Korolev's deputy Boris Chertok, a sweeping and riveting account of the Soviet space program from its origins in the postwar years to the end of the Cold War. Well-informed and well-told, these memoirs, nonetheless, are written entirely from the perspective of Korolev's engineering team.[77] In Russia, the reverence for such patriarch figures and the trust in their personal accounts reach extremes. The recent fundamental, 750-page-long Russian *Encyclopedia of Human Spaceflight* often draws on memoirs as a major source for its articles. For example, the entry on the Soyuz 15 mission is based largely on an extended quote from Chertok's memoirs.[78] In 1974, Soyuz 15 failed to dock with the Salyut 3 space station, and an internal controversy erupted over equipment malfunctions

---

Soviet Experience," *Kritika: Explorations in Russian and Eurasian History* 3:4 (Fall 2002): 577–610; and Barbara Walker, "On Reading Soviet Memoirs: A History of the 'Contemporaries' Genre as an Institution of Russian Intelligentsia Culture from the 1790s to the 1970s," *Russian Review* 59:3 (2000): 327–352.

75. See Aleksei Ivanov (Oleg Ivanovskii), *Pervye stupeni: Zapiski inzhenera* (Moscow: Molodaia gvardiia, 1970); Ivanov (Ivanovskii), *Vpervye: zapiski vedushchego konstruktora* (Moscow: Moskovskii rabochii, 1982); Oleg Ivanovskii, *Naperekor zemnomu pritiazhen'iu* (Moscow: Politizdat, 1988); and Ivanovskii, *Rakety i kosmos v SSSR: Zapiski sekretnogo konstruktora* (Moscow: Molodaia gvardiia, 2005).

76. Ivanovskii, *Rakety i kosmos*, pp. 164-166.

77. NASA History Division has sponsored the translation of these memoirs into English under Asif Siddiqi's editorship. Siddiqi has provided an excellent running commentary to the English edition, which places Chertok's story in a wider context. See Asif A. Siddiqi, "Series Introduction," in Boris Chertok, *Rockets and People* (Washington, DC: NASA SP-4110, 2005), pp. ix-xix.

78. See Iurii M. Baturin, ed., *Mirovaia pilotiruemaia kosmonavtika. Istoriia. Tekhnika. Liudi* (Moscow: RTSoft, 2005), pp. 209-210.

and the actions of the crew in that incident. By letting an engineer tell his story unopposed, encyclopedia editors in effect presented a vary partial view of that controversy, placing the blame on the crew.[79] When a personal perspective is thus validated and becomes a major reference source, this "counter-memory" of a previously hushed-up episode literally turns into a new master narrative.

THE NOSTALGIC POETICS OF POST-SOVIET SPACE MEMORY

In today's Russia, which has lost its former Communist ideals and is still searching for a unifying "national idea," Gagarin's pioneering flight—the pinnacle of the Soviet space program—often stands as a symbol of history that the Russians could really be proud of, despite the trauma of losing the superpower status. "If we did not have Gagarin, we would not be able to look into each other's eyes. It seems, we blew everything that we could. But we still have Gagarin. We will never lose him," writes one Russian journalist. "Gagarin is the symbol of a Russian victory over the entire world. A symbol for ages to come. We don't have another one and perhaps never will. Gagarin is our national idea."[80]

Sociological studies confirm that the Russians today rank Gagarin's flight as their second proudest historical achievement (91 percent), right after the victory in World War II (93 percent), and followed by Sputnik (84 percent).[81] Other Soviet symbols of national pride are falling far behind: the Stalin-era creation of the atomic and hydrogen bombs, the Khrushchev-era Virgin Lands campaign, and the Brezhnev-era Baikal-Amur giant railroad construction are all tainted by various historic revelations that cast a dark shadow over the former showcase projects.

The Russian space program occupies such a prominent place in collective memory that any critique of its past or present is often viewed as unpatriotic. The deorbiting of the *Mir* space station in March 2001 caused a public outcry. The loss of *Mir* was portrayed in the media as a major blow to the national psyche. Radical Communist opposition viewed the destruction of *Mir* as part of a sinister Western plot to bring down Russia, and accused President Putin of bowing to Western demands. Street protests were held, with signs reading, "Send the government to the bottom!" and "If you drown *Mir*, we'll drown you!"[82]

---

79. For an alternative account by the Soyuz 15 crew see Mikhail Rebrov, "Gor'kii privkus slavy," *Krasnaia zvezda* (September 9, 1994): 2; for an English translation, see "Cosmonauts Unfairly Blamed for Failure of Soyuz-15 Flight," JPRS-USP-94-007 (October 5, 1994): 3.

80. Ivan Iudintsev, "Rossiia stremitsia v kosmos …na skripuchei telege proshlykh uspekhov," *HotCom.ru*, vol. 16 (April 12, 2001) (available at *http://www.hotcom.smi-nn.ru/main/art.phtml?id=5888*).

81. Russian Public Opinion Research Center, Press Release 612, January 18, 2007 (available at *http://wciom.ru/arkhiv/tematicheskii-arkhiv/item/single/3864.html*).

82. Vladimir Plotnikov, "Rubikon Prezidenta," *Sovetskaia Rossiia* (March 22, 2001) (photo of street protests available at *http://sumpaket.webzone.ru/listwka.html*).

President Putin presents Gagarin's 1961 portrait by Nikolai But to the Cosmonaut Training Center head Petr Klimuk, Star City, April 12, 2001. (Photo from the author's collection)

Both critics of the government and government officials appealed to the public sentiment about space history, each side trying to claim historical memory in support of its legitimacy. The new, post-Soviet political leadership appropriated the image of Gagarin as its own ideological symbol, an emblem of national pride and technological prowess, and an inspiration for a superpower status. On April 12, 2001, on the 40th anniversary of Gagarin's flight and just three weeks after the de-orbiting of *Mir*, President Putin visited the Cosmonaut Training Center in Star City and gave a speech before the cosmonauts. The Center personnel prepared a special backdrop for Putin's speech—a giant, full-wall-size portrait of Gagarin in full regalia—a not-so-subtle message to the President, reminding him of the appreciation of cosmonauts' achievements by previous governments. For his part, Putin also showed historical sensitivity: he assured the cosmonauts that April 12—the Cosmonautics Day that was established to memorialize the date of Gagarin's flight—was celebrated not only by the cosmonauts, but by the entire country. To boost the cosmonauts' morale, which was at a historic low after the *Mir* demise, Putin brought them a gift. Apparently he concluded that nothing could be more valuable to the cosmonauts than reasserting the symbolic meaning of space memory, and he presented them with another portrait of Gagarin. The cosmonauts, in turn, handed the President their own gift: a watch with Gagarin's portrait on its face, and Putin immediately put it on.[83] By exchanging gifts, the President and the cosmonauts in effect exchanged their memories.[84] Both sides seemed keen to avoid confrontation over the present-day *Mir* controversy by reaffirming their connection with space history. This co-remembrance of the celebrated past of the Soviet space program reasserted their common identity as Russian heirs to the Soviet glory.

In post-Soviet culture, space history becomes part of what the cultural critic Natalia Ivanova has termed "no(w)stalgia": neither condemnation nor idealization of the past, but its actualization as a symbolic language for discussing today's pressing issues. The "no(w)stalgic" audience turns into "a collective participant and a collective interpreter; a creator of a myth, a part of the myth, and a debunker of the myth; the living past and a trial of the past at the same time."[85] The cultural anthropologist Serguei Oushakine has argued that the main task of "the postsocialist poetics of nostalgic clichés" is "to produce an already known and previously encountered effect of recognition, to evoke a shared experience, to point toward a common vocabulary of symbolic gestures"

---

83. V. Davydova et al., "40 let pervomu poletu cheloveka v kosmos!" *Novosti kosmonavtiki*, no. 6 (2001) (available at *http://www.novosti-kosmonavtiki.ru/content/numbers/221/01.shtml*).

84. On the Soviet tradition of gift-giving, particularly on gifts to political leaders, see *Dary vozhdiam / Gifts to Soviet Leaders*, edited by Nikolai Ssorin-Chaikov (Moscow: Pinakoteka, 2006).

85. Natalia Ivanova, *No$tal'iashchee: Sobranie nabliudenii* (Moscow, 2002), p. 62. See also Natalia Ivanova, "No(w)stalgia: Retro on the (Post)-Soviet Television Screen," *The Harriman Review* 12:2–3 (1999): 25–32.

and thus to overcome "a peculiar post-Soviet stylistic block, a particular expressive deficiency of postsocialism."[86] Old symbols become frames for entirely new meanings. When President Putin and the cosmonauts have to find a common language, both sides resort to nostalgic images of the past—Gagarin's portraits—to convey their messages.

The Gagarin iconography was no longer tied to the specific meanings attached to it in the Soviet era; it became a shared language that could express a wide range of new meanings. In the early 1990s, youth culture appropriated space iconography for the widely popular "Gagarin Parties," rave dance extravaganzas held at the Cosmos Pavilion in the famed Soviet Exhibition of People's Economic Achievements in Moscow. Giant make-ups of rockets and spacecraft hung from the ceiling, an enormous portrait of Gagarin was specially produced to adorn the festivities, and real cosmonauts were invited to have drinks at the bar and to mingle with the crowd. Placing old Soviet memorabilia into a youth party context had a strange liberating effect: space symbols were no longer perceived as ideologically loaded emblems of Soviet propaganda or perestroika revisionism. "The juxtaposition of Soviet symbols with rave symbols, which may seem ironic and absurd," writes the cultural anthropologist Alexei Yurchak, "in fact freed the symbolic meanings attached to Gagarin and the space program from their Soviet pathos and reinvented them, making them accessible for the new cultural production."[87] Yurchak has suggested the metaphor of "sampling" to express the (re)use of Soviet symbolism in the post-Soviet culture. "As with house music— which is continuously remixed, sampled, and quoted in new contexts—here, former official symbols were also *remixed* and presented in new contexts and in a fresh, nonlinear format," he writes. "Thus, the new 'symbolic samples,' containing quotes from past and recent Soviet meanings, were placed into a dynamic new context."[88]

RUSSIAN CAPITALISM AND THE SEMIOTICS OF SPACE

In the post-Soviet era, discourses of the past and of the present interact in complex ways. As the historian Martin Collins points out, the Global Age that we live in has both changed the cultural perception of spaceflight and shifted priorities for the Space Age. The meta-narrative of exploration no longer dominates the public image of spaceflight, and new large-scale space projects tend to involve global satellite communication systems, rather than ambitious

---

86. Serguei Alex Oushakine, "'We're Nostalgic but We're not Crazy': Retrofitting the Past in Russia," *The Russian Review* 66:3 (July 2007): 469, 481.

87. Alexei Yurchak, "Gagarin and the Rave Kids: Transforming Power, Identity, and Aesthetics in the Post-Soviet Night Life," in *Consuming Russia: Popular Culture, Sex, and Society Since Gorbachev*, edited by A. Baker (Durham, NC: Duke University Press, 1999), p. 94.

88. Ibid., p. 95.

human spaceflight endeavors. Instead of leading humanity away from Earth into the enchanting Unknown, space projects now connect disparate parts of Earth, changing the very terms in which we discuss culture in general and Space Age culture in particular.[89]

Collins draws our attention to the semiotic nature of new discursive regimes: cultural symbols do not simply represent things, they act. They create a "second nature" environment in which new identities emerge and a new form of cultural power competes with and reshapes old political and institutional structures. Thus culture cannot be seen as a mere gloss on the rough surface of the crude machinery of technological innovation, economic pressures, and political decision-making. Culture is an actor in its own right—an instrument of innovation, a tool of profit-making, and the stuff politics is made of.

Both capitalism and communism manipulated with symbols: capitalism made semiotics an essential part of marketing, while communism incorporated it into daily ideological indoctrination. Both generated mass production and mass consumption of symbols; any public representation sold something, be it a product or an ideological dogma. Communist propaganda officials dealt with some of the same issues as corporate marketing executives.

In post-Soviet Russia, the cultural heritage of the decades of the communist rule clashes with the newly developing capitalist culture. Russian advertising campaigns today often skillfully combine old Soviet symbolism with "new Russian" capitalist values. To what Collins has called the "mix of semiotics, capitalism, spaceflight, and the global and the local" they add the spectacularity of space symbols of the Soviet superpower, which are fashionable among the young and nurture the nostalgic feelings of the old. In the summer of 2006, the cell phone provider MTS launched a billboard campaign in Moscow, promoting its new "Number One" calling plan. The billboard depicted a cosmonaut in a spacesuit happily using a cell phone in space. Accompanied by a television advertisement with the slogan "Be Number One!", this blunt attempt to brand the company as the industry leader drew on the popular Russian association of the cosmonaut image with Gagarin, the "Number One" cosmonaut. In a truly postmodern fashion, the billboard message also had a self-mocking twist: the cosmonaut was wearing space gloves, which of course made it impossible to punch keys on the phone. Thus the advertisement pretended not to be an advertisement at all, but rather an invitation to the viewer to play a semiotic game, sorting out contradictory signifiers.

The mixed feelings of pride for the glorious space achievements of the past, shame for losing the superpower status, and the mockery of both pride and shame as ideological constructs provided a fertile ground for the semiotic interplay of past/present, reality/simulation, and truth/advertising. The

---

89. See Collins's article in this volume.

A billboard advertisement of the "Number One" cell phone calling plan by the MTS
company in the streets of Moscow, June 2006. (Photo from the author's collection)

ostentatious self-awareness of the simulated reality of advertising was taken to a
new level in a series of MTS television ads that followed the "Number One"
billboard campaign. Those ads first depicted a cosmonaut talking on a cell phone
during preparations for a takeoff, but then a wider camera shot gradually revealed
that the action was actually happening at a movie set being prepared for shooting
a takeoff scene.[90] In a sly reference to the popular conspiracy theories about
entire space missions staged on a movie set, these ads again invited the viewer to
blur the boundary between reality and simulation, between an advertisement
and a game, and between space history and today's marketplace.

Global satellite communication and positioning systems are increasingly
integrated into the Russian economy, but their political and cultural ramifications
remain peculiar to Russian society and are burdened with the remembrance of
the Soviet past. As late as 1999, there still was no legal framework for using
global positioning systems in Russia. In 1998, a batch of Volkswagen cars was
reportedly not permitted for sale in Russia, because they were equipped with

90. See Dmitrii Kozlov, "MTS: O iaitsakh, tarifakh, sovetskoi simvolike i butaforskikh kosmonavtakh,"
Reklamnye idei, no. 5 (2006) (available at *http://www.advi.ru/page.php3?id=287*, including one of
the television ads).

GPS receivers.[91] In 2001 the Russian authorities decided to build a Russian rival to GPS, and they revitalized the stalled military project called GLONASS (GLObal Navigation Satellite System), now broadening its use for civilian purposes. In May 2007, President Putin signed a decree authorizing free and open access to the civilian navigation signals of the GLONASS system to both Russian and foreign customers.[92] After adding three satellites in December 2007, GLONASS would soon provide almost complete coverage of the Russian territory. According to the planners, GLONASS should reach global coverage by 2010. The Russian authorities counted that foreign consumers, especially in the Middle East and South East Asia, would be interested in having access to an alternative to the U.S.-controlled GPS.[93]

Instead of fostering a sense of global unity, satellite navigation systems in the Russian context are becoming a subject of international technological competition, a tool of political influence, and a vehicle for boosting national pride. U.S.-Russian negotiations on achieving technical compatibility and interoperability between GPS and GLONASS progress very slowly. In the meantime, the Russian Ministry of Industry has proposed limiting the sales in Russia of GPS receivers that were not compatible with GLONASS.[94] Official policies toward global navigation systems in Russia seem to fall back on the old Soviet stereotype of national isolationism. In March 2007, Putin held a meeting of the State Council in Kaluga, the town nicknamed "the birthplace of cosmonautics" where Tsiolkovskii spent most of his life and produced his most important works. Having reestablished historical links with Tsiolkovskii's visions of space exploration, Putin instructed the Council members that GLONASS "must work flawlessly, be less expensive, and provide better quality than GPS." He expressed his confidence that Russian consumers would show "healthy economic patriotism" and prefer GLONASS over GPS.[95] In December 2007, the first batch of dual-signal GPS/GLONASS traffic navigators was quickly sold out in Moscow stores at $570 a piece, several months before the customers could take full advantage of GLONASS capabilities.[96]

91. V. Koliubakin, "'Iridium'–presentatsiia v Sankt-Peterburge," *Tele-Sputnik*, no. 3(41) (March 1999) (available at *http://www.telesputnik.ru/archive/41/article/40.html*).

92. Novosti Russian News and Information Agency report, May 18, 2007 (available at *http://rian.ru/technology/innovation/20070518/65722212.html*).

93. Novosti Russian News and Information Agency report, December 26, 2007 (available at *http://www.rian.ru/technology/connection/20071226/94147340.html*).

94. Anton Bursak, "Minprom zashchitit GLONASS, ogranichiv vvoz GPS-ustroistv," *RBK Daily*, February 22, 2007 (available at *http://www.rbcdaily.ru/print.shtml?2007/02/22/media/266488*).

95. Viktor Litovkin, "GLONASS ishchet oporu na zemle," FK Novosti Information Agency report, April 2, 2007 (available at *http://www.fcinfo.ru/themes/basic/materials-document.asp?folder=1446&matID=134457*).

96. PRIME-TASS Business News Agency report, December 27, 2007 (available at *http://www.prime-tass.ru/news/show.asp?id=746309*).

For individual Russian users, an "eye in the sky" often evoked Soviet-era cultural memories of total surveillance. In October 2007, General Nikolai Patrushev, the head of the FSB (the successor to the KGB), announced plans for a nationwide system of traffic control. Under the banner of fighting terrorism, the FSB intended to implement a system of monitoring individual motor vehicles on the Russian territory. Technical details of the new system were not revealed, but it was implied that it might involve the use of satellites for positioning and communication. Journalists quickly gathered initial negative reactions to the news: "it's an invasion of privacy"; "this smells like a violation of constitutional rights of citizens"; and "any surveillance brings up bad memories of Stalin's totalitarian system."[97] At the same time, individual users seemed quite willing to use GPS devices to track the movements of their own children.[98]

A shift in priorities from space exploration to satellite applications is clearly reflected in the Russian public opinion. In an April 2005 poll, the highest number of respondents (52 percent) said that scientific research and the development of advanced technologies should be a top priority of the Russian space program, and 44 percent supported defense applications. 17 percent mentioned the importance of space achievements for international prestige, and only 1-4 percent prioritized missions to the Moon and Mars, search for extraterrestrial civilizations, and space tourism.[99] Ambitious projects of space exploration serve as a token of memory, an emblem of the "no(w)stalgic" past, but they no longer dominate the cultural production of the present.

CONCLUSION

The Space Age both reinforced cultural boundaries—through the Cold War imagery and rhetoric—and blurred them through the emerging sense of the global. It produced vivid memories and engaging stories; individual retelling of these stories and collective propaganda projects of remembrance gradually turned historical events into mythological epics, shaping the identity of generations. The "Sputnik generation" of Russian citizens, who grew up in the 1950s, in recent interviews acknowledged the formative role of the key events of the Space Age, but had little personal recollection of their reaction

---

97. Andrei Kozlov, "Voditeli popali pod podozrenie," *Vzgliad*, October 16, 2007 (available at *http:// www.vz.ru/society/2007/10/16/117887.html*).

98. A. Kuznetsov, Report on testing the S-911 Personal Locator (available at *http://gps-club.ru/gps_think/detail.php?ID=8057*).

99. Russian Public Opinion Research Center, Press Release 187, April 11, 2005 (available at *http://wciom.ru/arkhiv/tematicheskii-arkhiv/item/single/1181.html*).

to Sputnik or Gagarin's flight.[100] In order to remember, we have to create our memories. And we create them out of the myths and symbols of our culture.

Cultural myths should not be seen merely as distorted memories. It is precisely these "distortions," cultural adaptations and appropriations of symbols, that give cultures their individuality, their unique character, and distinct perspective. Just as one's personal memories reveal more about one's current identity than about one's past, historical myths provide a valuable insight into the culture that produces them. At the intersection of space history and cultural history, the semiotics of Space Age remembrance ties together individual memory and collective myth, the materiality of objects and the pliability of symbols, the authenticity of fantasy and the deceptive nature of truth.

There can be no "true" memory, as any act of recollection reconstitutes our memories. As different cultures remember the Space Age, it keeps changing, revealing new symbolic meanings and providing an inexhaustible source of study for historians. By shifting the focus from debunking myths to examining their origins and their constructive role in culture, we can understand memory as a dynamic cultural force, not a static snapshot of the past.

---

100. Donald J. Raleigh, tran. and ed., *Russia's Sputnik Generation: Soviet Baby Boomers Talk about Their Lives* (Bloomington, IN: Indiana University Press, 2006).

CHAPTER 13

THE MUSIC OF MEMORY AND FORGETTING: GLOBAL ECHOES OF SPUTNIK II[1]

Amy Nelson

> In times when history still moved slowly, events were few
> and far between and easily committed to memory. They
> formed a commonly accepted *backdrop* for thrilling scenes of
> adventure in private life. Nowadays, history moves at a brisk
> clip. A historical event, though soon forgotten, sparkles the
> morning after with the dew of novelty.
>
> — Milan Kundera[2]

> It's been four long days since we first started experimenting
> on the dearly departed
> soon she won't communicate anymore.
>
> — Amoree Lovell[3]

While Americans' memory of the "Evil Empire" might be fading,[4] the Cold War continues to inform an increasingly diverse and interrelated global popular culture in often surprising ways. Among these, the enduring

---

1. The research for this essay was supported by a Summer Humanities Stipend from Virginia Tech, the Summer Research Laboratory on Russia, Eastern Europe, and Eurasia at the University of Illinois, and a Faculty Research Grant from Virginia Tech's College of Liberal Arts and Human Sciences. For assistance tracking down musical and poetic tributes to space dogs I am grateful to Karl Larson, Tom Ewing, Mark Barrow, Robert Stephens, Erik Heine, Andrew Jenks, and especially Evan Noble. I am indebted to Brian Britt and Greta Kroeker for their help translating lyrics in languages I wish I knew better. Some material from this essay also appears in Amy Nelson, "Der abwesende Freund: Laikas kulturelles Nachleben," in Jessica Ullrich, Friedrich Welzien, and Heike Fuhlbrügge, eds, *Ich, das Tier. Tiere als Personlichkeiten in der Kulturgeschichte* (Berlin: Reimer Verlag, 2008), pp. 215–224.

2. Kundera, Milan. *The Book of Laughter and Forgetting*, trans. Michael Henry Heim (Middlesex, England: Penguin Books, Ltd, 1980 [1979]), pp. 7–8.

3. Lines from the song "Laika: an Allegory," *Six Sadistic Songs for Children* (2005).

4. In its annual assessment of the attitudes of today's youth, Beloit College's "Mindset List for the Class of 2010" notes that for today's college students "the Soviet Union has never existed and therefore is about as scary as the student union." "Beloit College Mindset List," *http://www.beloit. edu/~pubaff/mindset/2010.php* (accessed January 20, 2008).

celebrity and complex historical memory surrounding "Laika," the mixed-breed dog that became the first living being to orbit Earth in November 1957, is certainly one of the most intriguing examples. Instantly famous as evidence that the Soviets led the race to conquer space, Laika joined a small group of animals who are celebrities in their own right. But while the fame of other creatures in this cohort often derives from humans' shared assessment of their symbolic importance—as an emblem of grit and courage in the case of a depression-era racehorse such as Seabiscuit or as an exotic token of national rivalry in the case of P. T. Barnum's giant pachyderm, Jumbo, (purchased in 1882 from the London zoo for a then-record sum of $10,000),[5] Laika's celebrity was more controversial at the outset and remains more complicated 50 years after the flight of Sputnik II. By examining the ongoing resonance of the first space dog in global popular culture, this essay shows how a defining episode of the early Space Age has been remembered even as its specific historical circumstances have been effaced. This contradictory legacy has much to say about the shifting, mutable nature of social frames of memory (and, by extension, forgetting), and about the complex ways that humans engage, imagine, and remember the life and death of an individual dog.

Speculation about Laika's fate and the significance of her voyage served as the crux of the initial controversy. As Susan Buck-Morss and David Caute have recently noted, the fierceness of the cultural Cold War derived, somewhat ironically, from the superpowers' shared Enlightenment heritage and the fact that both sides largely agreed on cultural values, including a faith in progress, a veneration of science and technology, and a determination to harness nature to human ends. The space race, inaugurated a short month before Laika's voyage with the launch of the first artificial satellite, tapped into all of these concerns while also serving as a proxy for armed conflict. Caute's bemused assertion that "a Soviet dog orbiting in space caused all American dogs to howl" highlights the international drama precipitated by Laika's flight. As ordinary citizens scanned the night sky and amateur radio operators tracked the satellite's radio signal, world headlines confirmed the Soviets' latest victory in the space race—a competition of scientific, engineering, and industrial might that was both more threatening and more fascinating than conventional warfare.[6]

Sending a dog into orbit further undermined Western confidence already shaken by the launch of Sputnik I. At the same time, this bizarre, public form of animal experimentation outraged animal welfare groups. For although Laika's

---

5. Laura Hillenbrand, *Seabiscuit an American Legend* (New York, NY: Random House, 2001); Harriet Ritvo, *The Animal Estate. The English and Other Creatures in the Victorian Age* (Cambridge, MA: Harvard University Press, 1987), pp. 220, 232-233.

6. David Caute, *The Dancer Defects. The Struggle for Cultural Supremacy during the Cold War* (Oxford, UK: Oxford University Press, 2003), pp. 4, 38-39; Susan Buck-Morss, *Dreamworld and Catastrophe. The Passing of Mass Utopia in East and West* (Cambridge, MA: MIT Press, 2002).

space capsule had food, water, and a climate control system designed to support her for several days, it was not engineered to be retrievable, so the dog's death was a certainty from the outset. For 40 years the Soviets maintained that Laika had died painlessly after several days in orbit, revealing only in 2002 that she succumbed to overheating and panic a few hours after launch.[7]

Sacrificed in the quest to make spaceflight a reality for humans, Laika the dog provoked intense reactions from people who regarded her variously as an "experimental animal," a "brave scout," a "faithful servant," or an "innocent victim."[8] At one level, these responses mirrored contradictory attitudes, common in their main contours across many cultural and national contexts, of people toward dogs. As such, conflicting human perspectives on the first space dog drew on and intensified more generalized tensions generated by the intertwined nature of domestic dog and human ecologies.[9] They also tapped the excitement and apprehension occasioned by the advent of the nuclear era and the Space Age, which suggested the compelling attractions as well as the tremendously destructive potential of technological and scientific advances.

Over the last 20 years or so, the multivalent echoes of Laika's immediate celebrity have inspired an array of creative endeavors, including Lasse Halström's film, *My Life as a Dog* (1985) and extending to a number of recent literary undertakings, an array of Web sites, and, most remarkably, a diverse and expanding corpus of music emanating from various points around the Northern hemisphere and the transnational arena of cyberspace. Since the mid–eighties, music groups in Scandinavia, Spain, Germany, Japan, the United States, and the United Kingdom have dedicated songs to Laika, and three have adopted her name as their own. This represents considerable name recognition. Indeed, in the musical arena of commercial cyberspace, the first space dog seems to have more currency than the first space man or even the founder of the Soviet state.[10] Nearly 50 short pieces are named after Laika or have lyrics referencing

---

7. David Whitehouse, "First Dog in Space Died within Hours," BBC News Online October 28, 2002, *http://news.bbc.co.uk/1/hi/sci/tech/2367681.stm* (accessed January 25, 2008).

8. These overlapping but often contradictory perspectives on Laika might be explained in terms of the sociological concept of the "boundary object." See Anita Guerrini, *Experimenting with Humans and Animals. From Galen to Animal Rights* (Baltimore, 2003), p. x; Susan Leigh Star and James R. Griesemer, "Institutional Ecology, 'Translations' and Boundary Object: Amateurs and Professionals in Berkeley's Museum of Vertebrate Zoology, 1907-39," *Social Studies of Science* 19 (1989): 387-420.

9. On the extent to which the destinies of humans and domestic dogs are inextricably linked by forces of nature and culture, see Raymond Coppinger and Lorna Coppinger, *Dogs. A New Understanding of Canine Origin, Behavior, and Evolution* (Chicago, IL: The University of Chicago Press, 2001); Donna Haraway, *The Companion Species Manifesto. Dogs, People, and Significant Otherness* (Chicago, IL: Prickly Paradigm Press, 2003); and Susan McHugh, *Dog* (London, UK: Reaktion Books, 2004).

10. A search across all genres in the Itunes store in January 2008 yielded 23 pieces with Yuri Gagarin's name in the title, 27 pieces named after Lenin, and 43 referencing Laika.

her story. Nineteen of these are exclusively instrumental, and the majority of those are electronica in the tradition of the "space music" popularized for the last 20 years or so by Stephen Hill in his syndicated program "Hearts of Space." Laika also has served as muse for classically trained musicians, including Max Richter ("Laika's Journey," 2002) and Ulrike Haage, whose "Requiem for Laika" (2005) interweaves vintage Soviet radio broadcasts and narration in German with sung portions of the Mass for the dead (with the sacrificial *agnus dei* recast as a wolf).

Given this prominence, one might expect Laika to provide an important bridge to the popular memory of the space race, the Cold War, and the Soviet past. But while Laika's initial celebrity depended heavily on the politically charged and highly publicized circumstances under which she was sent into space, her ongoing resonance derives more from her appeal as a symbol of the timeless human concerns of sacrifice, experimentation, alienation, and loss. Indeed, an analysis of the recent musical tributes to her suggests that the contemporary popular memory of the first space dog has become somewhat uncoupled from the history of Sputnik II.

To explain this paradox, we must note that while the realms of "memory" and "history" partially overlap, they also differ in important ways. Historians use many different kinds of evidence—including qualitative sources such as memoirs, diaries, and oral histories—to gain insight on the events of the past. But like other scholars in the behavioral sciences and the humanities, they distinguish between the act of remembering and the historical events being remembered. Recent research in this area reminds us that for individuals and societies as a whole, memory is an active, iterative process. Our recollection of events is not a literal recall of a fixed or imprinted image or experience, but rather a construction or reconfiguration of what happened.[11] That the democratizing impulses fuelling the "unofficial knowledge" of popular memory often run counter to the empirical and sometimes arcane preoccupations of the professional historian has been well-documented, even as recent scholarship has focused on understanding the current obsession with "memory" among scholars and laypeople.[12]

Intended as a satirical observation on the perversity of Czech communism, Milan Kundera's assertion that "nowadays, history moves at a brisk clip," while events themselves are "soon forgotten," offers a telling comment on how time

---

11. David Gross, *Lost Time: On Remembering and Forgetting in Late Modern Culture* (Amherst, MA: University of Massachusetts Press, 2000), p. 4.

12. On the significance of amateur collectors and preservationists to the construction and perpetuation of popular memory. see Raphael Samuel, *Theatres of Memory. Past and Present in Contemporary Culture* (London: Verso, 1994). For a recent attempt to historicize discourses of memory and modernity, see Alon Confino and Peter Fritzsche, *The Work of Memory. New Directions in the Study of German Society and Culture* (Chicago, IL: University of Illinois Press, 2002).

seems to have accelerated since the end of the last World War while at the same time historical memory has become less stable and, in many contexts, less valued. Where the cultural legacy of the first space dog is concerned, the inherent atemporality of the media-mediated images, sounds, and messages that have played an increasingly dominant role in framing social and cultural memory over the last 50 years seem to be critically important.[13] So, too, are the converging influences of globalization and the digital technologies that have transformed the production, distribution, and consumption of music since the late nineties.[14] With the rise of relatively small digital audio files, such as the MP3, the global internet became the ideal forum to facilitate the exchange and distribution of music, a creative medium uniquely suited to conveying the emotional charge of the Laika story. As the song cited for this essay's second epigraph suggests, by 2005 that story might sound more like a funeral for a friend than an early episode of the quest to send humans into space.

Laika's current visibility in various aesthetic and creative realms extends and expands on the celebrity status accorded her in the early years of the space race. Like several other dogs sent into space by the Soviets between Laika's voyage and Yuri Gagarin's manned flight in 1961, Laika became the subject of a sophisticated, anthropomorphized celebrity.[15] Photographs of the canine cosmonauts were printed on front pages around the world. Reporters flocked to their "press conferences," and millions tuned in to hear their barks transmitted on radio "interviews." Fame was fleeting for most of these dogs as the world's attention quickly shifted from their exploits to the more compelling drama of human space travel and exploration. Laika, however, proved to be the exception. The significance of her voyage and the fact that she was deliberately sent to her death inspired a number of commemorative projects in the Soviet Union and other countries as well.

Soviet tributes to the canine pioneer began within a year of her journey. Soon after her flight, a brass tag was attached to her kennel with the inscription translated here from the Russian: "Here lived the dog Laika, the first to orbit our planet on an earth satellite, November 3, 1957."[16] In keeping with the tradition of commemorating historic events and individuals, the Soviet mint issued an enamel pin of "The First Passenger in Space," showing the dog's head and a rocket hovering over Earth on a field of stars. Official commemorations in other countries soon followed as stamps bearing the dog's likeness were issued in

---

13. Gross, *Lost Time*, p. 123.

14. Timothy D. Taylor, *Strange Sounds. Music, Technology & Culture* (New York, NY: Routledge, 2001).

15. I examine the history of the space dog program and the dogs' celebrity in: Laikas Vermächtnis: Die sowjetischen Raumschiffhunde" in *Tierische Geschichte: Die Beziehung von Mensch und Tier in der Kultur der Moderne*, eds. Dorothee Brantz and Christof Mauch (Paderborn: Schöningh, in press).

16. A. Golikov and I. Smirnov, "Chetveronogie astronavty," *Ogonek* 49 (1960): 2.

Romania (1957), Albania (1962), Sharjah/Mongolia (1963), and Poland (1964).[17] In the fall of 1958, the Soviet Union began to market its first filtered cigarette, using Laika's name and image on the wrapper, and initiating a now 50-year-old process of commodification and "branding" of the space dog.[18] The high-relief at the base of the monument "To the Conquerors of Space" (dedicated in 1964) at the Exhibition of Achievements of the National Economy[19] (VDNKh) includes an alert, larger-than-life Laika, whose capsule provides the foundation for a rocket guided by the muscular male arms of an anonymous socialist-realist human.

While the pins, stamps, and monuments of the 1950s and 1960s might be fairly straightforward commemorations of a significant event or individual, other tributes to Laika were more complex.[20] Outside the Soviet Union, at least two musical memorials addressed the main concerns raised by Sputnik II— American preoccupation with the specter of Soviet domination and widespread shock over sending a dog to its death in space.

"Sputniks and Mutniks," recorded by Ray Anderson and the Homefolks in 1958, playfully captured the sensationalism and insecurity Laika's flight generated in the United States.[21] Jaunty and playful, the song's quick tempo and bluegrass style is at odds with the anxiety over the potential for weaponizing space expressed in the lyrics:

> Sputniks and mutniks flying through the air
> Sputniks and mutniks flying everywhere
> They're so ironic, are they atomic?
> Those funny missiles have got me scared.

While Anderson's song received relatively little distribution before Jayne Loader and Pierce and Kevin Rafferty identified it as a "must have" for the soundtrack of their satirical documentary *Atomic Cafe* (1982), the second song from this era, "Russian Satellite," enjoyed instant and enduring acclaim. As one of The Mighty Sparrow's three Carnival Road March Competition winners from 1958, the song helped catapult the "Sparrow" (born Slinger Francisco) to the forefront of the calypso world, where he has remained for nearly half a century. As in the case of "Sputniks and Mutniks," the lyrics and music of "Russian Satellite"

---

17. In the sixties, the Soviet Union and several other Eastern Bloc countries also issued stamps of other space dogs, especially Belka, Strelka, Chernushka, and Zvezdochka. Stamps of Laika were issued later in Hungary (1982) and North Korea (1987).

18. "Soviet Smokers Now Have Filters," *New York Times*, September 11, 1958.

19. In 1992, the title of this center was changed to the All-Russian Exhibition Centre, but it continues to be referred to by its previous acronym of VDNKh.

20. See for example, Leonid Vysheslavskii's poem, "Pamiati Laiki," *Zvezdnye sonety* (Moscow: Sovetskii pisatel', 1962), p. 71.

21. *Atomic Cafe Soundtrack* (Rounder Select, 1994).

work against each other for ironic effect. But whereas the appeal of "Mutniks and Sputniks" derives from its disarming simplicity, "Russian Satellite" exploits a hallmark of calypso style, setting deftly pointed social commentary against a bright, syncopated melody. "Murder, murder everywhere," begins the song, which goes on to examine one of the many widespread myths about Laika's demise: "Over a thousand miles in space . . . They poison the food for the poor puppy / Oh Lord, this is more than cruelty." In 2002, The Mighty Sparrow reminded fans that he is a "multi-faceted" individual whose concerns about social justice still extended past the human community: "I can remember when the Russians sent a satellite in the sky, with a dog in it. I was the only one who came out and said that I was sorry for the dog."[22]

Given that rock and roll music developed in tandem with the space race and the heyday of science fiction, the pervasiveness of space themes throughout rock's history is hardly surprising. Indeed, as Ken McLeod has recently noted, "the association of space and alien themes with rock'n'roll rebellion is found throughout rock's history and has had an impact on nearly all its stylistic manifestations."[23] But while any number of examples can be mustered to demonstrate the fertility of this connection from the 1960s on (i.e., David Bowie's *Space Oddity* [1969] and his glam rock alter ego "Ziggy Stardust," Pink Floyd's *Dark Side of the Moon* [1973], George Clinton's *Mothership Connection* [1974], etc.), the flight and plight of the first space dog seems to have found minimal resonance between the late 1950s and the era of glasnost. Beginning in the mid-1980s, however, a diverse assortment of filmmakers, musicians, artists, and authors began turning to Laika for inspiration. Most of the resulting creative work has originated outside the former Soviet Union, although statistical evidence suggests that the memory of Laika thrives in her homeland as well. The majority of Russians surveyed in 1994 could identify Laika more accurately than they could other major events from the post-war period, including the Cuban missile crisis, the 20th Party Congress, or the publication of *One Day in the Life of Ivan Denisovich.*[24]

The starting point for this renewed interest in the first space dog was the 1985 film *My Life as a Dog*. Set in Sweden in the late '50s, Lasse Halström's drama charts the coming of age of a boy named Ingemar, who copes with his mother's failing health and her inability to care for him by reminding himself of Laika's plight. He worries that Laika starved to death, identifies with her helplessness, and laments her physical isolation in an effort to gain perspective

22. "Sparrow, the Concerned Caribbean Villager," *The Jamaica Gleaner*, November 27, 2002, http://www.jamaica-gleaner.com/gleaner/20021127/ent/ent1.html (accessed January 25, 2008).

23. Ken McKleod, "Space Oddities: Aliens, Futurism, and Meaning in Popular Music," *Popular Music* vol. 22, no. 3 (2003): 340.

24. Howard Schuman and Amy D. Corning, "Collective Knowledge of Public Events: The Soviet Era from the Great Purges to Glasnost." *The American Journal of Sociology* 105, no. 4 (2000): 913-956.

on his own abandonment and loss, which culminates in the death of his mother and his own beloved dog, Sickan. As the film's title and these lines suggest, Ingemar reaches across the boundary of species to shore up his own identity and resolve:

> I can't help thinking about Laika. She had to do it for human progress. She didn't ask to go . . . she really must have seen things in perspective. It's important to keep some distance . . .

Halström's film garnered critical acclaim at film festivals in Berlin and Toronto before making headlines in the United States, where box office sales ultimately topped eight million dollars. Nominated for a raft of awards and winner in the Best Foreign Film category for both the New York Film Critics Award (1987) and the Golden Globe Awards (1988), *My Life as a Dog* inspired a new wave of (mainly musical) tributes to the first dog in space.

The first of these came in 1987 from the Spanish punk rock group, Mecano. Part of "La Movida," the counter-cultural movement that mobilized Spanish youth in the 1980s, Mecano found commercial success in France, Italy, and Latin America, as well as in Spain. Still readily accessible on YouTube, Mecano's song, "Laika," tells the story of a "normal Russian dog" and speculates about her "thoughts" as she looked down on Earth through the window of her space capsule.[25] Like many bands to follow, Mecano laments sacrificing a dog to human ambition and curiosity, and in so doing elevates Laika to realms normally reserved for humans: "We have to think that on earth there is one little dog less / and in heaven there is one star more."

The most long-standing musical group to appropriate Laika's name was also founded in 1987. Ironically retro in conception, Laika and the Cosmonauts offered updated instrumental surf rock in the 1960s tradition of Dick Dale, complete with loud reverberating solo guitar and lots of fast double picking. The irony here derived from the former studio musicians' Finnish citizenship. Like their compatriots, The Leningrad Cowboys, Laika and the Cosmonauts parodied Finland's ambivalent stance toward the Cold War superpowers by choosing a Soviet-themed name and adopting a quintessentially American style. The group rode the wave of the instrumental surf rock revival set off by the release of the surf documentary, *Endless Summer II* and the inclusion of Dick Dale's "Miserlou" on the soundtrack of the film *Pulp Fiction* in 1994. Reviving the connection between rock music and space themes dating back to the early 1960s, Laika and the Cosmonauts paid explicit homage to the Space Age with their first album and title hit, *C'mon do the Laika* (1988) and their 1996 compilation, *Zero Gravity*. Besides offering covers of surf-rock classics and

---

25. "Mecano-Laika," *http://www.youtube.com/watch?v=AgHkv1XPPis* (accessed January 25, 2008).

themes from '60s movies and television shows (including *Psycho*, *Vertigo*, and *Mission Impossible*), the group composed its own music with a sound one critic described as bouncing "between endless summer, lurching polka, spy flick, and spaghetti western themes. Sometimes moody, sometimes trippy . . . Party music supreme."[26] The quartet has released six albums since its founding in 1987 and counts *Pulp Fiction* director, Quentin Tarantino, among its diehard fans.[27]

With a career that spanned the transition from the end of Soviet communism to the age of the global electronic village, Laika and the Cosmonauts were among the first musical ensembles to tap the appeal of campy nostalgia for things (formerly) Soviet to a range of audiences. Others who mastered the appropriation of Soviet symbols and themes included Rasputin Stoy, whose German synth-pop band CCCP found an enthusiastic following in alternative dance clubs. Along with several homages to the Soviet space program, the band's 1996 album, *Cosmos*, includes a cut called "Laika, Laika" with the enthusiastic participation of the Russian Army chorus.

Over the last decade, however, explicit references to the Soviet past have become vaguer, focusing instead on a fairly generic nostalgia for the early space race or on the figure of the first space dog herself. For example, the American indie rock-power pop band Sputnik dedicated a smoky, strummed guitar ballad to Laika in 2004, but the other tracks on its debut album *Meet Sputnik* make little or no reference to the space race. Following the lead of the Leningrad Cowboys, the virtual band Gorillaz titled their hit remix album of 2002 *Laika Come Home*, combining the name of the Soviet space dog with the title commonly associated with the Anglo-American canine hero Lassie. While the album art evokes the glory days of dogs and chimps in space, the music consists of re-mixes of the group's first (eponymous) album in reggae and dub style.

Clearly for musicians, Laika's association with the creative possibilities and costs of innovation continues to serve as a compelling touchstone. The most explicit homage to the space dog and her legacy belongs to the eclectic British quartet, Laika, which uses sampling and electronics to achieve a celestial, innovative sound and features an image of the dog on all of its album covers. Founded in 1994, the group released five albums before "taking a break" in November 2007, the 50th anniversary of the launch of Sputnik II. While the group's "classic" sound is best exemplified in collections such as Silver *Apples of the Moon* (1995) and *Sounds of the Satellites* (1997), the incorporation of blues elements in *Good Looking Blues* (2000) followed from the group's determination to confound expectations. According to their Web site: "They're not a rock band, but they play guitars. They're not an 'electronic' group in the usual sense

---

26. Andy Ellis, "The Amazing Colossal Band," *Guitar Player* 29, no. 5 (May 1995): 129.

27. "Laika and the Cosmonauts," *http://www.laikaandthecosmonauts.com/news/index.php3* (accessed January 25, 2008).

of the term, yet they meld and twist samples with the best of them."[28] As for the name, Margaret Fiedler and Guy Fixsen explain their choice as follows:

> [W]e liked the sound of the word and we liked the asso-
> ciation with being "out there" in terms of experimentation
> while at the same time being a warm furry organic thing . . .
> The other reason we like the name is that it was probably the
> most high profile animal experiment ever—Laika died up
> there in her capsule—and we are strong believers in animal
> rights and things that seem kind of obvious to us, like not
> eating them.[29]

While innovation represents an essential component of artistic originality, concern about animal experimentation and sacrifice emerges as a recurring theme in Laika-themed songs across several genres. For example, American folk singer Kyler England uses phrases from the beloved nursery rhyme "Twinkle, Twinkle, Little Star" to frame an almost maudlin tribute to a brave dog sacrificed for human ends: "like a diamond in the sky / gave your life for humankind / what a view it must have been."[30] In the hands of Amoree Lovell, the Portland-based rocker cited in this essay's second epigraph, the same material gets an almost silly gothic twist, replete with rollicking arpeggios, cello counterpoint, and moaning bass chorus background. Others, such as the retro rock group Sputnik, Eurodance star Ice MC, and the grunge rock group Pond, denounce the human forces behind Sputnik II with little or no trace of irony. The lyrics of Pond's "My Dog is an Astronaut," for example, expresses this wish for Laika:

> I hope she sails on and on across the universe
> finds there some new world where she'll be safe from man's
> experiments
> that don't have come home parts

In many of these songs, Laika is no longer a stray dog captured for laboratory research, but rather an abused or abandoned pet. Since most people more easily relate to dogs as pets or companions than as research subjects, this slippage facilitates an emotional connection with Laika's experience even as it obscures the reality of her life. Other kinds of identity ambiguity in musical tributes to the space dog involve the performer appropriating a canine identity

---

28. "Laika," *http://www.laika.org/index_main.shtml* (accessed January 25, 2008).

29. "Laika," *http://www.laika.org/index_main.shtml* (accessed January 25, 2008).

30. Kyler, "Laika," *A Flower Grows in Stone* (Deep South, 2004).

or blurring the human-dog boundary in the vein of George Clinton's "Atomic Dog" or the hip hop artist Snoop Dogg. The most explicit example of this is probably Ice MC's Eurodance hit from 1990 in which the rapper announces:

> I'm a dog
> my name is Laika
> my ambition is to be like a f---in' astronaut
> and see Mars[31]

In other cases, the boundary between human and animal and the ethical perspective of the artist are unclear, as in Moxy Früvous's "Laika," which appeared on the Canadian group's 1994 smash hit album, *Bargainville*. The point of view shifts numerous times throughout this witty meditation on coming of age in the age of flying dogs. Like Ice MC and the death rock group Massacre (which speculates that Laika had a fear of heights),[32] Moxy Früvous projects human aspirations and feelings onto Laika with excellent ironic results ("Hey darling, throw this space pup a bone").

An even more arresting ambiguity surfaces when humans incorporate Laika into human pantheons. A physical example of this is the monument to fallen cosmonauts erected outside Moscow in 1997 that includes a likeness of Laika peering up at the faces of the humans who also died in the conquest of space. In the musical realm, we have a brilliant send up of real and artistic spectacular demises by British singer Neil Hannon. In the title cut of the 2004 album *Absent Friends*,[33] Hannon flanks a witty toast to Laika with tributes to the suicidal actress Jean Seberg, the World War I chaplain "Woodbine Willy" (who distributed cigarettes to doomed and dying soldiers), the persecuted Oscar Wilde, and the king of cool Steve McQueen (as "Hicks" in *The Great Escape*).

The flirtation with self-destruction in "Absent Friends" finds more direct expression in the song "Neighborhood #2 (Laika)," a ballad by the Montreal-based indie rock sensation Arcade Fire. While themes of death and loss run throughout the album (appropriately entitled *Funeral*), "Neighborhood #2" invokes Laika's name as the definitive marker of betrayal and rejection:

> Alexander, our older brother,
> set out for a great adventure.
> He tore our images out of his pictures,

---

31. "Laika," *Cinema* (Xyx, 1991).

32. "Laika, se va," *Aerial* (1998?).

33. The Divine Comedy, *Absent Friends* (Parlophone, 2004).

he scratched our names out of all his letters.
Our mother shoulda just named you Laika![34]

The music video for this piece shows a book of "memories" being pulled from the family bookshelves during the singing of the third line. As the last line is sung, a shell labeled "Laika" blasts out of a cannon. In this song and in other examples, the elision of canine-human identity facilitates a reversal of the original inflection of Laika's story. The historical Laika might still be a victim or a pioneer, but contemporary Laikas can also be agents of betrayal (as in the case of "Neighborhood #2) or emblems of lost causes.[35] When the specifics of Sputnik II are invoked, the ending of the story is subject to considerable revision: Laika might survive, return to Earth, or reappear in another realm. For example, in Niki McCretton's recent theatrical production, "Muttnik, the First Dog in Space," the British solo stage performer portrays Laika as a "canine adventurer" whose "rags to riches story" appeals to audiences of "Children and Childish Adults." [36]

As the Soviet particulars of Laika's story recede from the popular consciousness, musicians seem increasingly inclined to link her to more universal human concerns and struggles. The clearest example of using Laika's name without any reference to the circumstances surrounding her story is a dreamy, half-intelligible song about lost love and self-effacement by Damon and Naomi, the folk-duo, peace activist sponsors of Exact Change publishers.[37] Other songs, such as Massacre's "Laika, se va" or Blipp!'s "Laika," use selected elements of the Sputnik II story to frame meditations on a (human) longing to return home or the isolation of an endless journey.

In addition to the musical compositions discussed here, a number of recent literary endeavors refer to or are inspired by Laika as well. Among these are children's books and science fiction works, as well as more serious explorations of loneliness and alienation such as James Flint's novel *Habitus* (2000) and *Sputnik Sweetheart* (1999) by Haruki Murakami. Nick Abadzis's graphic novel *Laika* (2007) intertwines fact and fiction to examine the nature of trust and the implications of technological advances for what it means to be human. In Jeanette Winterson's recently published, *Weight* (2005), the first space dog appears as a grateful companion to a world-weary Atlas in a witty retelling of a classic myth-cum-meditation on choice, freedom, and coercion. In

---

34. Arcade Fire, "Neighborhood #2 (Laika)," *Funeral* (Merge Records, 2004).

35. For a recent example of this usage, see the comic strip "Get Fuzzy" from November 7, 2006.

36. "Muttnik the First Dog in Space," *http://www.angelfire.com/stars4/nikimccretton/cgi-bin/MuttnikShowDetails2006.pdf* (accessed January 25, 2008).

37. "Laika," *More Sad Hits* (Shimmy Disc, 1992); "Exact Change: Classics of Experimental Literature," *http://www.exactchange.com/frame/frame.html* (accessed January 25, 2008); "Damon and Naomi," *http://www.damonandnaomi.com/frameset/frame.html* (accessed January 25, 2008).

keeping with dogs' powerful role as mediators of realms in various mythic and legendary settings, Winterson's historical Laika helps the mythic hero negotiate his unbearable burden even as he saves her from the solitude of outer space.[38]

Additional evidence of Laika's continued resonance is found in the astonishing number of Web sites devoted to the dog. These range from a "rainbow bridge" memorial that places Laika in the sentimentalized cosmology of grieving pet owners, to sites concerned primarily with space history, stamp collecting, or vending space dog memorabilia.[39] A rescue organization for homeless animals in Moscow chose Laika for its Web site logo because, "she represents for us the plight of homeless animals everywhere—abandoned or exploited, but rarely treated with the respect and compassion which all living creatures deserve."[40]

And then there is Akino Arai's song "Sputnik," which appeared on her *Raining Platinum* album in 2000. In a manner perhaps befitting a famous anime singer, the real, the imaginary, and the fabricated are interwoven in this song of lost (human?) love. The lyrics refer to "the Laika dog on Sputnik II," but then conflate the historical Laika with "Kloka," a space dog fabricated by the Spanish artist Joan Fontcuberta for an installation called "Sputnik: The Odyssey of Soyuz 2." First exhibited in Madrid in 1997, "The Odyssey of Soyuz 2" used manipulated digital photos to present an elaborate, completely fabricated history of a fictional cosmonaut who allegedly vanished (along with his canine companion) in 1968.[41]

A song of human longing that invokes a fictitious dog to commemorate a real one might be the ultimate tribute to a global celebrity whose entire history is built on irony. For not only is Laika the dog a more meaningful figure—at least in the popular imagination—than the many human forces associated with her voyage, but, even more paradoxically, it seems that by perishing in space, she has become eternal. Laika endures as a symbol of futuristic adventure, sacrifice, and experimentation, as a foil for human anxieties about abused animals and pet dogs, and as a timeless echo of a unique historical moment. But in today's popular culture, the particulars of that moment seem to have been far easier to metabolize than the reality of sending "man's best friend" on a one way trip to

---

38. On dogs as negotiators of human identities and boundaries, see McHugh, *Dog*, pp. 47–48.

39. "Memorial to Laika," *http://www.novareinna.com/bridge/laika.html* (accessed January 25, 2008); Ted Strong, "Laika the Russian Space Dog!," *http://tedstrong.com/laika-trsd.shtml* (accessed January 25, 2008); Sven Grahn, "Sputnik-2, More News from Distant History," *http://www. svengrahn.pp.se/histind/Sputnik2/sputnik2more.html*, accessed January 25, 2008; Sven Grahn, "Sputnik-2, Was it Really Built in a Month?," *http://www.svengrahn.pp.se/histind/Sputnik2/ Sputnik2.htm* (accessed January 25, 2008); Melissa Snowden, "Russian Space Dogs," *http:// www.silverdalen.se/stamps/dogs/library/library_space_dogs_russian.htm* (accessed January 25, 2008).

40. "Moscow Animals," *http://www.moscowanimals.org/index.html* (accessed January 25, 2008).

41. Catherine Auer, "Ground Control to Comrade Ivan," *The Bulletin of the Atomic Scientists* vol. 58, no. 2 (2002): 10–12.

outer space. Like Oscar Wilde, a figure synonymous with wit and gay identity, Laika has become an iconic figure largely divorced from historical specifics. Her continued presence in the human imagination depends on her absence, on the bizarre and public circumstances of her demise, and on the contradictions between the grim realities of her life and people's idealized conceptions of dogs. In contemporary global culture, the memory of the first space dog remains vibrant, even as the historical particulars surrounding her place in the Space Age begin to fade.

LAIKA SINGLES WITH LYRICS
(Title, Artist, Album, Year, Genre, Artist's Country)

"Sputniks and Mutniks," Ray Anderson and the Homefolks, NA, 1958, country, United States

"Russian Satellite," The Mighty Sparrow, NA, 1959, reggae, Trinidad

"Laika," Mecano, *Descanso Dominical*, 1987, alternative, Spain

"Laika," Ice MC, *Cinema*, 1990, Eurodance/hip hop, United Kingdom

"Laika," Äge Andersen, 1991, folk rock, Norway

"Laika," Damon and Naomi, *More Sad Hits*, 1992, alternative / indie rock, United States

"Laika," Moxy Früvous, *Bargainville*, 1994, folk, Canada

"Laika, Laika," CCCP, *Cosmos*, 1996, rock, Germany

"My Dog is an Astronaut, though," Pond, *Rock Collection*, 1997, indie rock, United States (Oregon)

"Laika, se va," Massacre, *Aerial*, 1998, death rock, Argentina

"La Ballata Di Laika," Daisy Lumini E Beppe, *El Paese Dei Bambini con la Testa*, 1999, folk / acoustic, Italy

"Sputnik," Akino Arai, *Raining Platinum*, 2000, alternative, Japan

"Laika," Gionata, *L'uomo e lo Spazio*, 2002, alternative, Italy

"Laika," Kyler England, *A Flower Grows in Stone*, 2003, folk/indie rock, United States

"Neighborhood #2 (Laika)," Arcade Fire, *Funeral*, 2004, indie rock, Canada

"Absent Friends," Divine Comedy, *Absent Friends*, 2004, alternative, United Kingdom

"Laika," Little Grizzly, *When it comes to an end I will stand alone*, 2004, indie rock, United States (Texas)

"Sputnik (Song for Laika)," Sputnik, *Meet Sputnik*, 2004, rock, United States

"Laika," Blipp! *Impulser*, 2005, electronic / alternative, Sweden

"Laika," Per Bonfils, *Exotic Fruits*, 2005, electronic, Denmark

"Ultra Laika," Per Bonfils, *Exotic Fruits*, 2005, electronic, Denmark

"Laika: an Allegory," Amoree Lovell, *Six Sadistic Songs for Children*, 2006, gothic rock, United States (Oregon)

"Laika," Built by Snow, *Noise*, 2007, indie rock, United States (Texas)

"Laika," Handshake, *World Won't Wait*, 2007, folk, United Kingdom (London)

"Laika In Space," The Antecedents, *Letters from Rome*, 2007, indie rock/pop, United States (Oregon)

"Laika," Team Robespierre, *Everything's Perfect*, 2008, punk/dance, United States (New York)

Instrumental Singles Named After Laika
(Title, Artist, Album, Year)

"Laika," Honey B. & The T-bones, *On the Loose*, 1990

"Laika," The Cardigans, *The Other Side of the Moon*, 1997

"Laika," Those Norwegians, *Kaminzky Park*, 2003

"Like Armstrong + Laika, Tied and Tickled Trio, *Observing Systems*, 2003

"Laika," Alias, *Instrument No. 4*, 2004

"Laika," Ghost 7, *New Directions in Static*, 2004

"Laika's Theme," The Divine Comedy, *Absent Friends*, 2004

"Laika," Walnut Grove Band, *Black Walnut*, 2005

"Laika," KDream, *Spacelab*, 2005

"Flight of the Laika," Gabber Nullification Project, *Gabber Nullification Project*, 2006

"Laika Goes Techno," Deliens, *Impacts*, 2006

"Laika," Ratasseriet, *Beyond*, 2006

"Laika," The Take, *Dolomite*, 2006

"Laika," Jah on Slide, *Parole de Rude Boy*, 2007

"Neighborhood #2 (Laika), *Vitamin String Quartet*, 2007

"Laika," Tony Corizia, *Basswoodoo*, 2007

"Laika (Part 1), CNTR, *Northern Deviation*, 2007

"Laika (Part 2), CNTR, *Northern Deviation*, 2007

"Laika," Juri Gagarin, *Energia*, 2008

CHAPTER 14

FROM THE CRADLE TO THE GRAVE: COSMONAUT NOSTALGIA IN SOVIET AND POST-SOVIET FILM

Cathleen S. Lewis

"The Earth is the cradle of humanity, but mankind cannot stay in the cradle forever."
—Tsiolkovskii

"Of all the arts, for us the most important is cinema."
—Lenin

INTRODUCTION

Soviet film has featured space travel since its beginning. The first Soviet cinematic blockbuster drew on a contemporary science fiction novel about a pair of travelers to Mars. Since that time point, the popular images of human spaceflight and films in Russia and the Soviet Union have had a long, intertwined history that spanned a century. Over that period, the image of the cosmonaut changed along with political sensibilities. Prior to the revolution, the literary image of the cosmonaut began to take form when Russian writers began to explore the possibility of flying into space through the means of science fiction. As revolution approached, these writings took on ideological overtones, combining the ideas of spaceflight with concepts of utopia and revolution. After the Bolshevik revolution, the government undertook the reconstruction of the Russian film industry that had flourished during the years prior to the revolution. About the same time, recognizing the propaganda potential of the media, Lenin declared it a priority in the economic reconstruction of the country that followed the civil war. Over the next decade, a handful of movies treated the idea of space travel, each one conforming increasingly closely to predominant ideological mores about the demeanor and messages of space travelers should carry on their missions. The most popular media in the Soviet Union and the most popular and celebrated event in Soviet history combined to create a national memory and understanding of spaceflight.

In a conversation with Soviet Commissar of Enlightenment Anatolii Lunacharsky in the years immediately after the 1917 revolution, Lenin said,

"Of all the arts, for us the most important is cinema."[1] Whether Lenin referred to the propaganda potential of the media or its ability to satisfy the country's need for entertainment is unclear. Nonetheless, during the course of rebuilding the country after war and revolution, the new Soviet state went to great effort and expense to develop this young art form. Two of the earliest and most artistically innovative films of this era featured space travel and were adaptations of a Soviet science fiction novel that promoted the idea of interplanetary socialist revolution.[2] The reopening of Soviet cinemas and the first portrayal in spaceflight in film coincided with the cinematic production of Aleksei Tolstoy's *Aelita* in 1924. Months later, a team of animators created their own version of Tolstoy's tale, replicating the ambitious tone of revolutionary fervor of the time. By the end of the decade, Stalin had redirected that fervor internally towards transforming the U.S.S.R. into an industrialized country. Ideologically, transforming nature and political loyalty replaced the concept of exporting revolution. Man and machine traveling through space matched the prevailing political metaphor of the time of man using technology to master nature. Science fiction that emphasized man's ability to engineer mastery over nature and political and personal loyalty gained favor during that brief period when officials tolerated speculative literature.

---

1.   Translated and quoted in: *The Film Factory: Russian and Soviet Cinema in Documents, 1896-1939*, 1988, trans. Richard Taylor, ed. Richard Taylor and Ian Christie, paperback (London: Routledge, 1994), p. 56 from the original citation in G. M. Boltyanskii (ed.), *Lenin i kino* (Moscow/Leningrad, 1925), pp. 16-19. Although many historians cite Lenin's quotation, there is thin evidence that Lenin actually said precisely those words. In the introduction to Josephine Woll's book on the cinema of the Thaw era, Richard Taylor describes the quote thus, "Cinema has been the predominant popular art form of the first half of the 20th century, at least in Europe and North America. Nowhere was this more apparent than in the former Soviet Union, where Lenin's remark that 'of all the arts, for us cinema is the most important' became a cliché and where cinema attendances were until recently still among the highest in the world." Josephine Woll, *Real Images: Soviet Cinema and the Thaw*, Kino: The Russian Cinema Series, ed. Richard Taylor (London: I. B. Tauris Publishers, 2000), p. vii. Denise Youngblood casts doubts on whether Lenin actually made the statement, but does support the idea that Lenin had the intention to promote cinema as a means to propaganda, Denise J. Youngblood, *Movies for the Masses: Popular Cinema and Soviet Society in the 1920s* (Cambridge, UK: Cambridge University Press, 1992), p. 35. Peter Kenez discusses the likelihood that the words were consistent with Lenin's actions, Peter Kenez, *Cinema and Soviet Society: From the Revolution to the Death of Stalin*, Kino: The Russian Cinema Series, ed. Richard Taylor (London: I. B. Tauris Publishers, 2006), p. 22.

2.   The two films were adaptations of Aleksei Tolstoy's novel, *Aelita*. Aleksey Nikolayevich Tolstoy, *Aelita*, trans. Antonnia W. Bouis, ed. Theodore Sturgeon, Macmillan's Best of Science Fiction (New York, NY: Macmillan, 1981). The first was Yakov Protazanov's film by the same name: Yakov Protazanov, *Aelita: Queen of Mars*, Kuinzhi, Valentina; Tseretelli, Nikolai; Eggert, Konstantin; Solntseva, Yulia; Zavadsky, Yuri; Ilinsky, Igor; Batalov, Nikolai (Mezhrabpom-Rus, 1924), 120 minutes. The second was an animated version: Nikolai Khodataev, Zenon Komisarenko, and Yuri Merkulov, *Mezhplanetnaia revolutsiia (Interplanetary Revolution)*, animation (Biuro gosudarstvenno tekhnicheskogo kino, 1924), 7:40 min.

After Stalin's death, options for speculative expression began to reopen. Soviet science fiction reemerged in the late 1950s after the Soviets launched Sputnik in 1957 and Gagarin in 1961. During the early era of human spaceflight in the 1960s, filmmakers undertook a new effort at portraying spaceflight with ideological undertones similar to the previous era. This time, instead of demonstrating how the new technology was transforming the economy and society these movies reassured the public, combining documentary and theatrical components. The focus was on the present indicating that the era of science fiction and the present were one. After the collapse of the Soviet Union, spaceflight attracted new interest, this time without the inhibitions of Party ideology. These new, post–Soviet films were one component of a reexamination of the 1960s as a pivotal period in Soviet history.

While Soviet and Russian portrayals of spaceflight have been sporadic over the decade, they have been consistent in the way in which they reflect their contemporary ideological realities. Similar to the real cosmonauts, film cosmonauts carried the ideology of their nation into space.

SPACEFLIGHT GAINS IDEOLOGY

Whether or not Lunacharsky's memory of Lenin's statement on the importance of film to the young Soviet state was accurate, the new government indeed demonstrated a commitment to film production that made its importance clear. Movies had been popular in pre-revolutionary Russia. In 1913, St. Petersburg and Moscow had over one hundred movie theaters even though the Lumières brothers' invention of the motion picture camera and projector had only arrived in Russia in 1896, one year after its introduction in France.[3] Within five years of the first Russian film production, and at the onset of World War I, Russia was producing about ten percent of films that screened in nearly 1500 Russian movie theaters.[4] As was true with European audiences, the Russians preferred costume dramas and literary adaptations in this new medium.[5]

Film was a very expensive industry for the young U.S.S.R. What World War I did not destroy of the Russian movie industry, the Civil War finished off. Movie theaters and production companies, like most enterprises that were not essential to life, dissolved due to neglect and scavenging during the Civil War. New foreign films were far too expensive for the government to import during the 1920s into the few surviving theaters. and precious materials for domestic film production were beyond the means of the impoverished state. Promising

---

3. Kenez, *Cinema and Soviet Society*, pp. 10–11 and 34.

4. Kenez, *Cinema and Soviet Society*, p. 13, and Youngblood, *Movies for the Masses*, p. 2.

5. Youngblood, *Movies for the Masses*, pp. 2–3, and Kenez, *Cinema and Soviet Society*, pp. 13–18.

and experienced Russian directors had fled the country to Western Europe where filmmaking remained a viable career.[6]

Early Soviet attempts to reignite the film industry were not successful. A film industry was far more complex than a factory and relied heavily on foreign trade as much as artistic talent. It was only the implementation of the New Economic Policy (NEP) that materially changed the situation. The NEP allowed the formation of joint stock companies that could earn income, which was the fiscal solution that allowed movie houses to reopen and make profits from ticket sales. After several iterations, Sovkino, the Soviet film production company, was established as a corporation with shares owned by the Supreme Council of the National Economy, Moscow and Petrograd workers' councils, and People's Commissariat of Foreign Trade, which was the largest stockholder.[7] The resulting cooperation between Sovkino and the one remaining independent film studio, Mezhrabpom-Rus, solved the difficulty of marshalling resources to make films.[8] Mezhrabpom-Rus used the profits from Sovkino to pay for film production.

The NEP period not only marked a relaxed attitude towards the economy and business, but it also marked a period during which attempts were made to encourage the repatriation of Russian intellectuals who had fled the country during World War I or the Civil War. One such person was Iakov Protazanov, the Russian film director who had directed widely popular costume dramas before the war and had lived in exile in Paris and Berlin since 1917.[9] Today in the West, Protazanov's role in early Soviet cinema had been overshadowed by directors such as Vertov and Eisenshtein, but at the time, at age 41, this relatively old man of the cinema promised to reinvigorate Russian film.[10] Probably at the behest of Lunacharsky, Protazanov returned to Russia with the promise that he would be allowed to adapt Aleksei Tolstoy's *Aelita* to film with few expenses or

6. Youngblood, *Movies for the Masses*, pp. 3–5, and Kenez, *Cinema and Soviet Society*, pp. 16–21. Both authors recount the disassembly of the Russian film industry and the dispersal of its resources.

7. Kenez, *Cinema and Soviet Society*, p. 40.

8. "In addition, the NEP allowed the formation of the private joint-stock companies. Of these, the two most important were Rus and Mezhrabpom, which were later to form Mezhrabpom-Rus. Mezhrabpom was an abbreviation of International Workers' Aid, an organization established in Germany in 1921 by pro-Soviet and pro-Communist elements. Its original task was to help Soviet Russian fight famine." Ibid., p. 38.

9. "[Protazanov] made his directorial debut in 1912 with the production of The *Departure of the Great Old Man* ('Ukhod velikogo startsa'), an account of the final days of Lev Tolstoy. He made a star of Ivan Mozzhukhin in literary adaptations, such as *The Queen of Spades* ('Pikovaia dama') in 1916, based on Pushkin's short story, and *Father Sergius* ('Otets Sergei') after the novella by Tolstoy, made in 1918." David Gillespie, *Early Soviet Cinema: Innovation, Ideology and Propaganda*, Short Cuts: Introductions to Film Studies (London; New York: Wallflower, 2000), p. 10.

10. Jay Leyda, *Kino: A History of the Russian and Soviet Film*, 1960, Third (Princeton, NJ: Princeton University Press, 1983), p. 186.

resources spared. His allocation of film stock far exceeded the normal budgets of the time.[11]

Protazanov took advantage of his prestige and drew on then-dormant Russian artistic resources. He hired established Russian stage actors, such as Nikolai Tsereteli, and offered the first screen roles to new actors Igor Ilinsky, Iuliia Solnetseva, and Nikolai Batalov, all of whom later became Soviet film stars. Modernist artist Isaak Rabinovich designed the massive Constructivist sets for the Mars scenes. In addition, Alexandra Ekster designed the modernist Martian costumes just prior to her departure from the Soviet Union.[12]

As a completed film, *Aelita* was almost two hours long—very long for the standards of the time.[13] It was popular among the film going public. Rumors circulated that the director was unable to view the opening due to overcrowding.[14] Most importantly, Sovkino was able to distribute the movie throughout Europe, thus earning hard currency, improving the Soviet Union's balance in foreign trade, and making a profit for future productions. As a measure of his success, Protazanov went on to make ten more silent films in the next six years and continued to make movies until two years before his death at the age of 63 in 1945.[15] In spite of the taints of having returned from abroad after the revolution and producing an ideological suspect film as his inaugural post-Soviet film, Protazanov survived better than other, more revolutionary filmmakers.

As an adaptation of Tolstoy's novel, *Aelita* was the first Soviet science fiction film. It influenced subsequent and internationally better-known European

---

11. Protazanov's production of *Aelita* was clearly a priority for Sovkino, as the expense of the project revealed: "The production history of *Aelita* indicated that Protazanov prepared for his Soviet debut with great care and forethought, but without political foresight. Though schooled in the breakneck pace of pre-Revolutionary filmmaking, averaging more than ten films annually before the Revolution, he took over a year to complete *Aelita*. According to the handsome programs that was distributed at screenings of the picture, Protazanov shot 22,000 meters of film for the 2841-meter film (a 3:1 ratio was the norm) and employed a case and crew of thousands." Youngblood, *Movies for the Masses*, p. 109. Advertising for the film, too, was unprecedented. Almost a year prior to its release, Soviet film newspapers and journals reported on the status of the production. In the weeks leading up to the opening in Moscow, *Pravda* advertised teasers for the perspective Moscow audiences. Aleksandr Ignatenko, *"Aelita": Pervyi opyt sozdaniia blokbastera v rossii* (Sankt-Peterburg: Sankt-Peterburgskii gosudarstvennyi universitet kino i televideniia, 2007).

12. Gillespie, *Early Soviet Cinema*, p. 11. Although Ekster associated with the Constructivists, she considered herself to be an art nouveau designer as she did not adhere to the Constructivist tenets of utility. She immigrated to Paris in 1925. Christina Lodder, *Russian Constructivism* (New Haven, CT: Yale University Press, 1983), pp. 153-155 and 242.

13. Protazanov, *Aelita: Queen of Mars*.

14. Mike O'Mahony, "Aelita," in *The Cinema of Russia and the Former Soviet Union*, ed. Brigit Beumers, 24 Frames (London: Wallflower, 2007), p. 37.

15. Andrew J. Horton, "Science Fiction of the Domestic," *Central Europe Review* 2, no. 1. January 10, 2000: Kinoeye, February 7, 2007, *http://www.ce-review.org/00/1/kinoeye1_horton.html*, n.p.

science fiction films, such as Fritz Lang's *Metropolis* (1926).[16] However, *Aelita* was in fact not a close adaptation of Tolstoy's *Aelita*, consequently shifting the central theme of the film away from science fiction. In fact, Protazanov's liberties with the novel's plot turned the film into a fantasy melodrama that, while appealing to the public, drew harsh criticism from the ideologically strict party elite. In the movie version of *Aelita*, the heroes Los and Gusev do travel to Mars and precipitate a worker's revolution, but the travel takes place in Los's dream that he has because of his jealousy over his wife. The director turned the export of revolution into a fantasy. His liberties transformed space exploration from a revolutionary activity into the daydreams of an engineer with lingering bourgeois sentiments. Moreover, Protazanov's production dwelled on the corruption and hypocrisy of the NEPmen and pointed out that no one was above the corrupting influences of poverty.

The Martian sequences and Los and Gusev's travel to the planet are of particular interest for their style and design even though their significance to the story is diminished. In contrast to the Soviet-set portions of the movie, which were filmed largely in the streets of Moscow, Mars was represented entirely with the constructivist set. Modern-designed costumes and even the movements of the actors seem to follow the choreography of modernist dance in the manner in which Sergei Diagalev's *Les Ballet Russes* was popularizing in exile at the same time. Yet Protazanov used these modernist images to portray a dream fantasy of a feudal, slave-owning society. By doing so, he broke the intellectual link between the utopian ideal and modernist art that constructivist artists were demonstrating at the time.[17] Even though Protazanov disassociated the revolutionary notions from space travel, his version of *Aelita* nonetheless established a standard for fictional space travelers in Russian culture. Los, in his dream, discovered his true self through spaceflight, even though the flight was imaginary. He discovered that his dreams of spaceflight interfered with his acceptance of reality, much in the way that the NEP period had been a step back from revolutionary idealism.

The implications of Protazanov's inclusion of modernist and constructivist designs and sets in *Aelita* merits separate discussion. In many ways, the period

---

16. Ibid.

17. Art historian Christina Kiaer defines constructivism as "this concept of the 'socialist object' as Russian Constructivism's original contribution not only to the history of the political avant-garde art movements of the 20th century, but also to the theory of a noncapitalist form of modernity." Christina Kiaer, *Imagine No Possessions: The Socialist Objects of Russian Constructivism* (Cambridge, MA: The MIT Press, 2005), p. 1. "From this it may be concluded that the term 'Constructivism' arose in Russia during the winter of 1920-1921 as a term specifically formulated to meet the needs of these new attitudes towards the culture of the future classless society. Strictly speaking, the term should not be used with reference to those works of art which were made prior to the Revolution, completely free of any utilitarian content of social commitment on the part of the artist who produced them." Christina Lodder, *Russian Constructivism* (New Haven, CT: Yale University Press, 1983), p. 3.

of the NEP was the freest time of intellectual experimentation in the Soviet Union. Among the many experimental movements active at the time was Aleksandr Rodchenko's Constructivist group that emerged from the Moscow Institute of Artistic Culture (INKhUK).[18] The group sought to map out the role of material objects after the revolution eliminated the last vestiges of capitalism. Rodchenko and his group experimented with the modernist design of everyday objects, using geometric shapes and images of machines as the main themes of their designs. Their experimentation continued for some time into the 1930s, but it did not meet with any degree of success. Modernist preoccupation with stylistic innovation and machines contrasted with the central tenet of Socialist Realism that focused on the nature and concerns of the people.[19] Leaders within the architectural community favored neoclassicism since it appeared to resemble Russian national ideals.[20] The protracted competition for the design of the Palace of the Soviets is one example of the manner in which architects and designers pulled away from modernism.[21] These actions paved the way for the creation of monumental art works in which sculptors and painters collaborated with the architect.[22] Over the ensuing years, constructivist designs lost their associations with progress in the Soviet Union until the death of Stalin.

Protazanov's *Aelita* sparked an immediate ideological response over his portrayal of NEP Soviet society. It also sparked a cinematic response. The same year that *Aelita* came out, Soviet animators Nikolai Khodataev, Zenon Komisarenko and Yuri Merkulov released an eight-minute animated short, *Mezhplanetnaia revolutsiia* (*Interplanetary Revolution*).[23] This short, too, was loosely based on Tolstoy's *Aelita*. In this case, the revolutionary cosmonaut was Red Army Warrior Comrade Kominternov. His name is eponymous with the Communist International—the organization for the international spread of the revolution. The film began with the Bolshevik revolution that motivated the capitalists to flee Earth for Mars. Kominternov chased down the capitalists, following them on his own spacecraft. On Mars, he pursued the grotesque capitalists, emerged victorious, and then sent his message to an Earth receiving station decorated with a portrait of Lenin (see illustration). Khodataev's revolutionary message is not remarkable, but his techniques for portraying this

---

18. Kiaer, *Imagine No Possessions*, pp. 1–2.

19. Cynthia Simmons, "Fly Me to the Moon: Modernism and the Soviet Space Program in Viktor Pelevin's 'Omon Ra'," *Harriman Review* 12, no. 4 (November 2000): 4.

20. Arthur Voyce, "Soviet Art and Architecture: Recent Developments," *Annals of the American Academy of Political and Social Science* 303, "Russia since Stalin: Old Trends and New Problems" (January 1956): 107.

21. Vladimir Paperny, *Architecture in the Age of Stalin: Culture Two*, trans. John Hill and Roann Barris (Cambridge, UK: Cambridge University Press, 2002), pp. 1–8.

22. Voyce, "Soviet Art and Architecture,": 114.

23. Nikolai Khodataev, Zenon Komisarenko, and Yuri Merkulov, *Mezhplanetnaia revolutsiia (Interplanetary Revolution)*, animation (Biuro gosudarstvenno tekhnicheskogo kino, 1924), 7:40 min.

Kominternov declares victory in Mezhplanetnaia revoliutsiia (1924) Nikolai Khodataev, Zenon Komisarenko, and Yuri Merkulov. Mezhplanetnaia Revolutsiia (Interplanetary Revolution). Animation. Biuro gosudarstvenno tekhnicheskogo kino, 1924. 7:40 min. Redistributed in: Animated Soviet Propaganda: From the October Revolution to Perestroika. Films by Jove in Association with Soyuzmultfilm Studio. Executive Producer: Oleg Vidov. Director/Writer/Producer: Joan Borsten. Restored version (c) 2006. (Films by Jove)

message are startling in their ingenuity. The animation used a combination of hand drawn cells and cutout animation. Although the ideology of Khodataev's short differed from that of Protazanov, these were the last two portrayals of space travelers using modernist designs for nearly 40 years.

Although creative artists began the portrayal of utopian spaceflight in film in the 1920s, the science popularizers were about a decade behind in the production of science fiction films. *Kosmicheskii reis* (*Spaceflight*) was a 1936 film that was the brainchild of director Vasilii Zhuravlev, a young director who evaded the ideological controversies that plagued Protazanov.[24] In contrast to Protazanov and Khodataev, Zhuravlev's goal was to portray spaceflight

---

24. Vasili Zhuravlev, *Kosmicheskii reis (Space Flight)*, S. Komarov; K. Moskalenko; V. Gaponenko; V. Kovrigin; N. Feoktistov; (Gosudarstvennoe upravlenie kinematografii i fotografii (GUKF), 1936), 70 minutes. The English translation of the film title is sometimes referred to as *Cosmic Voyage*.

The spacecraft *Iosif Stalin* waits for launch to the Moon in Kosmicheskii reis (1936) Vasili
Zhuravlev. Kosmicheskii Reis (Space Flight). S. Komarov; K. Moskalenko; V. Gaponenko;
V. Kovrigin; N. Feoktistov; Gosudarstvennoe upravlenie kinematografii i fotografii (GUKF),
1936. 70.

realistically and to produce a technical science fiction film. Previously the
creator of educational scientific films, he called on the expertise of none other
than Konstantin Tsiolkovskii for technical advice.[25] Tsiolkovskii had been a
science popularizer, as well as an airship and rocket theorist. Even this late in
his life, Tsiolkovskii enthusiastically contributed to the project, sketching and
writing notes on his anticipation of the effects of spaceflight.[26]

Kosmicheskii reis is set in the futuristic year of 1946, and begins at the fictional Tsiolkovskii Institute for Interplanetary Communications. In contrast to
Protazanov's *Aelita*, the Earth scenes are modernist, resembling the art deco

25. Anatolii F. Britikov, *Russkii sovetskii nauchno-fantasticheskii roman* (Leningrad: Izdatel'stvo "nauka,"
1970), p. 27.

26. Ben Finney, Vladimir Lytkin, and Liudmilla Alepko, "Tsiolkovskii's "Album of Space
Voyages:" Visions of a Space Theorist Turned Film Consultant." 1997, in *Proceedings of the
Thirty-First History Symposium of the International Academy of Astronautics, Turin, Italy, 1997*,
ed. Donald C. Elder and George S. James, vol. 26, *History or Rocketry and Astronautics*, AAS
History Series (San Diego, CA: Univelt, 2005), pp. 3-16.

style that was popular in Hollywood films at that time. The film features aged astrophysicist, Pavel Ivanovich Sedykh, who bears a remarkable resemblance to Konstantin Tsiolkovskii. The film opens with Sedykh planning a spaceflight to the Moon on board his space rocket (see illustration), the *Iosif Stalin*, in spite of a previous failed test mission with a cat. After a dispute over issues of personal loyalty and bureaucratic interference, Sedykh balks at concerns over his health and insists on accompanying his assistant and an adolescent boy, Andrushka, on the flight. The three astronauts successfully land on the Moon, unfortunately losing fuel and their radio in the process.[27] While en route to the Moon, the trio experience weightlessness, and on the Moon, they experience diminished gravitational pull. In the process of making a visual signal for Earth about their successful arrival, they discover the cat from the previous mission has survived and that frozen remnants of the lunar atmosphere can be used as fuel for their return mission. Meanwhile on Earth, scientists plan to launch a rescue mission. Just as the launch is about to take place, the Iosif Stalin returns with the jubilant crew. Sedykh declares that they have "opened the path to space." During the late 1950s and early 1960s, Soviet cosmonauts, politicians, and journalists repeated that phrase again and again.

THE KOMSOMOL IN SPACE

Both science fiction film and literature diminished in prominence in the Soviet Union under Stalin. After *Kosmicheskii Reis,* there was not another space science fiction film in Moscow until 1958 when the East German film based on the Stanislav Lem science fiction story, *Der Schweigende stern* (*Silent Star*) opened in theaters in the Soviet Union under the Russian title *Bezmolvnaia zvezda.*[28] A new infusion of science fiction films, beginning with the prescient Soviet film *Nebo zovët* (The Sky Calls), followed in 1960. It predicted a space race between the United States and the Soviet Union to Mars.[29] Like *Kosmicheskii reis* before

---

27. Sedykh and Andrushka referred to themselves as "astronauts" and not "cosmonauts" throughout the film. This was the prevailing name of space travelers at the time, drawing from the Latin-based language of Verne. The decision to adopt the Greek-root cosmos for cosmonaut was deliberate and absolute in 1961. Morton Benson, "Russianisms in the American Press," American Speech 37, no. 1 (February 1962): 41-47.

28. Kurt Maetzig, *Der Schweigende Stern (The Silent Star), (Bezmolvnaia zvezda),* Tani, Yoko; Lukes, Oldrich; Machowski, Igancy; Ongewe, Julius (Deutsche Film (DEFA), 1959), 155 min.

29. Mikhail Kariukov and A. Kozyr,' *Nebo zovët (The Sky Calls),* Pereverzev, Ivan; Shvorin, Aleksandr; Bartashevich, Konstantin; Borisenko, Larisa; Chernyak, V.; Dobrovolsky, Viktor (Gosudarstvenii komitet po kinematografii (Goskino), 1960), 77 min. American producer Roger Corman purchased the rights to the film and hired a young Francis Ford Copolla to rework the movie. *Battle Beyond the Stars* was an American interplanetary war movie with no reference to Cold War competition. Jimmy T. Murakami, *Battle Beyond the Stars,* Thomas, Richard; Vaughn, Robert, Saxon, John (New World Pictures, 1980), 104 min.

Masha bids farewell to her crewmates from the Vega, Shcherba and Allan Kern, as they prepare to land on Venus to rescue their colleagues from the Serius in Planeta Bur' Klushantsev, Pavel. Planeta Bur' (Planet of Storms). V. Emel'ianov; Iu. Sarantsev; G. Zhzhenov; K. Ignatova; G. Vernov; G. Teikh, Leningrad Popular Science Film Studio, 1962. 83 min. (Courtesy Seagull Films of New York)

it, *Nebo zovët* took pains at demonstrating the effects of spaceflight through special effects and set design.

Two years after the release of *Nebo zovët* and within one year of Yuri Gagarin's historic flight, another Soviet science education film director, Pavel Klushantsev, presented his own fictional interplanetary tale, *Planeta bur'* (*Planet of Storms*).[30] The movie began with a crash. A meteor crashes into one of three Soviet spacecraft en

---

30. Pavel Klushantsev, *Planeta bur' (Planet of Storms)*, Emel'ianov, V.; Sarantsev, Iu.; Zhzhenov, G.; Ignatova, K.; Vernov, G.; Teikh, G. (*Lennauchfilm*, Leningrad Popular Science Film Studio, 1962), 83 min. Like its immediate predecessor, this film, too, had a second cinematic life in American theaters, first as the 1965 *Voyage to the Prehistoric Planet* and then in 1968 as *Voyage to the Planet of Prehistoric Women*. Director Peter Bogdanovich created the second American version. Curtis Harrington, *Voyage to the Prehistoric Planet*, Rathbone, Basil; Domergue, Faith; Shannon, Marc (Roger Corman Productions, 1965), 78 min. and Peter Bogdanovich, *Voyage to the Planet of Prehistoric Women*, Van Doren, Mamie; Marr, Mary; Lee, Paige (The Filmgroup, 1968), 78 min.

route to the planet Venus. The crews of the surviving spacecraft, "Vega" and "Serius," had to make a decision about exploring the planet while waiting for a third craft to join them. They decided jointly that in the name of the party and the Soviet Union, they would go ahead with the risky exploration, leaving a lone crewmate and the only woman, Masha, in orbit (see illustration). With them, the men took the robot and its designer, an American, to the surface. In this movie, the cosmonauts found themselves separated from their spacecraft on the planet Venus, fighting prehistoric animals and surviving erupting volcanoes en route back to their spacecraft. The men survived the mission by maintaining Komsomol discipline. The robot was not capable of sacrificing himself for the good of the collective; therefore, the cosmonauts abandoned the robot in a river of molten lava. Meanwhile, onboard the sole remaining spacecraft orbiting the planet, Masha struggled to maintain discipline and remain in orbit over her desire to commit a pointless act of heroism. In the end Masha overcame her emotions, obeyed orders, and aborted her rescue attempt to the planet's surface, thus leaving open the possibility of salvaging the mission.

Even though they were made a generation apart, *Planeta bur'* resembled *Kosmicheskii reis* in content and values. Both films relied heavily on the principles of science education for content, although the latter used wild fantasy in the Venus segments of the movie. The former presented an image of cosmonauts that announced the new age of human spaceflight to the world. *Planeta bur'* demonstrated that cosmonauts took the values of party discipline with them as they explored the solar system. The film also began Klushantsev's 1960s trilogy that included the movies *The Moon* and *Mars*. All three films combined science education with realistic portrayals of science fiction, even though the later two were hybrids of documentary and theatrical film, switching from scientific lectures and interviews to dramatic demonstrations of scientific principles. This new genre reinforced the cosmonaut message during the 1960s. Through party discipline, the Soviet Union was leading the way into space.

Although popular cinema never achieved the propaganda effect that Lenin and Lunacharsky had predicted during the revolution, it did remain a popular diversion from everyday life. In spite of ideological mandates and the international popularity of Modernist film directors such as Eisenshtein and Vertov, Russian audiences preferred a comprehensible story line and rational adventure. As a result, the earliest Soviet science fiction movies subordinated the ideological aspects of space travel to the fantasies about the appearance of other worlds. Subsequent films attributed a nation's ability to fly in space to personal loyalty and party discipline. This trend continued in the new Soviet science fiction films at the dawn of the Space Age. This style would continue in films and the portrayal of cosmonauts through the collapse of the U.S.S.R.

Post-soviet Reexaminations

After the dissolution of the Soviet Union in 1991, postmortem examination of the Soviet experience became a national pastime. Artists, writers, and filmmakers joined with journalists and common citizens to assess what the 75-year Bolshevik experiment had meant. There have been two recent Russian theatrical films, Aleksei Uchitel's *Kosmos kak predchuvstvie* (*Space as Premonition*, 2005)[31] and Aleksei Fëdorchenko's *Pervye na lune* (*First on the Moon*, 2005),[32] that address the legacy of the golden years of Soviet spaceflight in their own unique manner. Each film places spaceflight into the context of a specific period. Uchitel's film is set in the early 1960s and Fëdorchenko's begins in the 1930s. Both dissect the origins of the culture of real spaceflight.

Aleksei Uchitel's film takes a nostalgic approach in which the early Soviet space program provides the background for a story about the illusion of nostalgic optimism. The film takes place between the time of the launch of Sputnik and Yuri Gagarin's flight. The protagonist of the story is a hapless young man, Konëk, whose naïveté has benefited him. Unaware of the injustices around him, he is able to wonder through life unaffected by it. The main character is a cook whose real name is Viktor, but he goes by the nickname "Konëk" (Horsie). The story focuses on Konëk's relationship with a former sailor and dockworker, German. German, who is also known as "Lefty," is a former sailor who is trying to defect to the West. His persona allures Konëk, a man who is haplessly living with his mother and indecisive about committing to his girlfriend, Lara. In contrast, German is worldly and sophisticated. The men form bonds: both are war orphans and relish fights with sailors. To Konëk's mind, German is exotic and mysterious, possessing superior skills and knowledge about the world, as well as material possessions including an East German radio that picks up BBC. In spite of his seeming sophistication, German cannot articulate properly the English words to declare his intention to defect. Ironically, his new hapless friend demonstrates the ability to mimic the voices on BBC radio almost perfectly, although he has no ambitions for contact with the West and understands little of what he is saying.

Over the course of their relationship, Konëk begins to dress and act like German. He goes as so far as to practice swimming in the harbor. Konëk assumes that this activity is to improve his athletic performance; he is clueless that his

---

31. Aleksei Uchitel,' *Kosmos kak predchuvstvie (Space as Premonition)*, Mironov, Evgenii; Pegova, Irina; Tsyganov, Evgenii; Liadova, Elena (Rock Film Studio, 2005), 90 minutes. Uchitel' won the "Golden St. George" award at the Moscow International Film Festival in 2005 for this movie.

32. Aleksei Fedorchenko, *Pervye na lune (First on the Moon)*, Vlasov, Boris; Slavnin, Aleksei; Osipov, Andrei; Otradnov, Anatolii; Ilinskaia, Viktoriia (Sverdlovsk Film Studio and Film Company Strana, 2005), 75 minutes. Ironically, this film won the "Best Documentary" award at the Venice Film Festival in 2005. The same year, it won "The Best Debut" prize at the Kinotaur Festival in Sochi, Russia.

friend intends to defect by swimming out to a foreign ship. Sailors eventually beat him up for his attitude and for flagrantly walking around town with the forbidden radio. As their relationship develops, German confesses that his mysterious secret assignment is to seek out the ten cosmonauts who are training for the first spaceflight in Kustanay in the Kazakh Republic. For a while, German insists to Konëk that soon men will travel to the Moon everyday, but, ultimately, he confesses again that he had been in prison in Kustanay, convicted for making wisecracks while in the Navy. German is last seen swimming toward a shipping vessel marked "Lake Michigan" as the ship moves away. Viktor (Konëk) marries Rima, Lara's sister, and they take a train to Moscow.

In *Kosmos kak predchuvstvie*, space is metaphor for hope. In the movie, Lara asks Konëk as a plea for reassurance if he can see *Sputnik* after German seduces her. A second use of the metaphor occurs during Konëk and Rima's trip to Moscow. While on the train, Konëk crosses paths with an equally unassuming young pilot named Gagarin whom the hero and audience believe to be the Yuri Gagarin. When speaking to Gagarin in the train, Konëk asks him if he is going to fly rockets. Gagarin responds by asking if he was referring to the predictions of Tsiolkovskii. Konëk replies, "No, German." Gagarin has not heard of that scientist, to which Konëk replies, "He is not a scientist, but he has already flown." When the pilot arrives at his stop, Konëk asks his name and notices that his shoelace is untied. Later Konëk recognizes Gagarin by this untied shoelace. By this time, Gagarin has made his flight and is walking down the red carpet to greet Khrushchev.

It is through this meeting that the director has tied the meaningless life of his hero to the equally unpurposeful mission of the space program. The experienced and knowledgeable character, German, is determined to escape the Soviet Union, even if it costs him his life. The more meandering of the two, Konëk, identifies most closely with Gagarin. One film reviewer has described the time between Sputnik and Gagarin's mission, "the two moments of Soviet triumph in space that, the contemporary audience knows, led nowhere and that provide the bookends of the film (the flights of Sputnik and of Gagarin)."[33] These two moments of triumph represent a memorable period that benefited the nation through their naiveté but provided no objective improvement in its circumstances.

Fëdorchenko's *Pervye na lune* is closer in tone to Viktor Pelevin's novel *Omon Ra* in its take on the space program.[34] Produced as a mock documentary or

---

33. Katerina Clark, "'Aleksei Uchitel,' Dreaming of Space [*Kosmos kak predchuvstvie*] (2005)," *KinoKultura*, no. October 2005, *http://www.kinokultura.com/reviews/R10-05kosmos.html* (accessed on August 11, 2006).

34. Pelevin's novella was an award-winning book in Russia in the early 1990s. The book is a modernist satire that tells the story of a boy, Omon, who wants above all things to become a cosmonaut. His journey to that goal takes him through the Byzantine depravations of a

mockumentary, this film fabricates the existence of a secret Cheka film archive of a Stalinist program to send men to the Moon during the 1930s. Where *Space as Premonition* parodies the unfulfilled potential of spaceflight weighed-down by a corrupt system, *First on the Moon,* portrays the cosmonauts as tragic victims of the state. The events of this film take place during the late 1930s and present footage of the selection and training of cosmonauts for a secret flight to the Moon. The purported documentary reports on the uncovered mission and the search for the survivor—cosmonaut Ivan Sergeevich Kharlamov—his journey from his crash site in Chile, and ultimate return to the Soviet Union. Although this is a parody, partly of the Stalinist Falcon's flights of the 1930s and partly of the space program of the 1960s, the treatment of the cosmonauts is deeply affectionate. They, too, are hapless and blameless in their efforts.

Ivan Sergeeivch Kharlamov started his career as an aviation pioneer along with Chkalov and Baidukov, aviators who were the first heroes of the Soviet Union. At some point in his career, he joins a team of Soviets and Germans who were cooperating on rocket development under the conditions of the 1922 German-Soviet Treaty of Rapallo. Among the finalists for the mission are four people: Kharlamov; a girl known as the Komsomol princess, Nadia; a central Asian, Kharif Ivanovich Fattakhov; and a midget. All endure the final testing, and the chief designer makes the final selection of Kharlamov for a launch in 1938. After the launch, the ground control loses all contact with the spacecraft, throwing the entire program into turmoil. The chief designer commits suicide. Mysterious men sedate and kidnap the remaining cosmonaut-candidates from their barracks and destroy all evidence of the mission, saving only a scale model of the spacecraft.

What could have been a complete coverup of the program's existence was, however, imperfect. In March 1938, Chilean peasants report seeing a fireball in the sky. This episode refers to a real event in history of a meteor landing in the country. In Fedorchenko's film, it is not a rock, but Kharlamov's spacecraft that lands in Chile. He has survived his mission. However, without official status, Kharlamov has no resources with which to return home and has to become personally resourceful to do so. The Cheka interviews trace his steps in the return home. He travels through Mongolia during the Russo-Japanese War, after crossing the Pacific Ocean via boat and through China. Speaking only gibberish upon his return to the U.S.S.R., Kharlamov ends up spending time in a psychiatric hospital. It was this time that the People's Commissariat for Internal Affairs (NKVD) takes notice of his return to the U.S.S.R. Always not close enough on his trail, 30 NKVD agents gather information on his life as they follow him.

---

corrupt Soviet system that stages a mock-flight to the Moon. When Omon discovers the true level of deception about the program, and implicitly, the Soviet system, he flees. Viktor Pelevin, *Omon ra* (Moscow, Russia: Vagrius, 2001).

The bulk of the film contains interviews with those who knew or saw him. The materials that the NKVD agents salvaged include psychiatric footage of the administration of electroshock and insulin therapy that bear a remarkable resemblance to the flight-test footage. Agents also interview his wife after he leaves the psychiatric hospital. They later interview Fattakhov and the midget. The interviews with Fattakhov about Kharlamov's background are particularly instructive. Fattakhov has become a builder of giant mechanical insects for children's museums. His profession makes the analogy to Franz Kafka's *Metamorphosis* obvious, especially because the insects are always on their backs except for one final scene with Fattakhov at a museum.

The NKVD's trail of Kharlamov goes cold after they tracked down the midget who returned to his original profession—a circus performer. At one point, Kharlamov joins him where, for a while, he plays the part of a circus version of Aleksandr Nevskii, repelling the Teutonic invaders.[35] Although the NKVD gives up pursuit of Kharlamov at this point and eventually opts to destroy all evidence of the lunar program, they neglect one thing. There is one remaining source of evidence that they cannot destroy. In the closing scenes, the movie takes the viewer to the natural history museum in Chile, near where the peasants had seen the fireball in 1938. This museum retains the footage from Kharlamov's lunar mission and the hardware from his flight. The movie closes with film footage of a lunar landscape and a lone, silent cosmonaut sitting inside his spacecraft (see illustration).

Pervye na lune goes far to dissolve the links between the hero cosmonaut and the Soviet state. Kharlamov was loyal to his country to the end, returning even after his existence had been denied. Even his final known role was that of the legendary and publicly manipulated Nevskii. His reward for all this had been pursuit and abuse by the system that created him.

Both Uchitel and Fëdorchenko drew on the traditions of realistic science fiction in their films. In each case, spaceflight is technically accurate and not metaphysical. These films refer to the traditions that Zhuravlev and Klushantsev had pioneered. The difference between these post–Soviet filmmakers and their predecessors is that they have portrayed Soviet cosmonauts stripped of the either implicit (*Kosmicheskii reis*) or explicit (*Planeta Bur*) ideological discipline. Uchitel's Gagarin lived in morally reprehensible system that only the hapless can ignore. Fodorchenko's Kharlamov returned to a country whose state apparatus is determined to remove all evidence of his existence.

---

35. This scene was clearly homage to Sergei Eisenshtein's film version of the Nevskii story, completed in 1938 and withdrawn in 1939. In Fedorchenko's film, the invading midget Teutonic Knights bear the swastika-like crosses that Eisenshtein's attackers on Novgorod did. Sergei Eisenshtein, *Aleksandr Nevskii*, Cherkasov, Nikolai; Okhlopkov, Nikolai; Abrikosov, Andrei; Orlov, Dmitri; Novikov, Vasili (Mosfilm, 1938), 1:37. This is one of many allusions to Soviet films in the movie. At one point, the cosmonauts go to see Zhuravlev's *Kosmicheskii reis* during the course of their training.

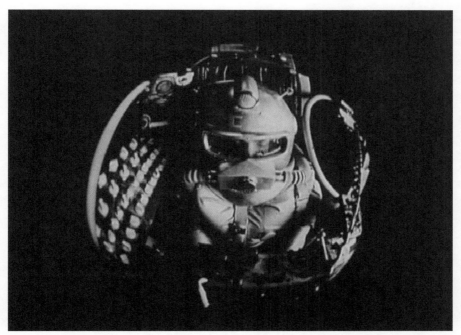

Cosmonaut Kharlamov en route to the Moon in Pervye na lune Fedorchenko, Aleksei. Pervye Na Lune (First on the Moon). Vlasov, Boris; Slavnin, Aleksei; Osipov, Andrei; Otradnov, Anatolii; Ilinskaia, Viktoriia. Sverdlovsk Film Studio and Film Company Strana, 2005. 75 minutes. DVD: ©2004 Prokatnoe upravlenie Sverdlovskoi konostudii. ©2005 OOO "SR Didzhital." (©Dizain Oblozhki OOO "SR Didzhital.")

CONCLUSION

The 20th century began with great expectations about spaceflight and communist revolution. By the end of the century, although spaceflight had become a reality, it did not meet the expectations of the early visionaries, and the revolution had become a bitter disappointment to most. In real life, Russian and Soviet cosmonauts have traveled barely higher than Yuri Gagarin did in his 1961 flight, escaping the cradle of Earth by only a few hundred kilometers. In film, they traveled to Mars, the Moon, and Venus, carrying with them bold messages of interplanetary revolution and party discipline. When Russians began to reexamine the Soviet experience, they did not spare the once celebrated experience of spaceflight. Spaceflight had been a central focus of mid-century Soviet propaganda as demonstrable evidence that the Soviet Union had exceeded recovery from World War II and was overtaking the United States. Although pre-Gagarin film directors could easily refit fictional characters with appropriate political awareness for the given situation, once the

Soviet Union had sent men and a woman into space, those individuals were forever bound to the Soviet ideology of the 1960s. For that reason, the post-Soviet reassessment of the early years of human spaceflight paid close attention to the cosmonaut's relationship with the state. One consequence was that these films extricated spaceflight and cosmonauts from their ideological burden by passing the judgment that what might have held cosmonauts back from fulfilling the spaceflight dreams of the 1920s. According to these directors, it had been ideology and not technology. As Tsiolkovskii had predicted, humans had emerged from the cradle of Earth, but it was from the gravesite of the Soviet Union that cosmonauts gained their freedom from ideological burdens.

CHAPTER 15

Examining the Iconic and Rediscovering the Photography of Space Exploration in Context to the History of Photography[1]

Michael Soluri

The exactly repeatable pictorial statement in its photographic forms has played an operational role of the greatest importance in the development of modern science and technology. It has become an essential to most of our industries and to all of our engineering.[2]

I. Preface

My earliest recollection of space travel was the star-size dot of the Soviet Union's Sputnik blinking on then off then on as it arced across the then starry sky of Niagara Falls, New York. And the first photograph that ignited my fantasies of space exploration (not withstanding motion pictures like *Forbidden Planet, This Island Earth*, and *The Day the Earth Stood Still*) was the dismaying launch-explosion of the Navy's Vanguard One. It was the United States' first effort to place a satellite into Earth orbit in December 1957. The result was a

1. I wish to thank Steve Dick and his committee (which included Asif Siddiqi) for having chosen me to participate in the "Remembering the Space Age Conference"—a dream come true. Thank you as well to Leslie Martin and the Aperture Foundation for their invaluable assistance in obtaining the artists' rights to some of the fine art photographs used in this paper. My gratitude to them for the inspiration I constantly receive from the world-class exhibitions, books, and lectures that have catalyzed my thinking for this paper. NASA Johnson Space Center's (JSC) Media Resource Center, Michael Gentry (and his dedicated team) was always available to find sought-after imagery, along with a wealth of information and insight on human spaceflight photography. Mary Ann Hager at the Lunar and Planetary Institute provided invaluable guidance on the flight photography of Project Apollo. Ed Wilson and Maura White of JSC's Information and Imaging Systems branch provided research and high resolution files from Apollo flight films. The Hubble SM4 crew of STS-125, whose wish to make more insightful images during their mission, inspired me to explore and share with them the photographic history of astronaut flight photography. And thanks to Loralee Nolletti, my wife and mother to our son, Gabriel, for her remarkable skill and patience in the editing of this paper.

2. William M. Ivins, Jr., *Prints and Visual Communication* (Cambridge, MA: The MIT Press, 1969), p. 179.

black-and-white image published over and over in various print media venues that, over the last five decades, has become a convincing visual report on the American effort to compete with the Soviets who already had two satellites in Earth-orbit. The Vanguard image is one of many now-familiar images that visually communicate the early days of the Space Age. As a result, it can be argued that this photograph has become iconic in the sense that it is a recognizable image whose familiarity is framed both by its historic relevance and its repeated publication in chronicling the first 50 years of space exploration.

In my process of rediscovering the iconic within these first 50 years of space exploration photography, I will first identify those images that I consider to be iconic, including the Vanguard launch explosion. Thereafter, I will discuss these photographs within the context of the history of photography, looking at both their technical evolution as record-keeping tools and their aesthetic appeal and importance as historic markers of American culture and beyond. Once this historical framework is established, I will move beyond the known iconic imagery to the emergence of other imagery. And why not consider new imagery? Aren't there other photographs beyond that one iconic photograph—the new and largely undiscovered photograph that tells the same story of an event? In fact, that new photograph may even offer a fresh perspective on the event. I will then critique these new and emerging photographs and juxtapose them to aesthetic markers in the history of photography, drawing from landscape, portraiture, documentation, photojournalism and fine art photography.

II. Beginnings

On a January night in 1958, not even 60 days after the Vanguard's explosive entrée, the U.S. Army launched its Jupiter-C version of the Redstone rocket, introducing the first successful orbiting of an American satellite: Explorer 1. This event was historically significant for both space science and photography because it created a new and emerging iconographic image: the launching of a rocket thrusting into space. In addition, the postflight news conference was memorialized in a photograph of the jubilant Wernher Von Braun, William Pickering, and James Van Allen holding a model of America's first satellite over their heads.[3] The photographs of Vanguard One, Explorer 1, and the postflight press conference were products of a distinctly American culture. With the Soviets not providing any immediate photographic evidence of their similar events, the American print media had ushered in a new photographic genre: space exploration photography. These photographs would be repeated again and again in both the print and electronic media for the next five decades. Moreover, images of a rocket explosion, a rocket launch and the people behind the scenes

---

3. The photographer of this image is unknown, according to Erik M. Conway, the Historian at the Jet Propulsion Laboratory.

Vanguard One Explosion at Cape Canaveral, Florida: December 6, 1957. (NASA)

would become immediate archetypes for those recording the exploits of future space exploration. These iconic images (and others that also eventually emerged) have had a profound impact on American culture, and the adaptation of American scientific and technological culture abroad.[4]

---

4. For a greater understanding on the association between space exploration and the American culture's space policy and history, see Howard McCurdy's *Space and the American Imagination* (Smithsonian Institution, 1997).

Explorer 1 Launch, Cape Canaveral, Florida: January 31, 1958. (NASA)

In examining the first 50 years of iconic imagery, space exploration photography has largely focused on the reporting of an event. According to William Ivins, photography actually has two possible outcomes: the recording of an event and the interpretation of that event:

> The flood of photographic images (since its invention) has brought about a realization of the difference between visual reporting and visual expression. So long as the two things were not differentiated in the mind of the world, the world's greater practical and necessary interest in reporting had borne down artistic expression under the burden of a demand that it be verisimilar (true or real), and that a picture should be valued not so much for what it might be in itself as for the titular subject matter which might be reported in it.[5]

It is my intent to explore these assertions by examining the notions of visual reporting and visual expression in the photography of space exploration in the context of the evolving history of photography. What follows is a brief summary of that history.

---

5. William M. Ivins, Jr., *Prints and Visual Communication* (Cambridge, MA: The MIT Press, 1969), p. 177.

Explorer 1 post-launch news conference, Washington, DC: February 1, 1958. (NASA)

III. An Abbreviated History of Photography

> The nineteenth century began by believing that what was
> reasonable was true and it wound up by believing that what
> it saw a photograph of was true—from the finish of a horse
> race to the nebulae in the sky. The photograph has been
> accepted as showing that impossible desideratum of the his-
> torian—wie es eigentlich gewesen—how it actually was.[6]

In 1836, more than 120 years before the dawn of the Space Age, the sciences
of chemistry and optics began to come together, resulting in the inevitable
invention of photography.[7] As early as 1826 in France, Joseph Nicéphore Niépce
was able to capture an image with an eight hour exposure of sunlight through
a camera obscura. He called his results, which eventually faded on bitumen
paper, a heliograph. What eluded Niépce, however, was the ability to fix and
chemically secure the image permanently. A few years later, in 1837, Louis-
Jacques-Mandé Daguerre discovered a process that captured and fixed an image
on a polished silver surface. Before exposure to sunlight through a camera, the
silver surface had to be exposed to the fumes of iodine. Once sensitized and
exposed, it was developed in a vapor bath of hot mercury. This process resulted
in a daguerreotype. Daguerreotypes, however unique and precious, were one-
of-a kind, small in size, fragile, and not reproducible.[8]

By 1835, William Fox Talbot, a British nobleman, was able to do what
scientists, alchemists, inventors, and artists had been unable to do: create exactly
reproducible pictorial images. He called his image reproduction a calotype.
With the calotype, Talbot figured out the basic principles of photography:
how to get images of things he saw in a camera obscura on paper and how to
make them permanent. Given both the competitiveness and thus simultaneity
of photography's invention, he put together the findings in a paper presented
to the Royal Society in London.[9] Talbot read his paper, "Some Account of the
Art of Photogenic Drawing, or the Process by which Natural Objects may be
to Delineate themselves without the Aid of the Artist's Pencil," on January 31,
1839—six months before Daguerre's official presentation.[10] When considering
a timeline, it was 119 years to the day until the launch of Explorer 1.

---

6. Ibid., p. 94.

7. Ibid., p. 116.

8. Ibid., p. 120.

9. Ibid., p. 122.

10. By the next decade, photographic experimentation moved across the ocean to America. In 1840,
Dr. John Draper of New York City was the first to make a 20 minute exposure of the Moon on
a daguerreotype. To mark the historical significance of Draper's image, consider that the date was
230 years after the 1610 publication of Galileo's drawings of the Moon in Sidereus Nuncius and

Fall of Richmond, Virginia: 1865. (Mathew Brady/Library of Congress)

And just two years into the Space Age in 1959 and approximately 95 years since Mathew Brady's exact and reproducible photographic images of the Civil War, the first still photograph from the backside of the Moon was radioed to Earth by the Soviet Union's robotic space probe, Luna III.[11] The resulting series of images, as crude as they were, added another element to the genre of space exploration photography: images captured and transmitted to Earth by robotic space probes of interplanetary objects and solar system phenomena.

Over the next 48 years since Luna III, the scope and scale of space exploration photography has evolved from the combination of interplanetary and Earth-orbiting robotic spacecraft, Earth-based telescopes, and human spaceflights. The methodical recording of space exploration technology, the engineering and construction of spacecraft, and the day-to-day operations of space centers and aerospace corporations (recorded by both NASA and its contractors) has augmented this robotic imagery. The sheer quantity of these images is considerable. Some of these photographs are classified or restricted and only available to engineers, scientists, and NASA Administrators. Others, depending on varying degrees of governmental policy, are classified, edited, catalogued, and released into the public domain for use by the media, academia, and the general public. Of these released images, they can be said to have become symbolic in the reporting and representation of space exploration.

---

128 years before the Apollo 8 astronauts became the first humans to photograph Earth rising over the Moon's surface in 1968.

11. Between 1861 and 1865, less than 25 years after the invention of the photograph, (not much more than the time between the launch of Sputnik and the first launch of the Space Shuttle), Mathew Brady's photographic record of the American Civil War—documentary images of people, battlefields, and the tools of war—became a first. His photographs eternalized the damage, carnage, and technology of that war for generations to come.

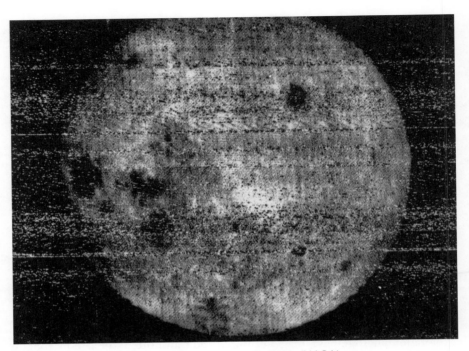

Back side of the Moon by the Soviet's Luna III: 1959. (NASA)

Indeed, a few of these photographs have even become iconic. These are photographic images that have transcended from the temporal to the iconic; they are now a part of our popular culture and photographic history.[12]

Examples of such iconic photographs are Neil Armstrong's 1969 snapshot-like portrait of Apollo 11's Buzz Aldrin at Tranquility Base and a series of self-portraits by Aldrin of his boot and resulting boot prints in the pristine lunar dust.[13] These color photographs are now almost mythic in the sense that they have become the definitive photographic default setting for "first men on the Moon" and/or "first human-Moon landing." In context, however, other lesser known images from Apollo 12 and video still-frame images from Apollo 11 emerge equally important in terms of technical proficiency and aesthetics. For example, the black-and-white series of surface photographs made by the Apollo 12 landing crew (Charles Conrad and Alan Bean) are remarkable in

---

12. See program cover photograph for:"Remembering the Space Age: 50th Anniversary Conference," NASA History Office Division, Office of External Relations, NASA Headquarters and Division of Space History of the National Air and Space Museum (NASAM), October 22-23, 2007.

13. See Apollo 11 images AS11-40-5874 through AS11-40-5880 at the Lunar and Planetary Institute Apollo Image Atlas, 70mm Hasselblad Image catalog, *http://www.lpi.usra.edu/resources/apollo/catalog/70mm/mission/?11.*

Unaltered Buzz Aldrin Apollo 11 portrait: July 20, 1969. (Neil Armstrong/NASA)

their variety, description, and aesthetics.[14] Specifically, Bean's photograph of
Conrad along side, the Surveyor 3 spacecraft with their lunar module in the
background documents the first evidence in one location, of human and robotic
exploration on a celestial body.[15] Bean's photograph of Conrad—their LEM in

---

14. The first four frames of Apollo 12's magazine "Y" by Alan Bean of Pete Conrad offers the most
distinctive (color) sequence of an Apollo astronaut descending the Lunar Excursion Module
(LEM) to the surface of the Moon of any of the six manned landings.

15. See Apollo 12 film magazine "X": AS12-48-7133 in the Apollo 12 flight photography at the
Lunar and Planetary Institute Apollo Image Atlas – 70mm Hasselblad Image catalog, *http://www.
lpi.usra.edu/resources/apollo/catalog/70mm/mission/?12*.

the background and Bean being reflected in Conrad's visor—is as captivating an image of man on the Moon as is the definitive Aldrin portrait.[16]

In addition to these photographs are the largely overlooked black-and-white photographic images of Apollo 11 made by Ed von Renouard, a video technician at the Honeysuckle Creek Tracking Station in Canberra, Australia.[17] During Armstrong's historic descent from the LEM, von Renouard took 35 mm black-and-white images off of his video monitor while it was receiving the first live downlink to Earth. The downlink was being transmitted from a remote slow-scan, black-and-white video camera that was attached to the Modularized Equipment Stowage Assembly (MESA) unit on the side of the LEM descent stage.

Von Renouard photographs captured the ethereal look of "live" black-and-white TV images mixed with the silver-grain textures of black-and-white film as Armstrong descended down the ladder. In some respects, both the aesthetics and the nature of how and where these black-and-white images were made complement Armstrong and Aldrin's Hasselblad surface photography. Moreover, they document the first human to descend to the surface of another celestial world on film through video. These "mixed media" images made it possible to photograph and participate in an historic event as it happened off of a live TV broadcast without even being there.

Up to this point, I have discussed several iconic photographs from the first 50 years of American space exploration. However, in compiling a short list, I would also include: the first photograph of Earth from the orbiting satellite Explorer 6 in 1959; the May 1961 Redstone launch of Mercury astronaut Alan Shepard; the February 1962 Atlas launch with John Glenn; the first EVA (Extra Vehicular Activity) ballet of Ed White from Gemini 4 in 1965; the first Earthrise as seen from the Moon by Lunar Orbiter 1 in 1966; Neil Armstrong's Apollo 11 portrait of Buzz Aldrin standing against a black sky on the surface of the Moon in 1969; the full "blue-marbled" Earth showing Antarctica as photographed by the crew of Apollo 17 in 1972; Voyager 1's first image of Earth and the Moon from seven million miles away in 1977; Bruce McCandless, II in his MMU (Manned Maneuvering Unit) floating away from the Space Shuttle in 1984; the in-flight explosion of the *Challenger* in 1986; the Hubble Space Telescope's photograph of M16, the Eagle Nebula's "Pillars of Creation," in 1995; the "pale blue dot" of Earth as photographed by Voyager 1 from four billion miles in

---

16. For further discussion on the altering of the iconic AS11-40-5903 of Buzz Aldrin by NASA, please refer to Eric Jones commentary from the Apollo 11 Lunar Surface Journal at: *http://history.nasa.gov/alsj/a11/a11-5903history.html.*

17. For an extensive description on both the role of the Honeysuckle Creek tracking station during the Apollo 11 and the resulting video and photography, see *http://www.honeysucklecreek.net/Apollo_11/index.html.*

Pete Conrad, Apollo 12: November 19, 1969. (Alan Bean/NASA)

1996, and a full planetary view of Saturn and its myriad of rings by the Cassini spacecraft in 2006.[18]

In further examining the causal relationship between the emergence of an image that becomes iconic and its ability to sustain itself, William Ivins's analysis of "an exactly repeatable" report of an event (in this case a photographic one) can provide some context. Ivins writes that "The role of the exactly repeatable pictorial statement and its syntaxes resolves itself into what, once stated, is the truism that at any given moment the accepted report of an event is of greater importance than the event, for what we think about and act upon is the symbolic report and

---

18. This short list is based on what I believe represents classic iconic space exploration imagery.

Armstrong descends the LEM. (Ed von Renouard/Honeysuckle Creek Tracking Station)

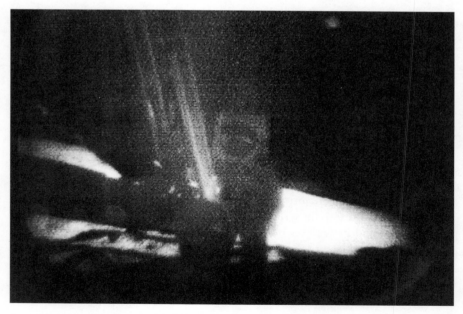

Armstrong Steps Off the LEM. (Ed von Renouard/Honeysuckle Creek Tracking Station)

not the concrete event itself."[19] As a result, the iconic images just cited and others like them—however compelling and important as markers in the visual chronology and history of space exploration—have become the symbolic reports. And as symbolic reports, these photographs have been seen and experienced to the point of being too familiar. They have been routinely published over the decades in academic and scientific journals, popular weekly and monthly magazines, newspapers and magazine supplements, posters, online journals and reports, and any number of photographically inspired books. As a result, these iconic photographs have become predictable markers leaving the viewer with few surprises.

However, there are alternative editing possibilities to draw from the first 50 years of space exploration photography. These alternatives lend themselves to aesthetic considerations worthy of examination as I have attempted with the iconic space exploration photographs already defined and discussed. If the scope of space exploration photography is to mean anything beyond its intended technical, scientific, and utilitarian (day-to-day) reporting, there needs to be an aesthetic framework for examining these photographs as well. Given the accessibility to NASA's photographic archives—and hopefully in the future from Russian, former Soviet, Chinese, European, Indian, and Japanese space agencies—it should be possible to reexplore the familiar and discover the unfamiliar with history and aesthetics in mind.

IV. THE AESTHETIC POSSIBILITIES OF PHOTOGRAPHIC DOCUMENTATION

His (T. H. O'Sullivan) primary aim was not to philosophize about nature, but to describe the terrain. The West was a place to span with railroads, to dig for gold and silver, to graze cattle, or perhaps sell groceries and whiskey. Occasionally—and remarkably—an especially extravagant sample of spectacular landscape would be set aside, sacrosanct, for the amazement of posterity, but this was neither the first function, nor the first interest, of the Surveys. [20]

In the last half of the 19th century, as the technology of photography continued to evolve—the size of cameras, sensitivity of glass plate negatives to light, and darkroom apparatus—it became less studio-dominated and more portable. Furthermore, photography's apparent verisimilitude resulted in opportunities for both American and European photographers to use the medium as a means of

---

19. William M. Ivins, Jr., *Prints and Visual Communication* (Cambridge, MA: The MIT Press, 1969), p. 180.

20. John Szarkowski, *The Photographer and the American Landscape* (New York, NY: The Museum of Modern Art, 1963), p. 3.

communicating to others what unexplored landscapes and native peoples looked like. Forty-five years after photography's invention, pioneers in America like T. H. O'Sullivan, William Henry Jackson, Alexander Gardner, J. K. Hillers, Edward Muybridge and Mathew Brady documented the territories of the United States unexplored by non-indigenous peoples. In numerous surveying expeditions between the Mississippi River and the Pacific (1867–1879), these photographers documented the "geographical and geologic" for the U.S. Government Survey.[21]

The aesthetic and historic examination of 19th and early 20th century landscape photography was first placed in context to the significance of the documentary photograph as art by John Szarkowski. He accomplished this in his seminal exhibition, and subsequent catalogue and book *The Photographer and the American Landscape*—by the Museum of Modern Art (MoMA) in New York in 1963.[22] Szarkowski, then the eminent curator of photography at the MoMA, examined the aesthetics of landscape photography by looking at the role of the photographer:

> The photographer-as-explorer was a new kind of picture maker: part scientist, part reporter, and part artist. He was challenged by a wild and incredible landscape, inaccessible to the anthropocentric tradition of landscape painting, and by a difficult and refractory craft. He was protected from academic theories and artistic postures by his isolation, and by the difficulty of his labors. Simultaneously exploring a new subject and a new medium, he made new pictures, which were objective non-anecdotal, and radically photographic.[23]

From the photographer's new role as explorer-in-the-wilderness, Szarkowski continued: "This work was the beginning of a continuing, inventive, indigenous tradition, a tradition motivated by the desire to explore and understand the natural site."[24] As a result, it is interesting to note that the recognition of landscape photography in the early 1960s paralleled the emerging human and robotic exploration of space. Perhaps these parallel photographic developments were no accident. They shared certain thematic roots, namely "incredible" landscapes and a certain challenge in the mechanics and labor of picture making.

Some of the still imagery captured by the 12 Apollo astronauts between 1969 and 1972 while on the low-angled, Sun-lit surface of the Moon, can be

---

21. Ibid., p 3.

22. Later in 1971, the George Eastman House in Rochester, New York, originated and exhibited "Figure and Landscape." This exhibition further explored the relationship between manmade objects, people, and landscape.

23. John Szarkowski, *The Photographer and the American Landscape* (New York, NY: The Museum of Modern Art, 1963), p. 2.

24. Ibid., p. 2.

Truckee Desert, Nevada: 1869. (T. H. O'Sullivan/U.S. Geological Survey)

likened to 19th and 20th century landscape photography. For example, the 19th century landscape photographer T. H. O'Sullivan's other worldly image of an ambulance covered wagon (containing his portable darkroom) and horses among the sand dunes of Nevada's Carson Desert can be compared to Apollo 12 astronauts Pete Conrad and Alan Bean's documentation of their Oceanus Procellarum landing site with the Surveyor 3 spacecraft in view.[25]

In this 1869 black-and-white image, O'Sullivan positions his camera to look back towards his footprints that lead to a team of horses attached to a covered wagon. The mis en scene gives a sense of scale of the wagon to the vastness of the pristine dunes and the washed out white sky. By comparison, Alan Bean's 1969 black-and-white photograph of the surface of the Moon includes his shadow looking toward the near distant and insect-like Surveyor 3 robotic spacecraft in its 1967 landing place. The abstraction of Bean's long shadow—postured in making this photograph—falls off against the desolate lunar landscape of the Surveyor crater and the stark blackness of space. Both photographs give life to an otherwise lifeless landscape.[26]

---

25. Apollo 12 black-and-white photograph AS12-48-7093 taken by Alan Bean.

26. Other comparisons to early landscape photography emerge from studying the photographs made by the Apollo 17 crew. During the Apollo 17 mission, Eugene Cernan and Harrison Schmitt

Shadow of Apollo astronaut Alan Bean at Surveyor III lunar landing site: 1969. (NASA)

The influence of 19th century landscape photography may extend beyond photography itself. Consider the work of the mid-20th century illustrator Chesley Bonestell. He was an illustrator of space exploration whose mixed illustrations often combined photography with pen, ink, and paint. His style of photorealism ultimately contributed to the American public's imagination of what space and

---

shot more than 3000 photographs. Schmitt and Cernan took a range of pictures documenting surface features and geologic formations in the Valley of Taurus Littrow with Earth floating above. The photographs showing a celestial body above the given horizon harkens back to Ansel Adams's "Moonrise over Hernandez, NM." This black-and-white photograph shows a turn-of-the-century adobe-like village set in the foreground with snowcapped mountains and a near full Moon set in a deep gray-black sky. Likewise, the Schmitt/Cernan photographs depict the blue-marbled Earth suspended in a black lunar sky and floating above the Moon's hills, craters, massive rocks and, occasionally, the landed LEM.

spaceflight might look like.[27] For example, Bonestell's photographic illustration of "Saturn from the surface of its Moon Titan" first published in 1944 offers an imaginary vision of what the exploration of our solar system might produce.[28] With Saturn and a few of its other moons, its rings suspended in the background and jagged mountains resting on an icy surface in the foreground, the sense of a mysterious and alien world permeates the work. Bonestell's technical and aesthetic process, however, was uniquely photographic. Like most of his illustrations, "Saturn from the Surface of Titan" was actually a carefully composed and artificially lit (to simulate the angle of the Sun) photograph of a model in which the artist constructed a mountainous landscape and painted the backdrop of Saturn, its rings, moons, and stars. Aesthetically, the constructed mountains of Titan may very well have been influenced by 19th century Western landscapes like those of the pioneer photographer J. K. Hillers.

Hillers worked on the documentation of the unsettled West during the Powell Survey (1870-1879) for the U.S. Geological Survey. His photographs, which often captured the monumental, depicted and celebrated geologic formations like the Grand Canyon and Yosemite Valley. However, it is Hillers's extensive documentation in Arizona's Canyon de Chelly that is likened most to Bonestell's imagined landscape of Saturn's moon Titan.[29]

The work of Edward Curtis offers a transition from the photography of landscape to the inclusion of people and their habitats. Curtis's documentation, mostly during the first quarter of the 20th century, is often attributed to be among the most remarkable portrayal of Native Americans and their customs and habitats before emerging gentrification and containment on reservations. His work is remarkable because of his ability to gain access to and trust of his subjects, which resulted in a quality of photography that suggested the subject's inner life. Curtis worked with the complexity of the era's photographic technology and adapted it to his personal style, allowing him to connect with and capture the humanity of his subjects and their sacred landscapes in Alaska's Northwest, the Great Plains and South West. Along the way, he developed a list of "twenty-five cardinal points" which outlined the ethnographic and anthropological details to accompany the captions in his photographs.[30]

A few of Curtis's landscape images can be compared to some of the photographs beamed to Earth from the surface of Mars by NASA's Viking Landers (1976), Pathfinder (1996), and most recently the Mars Exploration Rovers. The

---

27. A short list of other prominent illustrators of imagined space are Robert McCall, Pat Rawlings, Ron Miller, and David A. Hardy.

28. For a description of how Bonestell made this iconic image, see *http://www.bonestell.org/titan.html*.

29. The online site of the U.S. Geological Survey offers an impressive and comprehensive series of photo galleries regarding their 19th century pioneer photographers, including O'Sullivan and Hillers, at *http://libraryphoto.cr.usgs.gov/photo.htm*.

30. Alan Porter, "The North American Indian" *Camera* 52, no. 12 (December 1973): 4, 13-14, 23-24.

1944: "Saturn from the Surface of its Moon Titan." (©Bonestell Space Art)

two rovers have been scientifically studying, sampling, and photographing the surface of the Martian landscape since January 2004. Among the many startling photographs are the serene images looking back at the rover's wheel tracks and a hint of its solar panels, the lone trace of human ingenuity and technology amid a landscape of rocky debris on the wind-blown Martian sand. Some of these photographs recall Curtis's landscape documentation of Native American villages in the American Southwest. For example, Curtis took a photograph of a deserted Hopi Indian building in the village of Walpi, a 500-year-old village on a mesa in northern Arizona. The village rests at the near edge of an eroded rocky cliff protruding out into the desert. The aesthetics of this image and the warm-toned yellow hue, resulting from the gravure process that reproduced the original, can be compared to some of the rover Opportunity's panoramic photographs made in a location known as Meridiani Planum. In one reddish hued image, the curving wheel tracks from the rover show how it navigated around rock debris in the Meridiani Planum, a vast dry lakebed that may have once contained water, and the rim of Victoria crater on the horizon. The quality of light, the similarities in camera framing, and the feeling of desolateness offer a comparison between the geological evolution of Earth and

1870–1879: Canyon de Chelly, Arizona. (J. K. Hillers/U.S. Geological Survey)

Walpi Village v.12: 1907. (Edward S. Curtis/McCormick Library of Special Collections, NW University Library)

Mars. It also offers consideration to the erosion of a landscape that may have once nurtured some form of life.

The Curtis Hopi landscapes were part of a grand design: the photographer attempted to systematically document the indigenous peoples of North America, focusing on Native Americans west of the Mississippi and into Canada.[31] Consider Curtis's 1903 black-and-white portrait of a Zuni woman with a decorated ceramic bowl. The photographer's connection with his subject and the simplicity of his lighting and composition make this photograph a symbol of its era. The Zuni woman has a quality of the timeless, contributing to a sense of her profound dignity and humanity. Nearly a hundred years later, the relevance of Curtis's cardinal points and his aesthetic approach to documentary portraiture serve to influence my own photographic documentation of the people and place in space exploration. By comparison, my 2007 black-and-white portrait of an American astronaut, Megan McArthur, conveys a similar quality of lighting and composition. As with the Curtis image, I sought compositional

---

31. Ibid.

MER Opportunity on Mars at Victoria crater: 2008. (JPL/NASA)

simplicity, dramatic lighting, and eye contact that reveal McArthur's purpose, pride, dignity and humanness.[32]

While the photographers of the American frontier were drawn to the possibilities inherent in the open range, others who were captivated by life in urban and industrial centers were emerging. The evolution of photographic technology—glass plate to acetate-based negative film and more reliable handheld cameras—resulted in equipment that was less cumbersome, allowing the photographer to respond to situations and environments with greater spontaneity than previously possible.[33] This directly contributed to the emergence of industrial and urban landscape photography. Paul Strand was among a group of early 20th century American photographers who explored the contrasts between urban people and place. His documentation captured a moment in American urban history. As his work defines the urban landscape at the time, it may seem to be a commentary on American and even western civilization. Take, for example, Strand's black-and-white image "Wall Street 1915." The photograph depicts the side of an indifferent stone building with massive, black rectangular windows reigning over shadowed and silhouetted figures. The figures walk anonymously alongside the tall and seemingly impenetrable building. In an interview in New York City in 1973, Strand discussed his aesthetic:

---

32. In documenting the SM4/STS125 mission preparations to the Hubble Space Telescope, the author secured the first authorized portrait session of an astronaut crew in more than 25 years. McArthur's portrait is from that series which was photographed in black-and-white and color in the anechoic laboratory at JSC.

33. 20th century landscape photographers like Alfred Stieglitz, Edward Steichen, Edward Weston, Ansel Adams, Paul Caponigro, and Harry Callahan among others sought environments and subject matter that responded to their intellectual curiosity and idiosyncratic manner of combining lighting and composition—all of which influenced their approach to interpreting both the natural and the human made.

Zuni Woman v.17: c. 1903. (Edward S. Curtis/McCormick Library of Special Collections, NW University Library)

Hubble SM4 Astronaut K. Megan McArthur: 2007. (©Michael Soluri)

Paul Strand: Wall Street, New York, 1915. (©Aperture Foundation Inc., Paul Strand Archive)

It wasn't just that I was wandering around and I happened to see something like that. I went down there with a (hand-held 3-1/4 x 4-1/4 English Reflex) camera in order to see whether I could get the abstract movement of the counter-point between the parade of those great black shapes of the building and all of those people hurrying below.[34]

That same feeling of anonymity and imposing height extends 50 years later in an industrial photograph taken at NASA's former Lewis Research Center. There a man stands by a swinging valve door of a supersonic wind tunnel.[35] This uncredited photograph taken by an unknown space center photographer is among a significant body of NASA and aerospace industrial imagery that conveys the relationship between humans and their tools. More precisely, the

---

34. Jonathan Green, *The Snapshot* (Millerton, NY: Aperture, Inc.,1974), p. 47.

35. See Glenn Research Center GRC Image Net C1956-42070, "Swinging Valve for Supersonic Wind Tunnel," NASA GRIN database number: GPN-2000-0014474.

Swinging Valve for Supersonic Wind Tunnel: 1956. (NASA Lewis Research Center)

image depicts a nameless engineer in context to an imposing structure designed and built during the early years of space exploration.

In exploring the documentation of people and their industrial achievements, and focusing on the significant body of work made at the former NASA Lewis Research Center, I cannot ignore the work of the New York City photographer Eugene de Salignac. De Salignac was a New York City civil servant and the sole photographer of New York City's Department of Bridges/Plant and Structures from 1906 through 1934. During that period, he made over 20,000 glass plate negatives that documented the evolving infrastructure of

New York. Though the volume of de Salignac's work is impressive, the work had been largely ignored through the 20th century. Only recently were his photographs re-discovered, edited, and afforded a major exhibition at the Museum of the City of New York and a book in 2007.[36]

Made within a year of Strand's "Wall Street 1915," de Salignac's "Brooklyn Bridge 1914" portrait shows a group of painters randomly suspended along the numerous ascending cables of the bridge, appearing like musical notes in a composer's score. This visualization of random or musical placement reoccurs in a NASA Lewis Research photograph of two engineers working with an oscilloscope connected to a scale model supersonic aircraft inside a wind tunnel. In the black-and-white image random streaks of white light from the multiple flash systems that illuminate the engineers and the model aircraft bounce around the highly reflective tunnel and its three circled windows. Again, this is an uncredited photograph made in 1957 with what I suspect was a large format camera (typical for extensive detail).[37] The photograph evokes a mid-20th century feel for communicating "high-tech men-at-work." It does so by posing the men (as is often the case) and then asking the two engineers to do their work. In comparison to de Salignac's (apparent) found moments of (staged) randomness, the Lewis Center image is a visually compelling portrait of the emerging era of aerospace, just as the de Salignac photograph is a portrait of emerging urban infrastructure in America.

By the 1930s the evolution of the camera and the subsequent reduction in the size and quality of film revolutionized the look of news reporting and signaled the emergence of the picture magazine. Magazines like *Time*, *Life* and *Fortune* sprang up in competition to newspapers. Along with this, photographers were beginning to discover that they could access, experience, and interpret a range of events and situations in unobtrusive ways with handheld cameras.[38] The result was a new genre of photography called photojournalism—an approach in both news gathering production and photographic documentation. Small, handheld cameras like the 35 mm and 2 1/4 size were formidable tools with which to discover more fluid, intimate, and strikingly visual opportunities between people and place. The small camera created opportunities to tell a story, communicate a point-of-view, or report an unfolding event in one or even in a group of photographs (photographic essay). The small camera afforded the photographer a means to enter new social and industrial worlds without lengthy planning

---

36. See *New York Rises – Photographs by Eugene de Salignac* (New York, NY, Aperture Foundation and New York City Department of Records/Municipal Archives, 2007), with essays by Michael Lorenzini and Kevin Moore.

37. See Glenn Research Center GRC Image Net C1957-45670, "Engineers Check Body Revolution Model," NASA GRIN database number: GPN-2000-001473.

38. However, given the low grade quality of newsprint and photographic reproduction quality, large format cameras were still commonly used to obtain extensive detail and sharpness.

Brooklyn Bridge: 1915. (Eugene de Salignac/Courtesy NYC Municipal Archives)

and pretense. This new found spontaneity augmented the photographer's access with the given subject, thereby facilitating the telling of a story with depth and understanding. Photographic reporting and documentation—since the era of Curtis, Strand, and de Salignac—had now progressed to the point of actually capturing the immediate and even intersecting with the very fabric of life, war, industrialization and, on the not-too-far horizon, the exploration of space. At the same time, the increasing portability and decreasing cost of photography lead to the mass consumerization of home photography and the resulting cultural emergence of snapshot photography.

And with the snapshot, a new kind of photographic aesthetic evolved: a quality of imagery that has a kind of throwaway immediacy. Typically, they were images that were neither studied nor anticipated, but could be. For the casual amateur, the immediacy brought with it a freedom from the formal rules of photography. Since rules did not necessarily have to be adhered to, standards by which to measure quality changed. Hence evolved the notion that a good photograph is one that is not only technically correct, but easy to make. In all respects, it is the

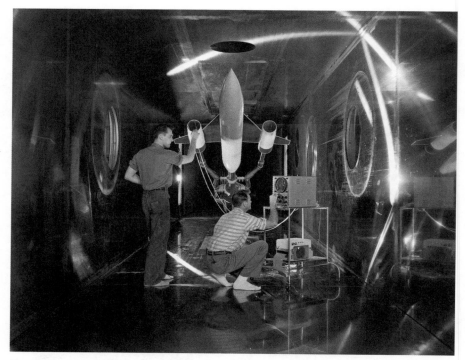

Engineers Check Body Revolution Model: 1957. (NASA)

essence of George Eastman's classic slogan: "You press the button and we do the rest." This idea has permeated American photography and photography itself for nearly a century. In his classic book, *The Snapshot*, Jonathan Green summed up the ambivalence toward and relevance of snapshot photography:

> It has been bandied about as both praise and condemnation. It has been discussed as both process and product. A snapshot may imply the hurried, passing glimpse or the treasured keepsake; its purpose may be casual observation or deliberate preservation. The snapshot may look forward in time to a chaotic, radically photographic structure, the appropriate equivalent of modern experience; or it may look backward to the frontal formal family portrait of a bygone age.[39]

39. Jonathan Green, *The Snapshot* (Millerton, NY: Aperture, Inc.,1974), p. 3.

One of the first snapshots taken in Earth orbit by John Glenn: February 20, 1962.
(NASA)

This brings us to what I would call the snapshot aesthetic in space exploration photography. In examining the snapshot aesthetic, I cannot ignore the photography of astronauts in-flight by the astronaut themselves. Consider John Glenn on the eve of America's first manned orbital mission in 1962. John Glenn went out to buy an Ansco Autoset 35 mm handheld camera (with a 55 mm lens) from a local store in Cocoa Beach.[40] With out formal photographic training, Glenn bought this modest camera because no one in the program at that time foresaw the necessity of an astronaut taking pictures. Glenn took the Ansco Autoset on board MA6. This resulted in a series of 48 snapshots of Earth taken from the window of his Mercury space capsule.[41] From Glenn's efforts and the resulting images, it can be argued that Glenn was the first human to record and take iconic snapshots in space.

---

40. Gary H. Kitmacher, "Astronaut Still Photography During Apollo," (Washington, DC: National Aeronautics and Space Administration), *http://history.nasa.gov/apollo_photo.html* (accessed March 6, 2008).

41. See John L. Kaltenbach, "A Table and Reference List Documenting Observations of Earth from Manned Earth Orbital and Suborbital Spaceflight Missions Including the Unmanned Apollo–Saturn 4 and 6 Missions" (Houston, TX, National Aeronautics and Space Administration Lyndon B. Johnson Space Center, December 1976), *http://eol.jsc.nasa.gov/sseop/metadata/Apollo-Saturn_4-6.html* (accessed March 18, 2008).

Of historic interest then is the in-flight photography by the Gemini crews (1965–66).[42] Here the snapshot-like photographs that James McDivitt made of Ed White during the first American EVA (from one of the open twin hatches of Gemini 4) is among the iconic images of American human spaceflight. However, what has not been uncovered are the photographs that White made with a 35 mm Zeiss Contarex camera mounted on his handheld mini propulsion system.[43] By in large, these photographs should document his perspective of the Gemini capsule with both capsule doors open and his snapshots of his colleague James McDivitt. Of aesthetic relevance (in a snapshot sense) are the EVA images from Gemini 9 through 12. Among these photographs is the partially sunlit and shadowed close up of Buzz Aldrin during an EVA. The photographs' setting is outside the open hatch of his Gemini 12 spacecraft. James Lovell took this snapshot. Lovell's photograph captures an intense look in Aldrin's eyes. This image is unique because it captured in a passing glance a quality of human vulnerability in the void of space.

Perhaps the most distinctive in-flight astronaut "snapshot" photography is the Apollo 7 crew's photography of themselves.[44] Each portrait of the three crewmembers (Wally Schirra, Donn Eisele, and Walt Cunningham) was initiated by Cunningham, who felt that they needed a souvenir from their mission. The photographs are framed using a handheld Hasselblad 70 mm camera with an 80 mm lens. Cunningham also used a handheld spot meter to measure the sunlight entering the cabin windows. The results are not just a series of technically accurate exposed images, but they are a series of exquisite snapshots made under controlled conditions: the same environment, camera, lens, and quality of sunlight. Cunningham was able to capture both a vulnerability and intensity of each of his two crewmates. In turn, Schirra was able to capture similar qualities in his photograph of Cunningham under the same conditions. When all three snapshots are grouped together, the square-framed tight close-ups of Schirra, Eisele, and Cunningham offer an unimagined glimpse—in the stark sunlight of outer space—of three men's faces within the tight confines of the first ever Apollo space mission.[45] These photographs are timeless "souvenirs" whose aesthetic relevance

---

42. Since Project Mercury, handheld cameras have accompanied crews into space. The cameras were initially recordkeeping tools to study Earth from space. From these handheld cameras there also resulted opportunities to capture the spontaneous moments during spaceflight, both within and outside of a spacecraft.

43. See John L. Kaltenbach, "A Table and Reference List Documenting Observations of the Earth from Manned Earth Orbital and Suborbital Spaceflight Missions Including the Unmanned Apollo-Saturn 4 and 6 Missions" (Houston, TX, National Aeronautics and Space Administration Lyndon B. Johnson Space Center, December 1976), *http://eol.jsc.nasa.gov/sseop/metadata/Apollo-Saturn_4-6.html* (accessed March 18, 2008).

44. See Lunar and Planetary Institute *Apollo Image Atlas–70mm Hasselblad Image catalog* for Apollo 7, *http://www.lpi.usra.edu/resources/apollo/catalog/70mm/mission/?7.*

45. See John L. Kaltenbach, "A Table and Reference List Documenting Observations of the Earth from Manned Earth Orbital and Suborbital Spaceflight Missions Including the Unmanned

are—referencing Green's consideration of the snapshot—a combination of both deliberate preservation and the appropriate equivalent of modern experience.

V. Confounding Expectations

At NASA, the elegance was in the design of the engineering systems rather than in the manners of the men.[46]

Some 38 years after the first publication of *Of a Fire on the Moon*, Mailer's astuteness on NASA's institutional culture during the era of Apollo 11 helps to suggest why most NASA generated photography from the first 50 years of space exploration tends to typically focus on the elegance and design of its engineering systems. Yet the results from some of its design and engineering systems are typically the first communication to the greater public. As a result, NASA releases pictures from weather satellites, robotic space craft flybys of the inner and outer planets, the era of Apollo 11, Skylab, the Space Shuttle, and the ISS programs. The elegance of its engineering systems is also celebrated in the manner of its rockets, rocket engines, guidance system avionics, the integration, testing and assembly of Earth orbiting satellites and space probes, interplanetary robotic spacecraft and so on. Yet confounding expectations in the need to report on the "elegance of its engineering systems" at often the expense of the "manners of men," are the photojournalists who, given the precious commodities of accreditation, access, and time work mostly from "behind the velvet rope" to capture these essential moments. Essential moments that are typically captured in the routine of rocket launches, press conferences, and guided media tours.

Photographic coverage by the print media (in daily newspapers and weekly news magazines) reached its zenith during the first decade and a half of the American space program. Photographers working for the wire services like AP (Associated Press), UPI (United Press International), and Reuters provided the American audience with a steady supply of rocket launches—manned and unmanned. More extensive storytelling in the form of photographic essays and written reportage typically appeared in weekly magazines like *Life*, *Look*, *Time*, *Newsweek*, *US News and World Report*, and the monthly *National Geographic*. In *Life* magazine's coverage of the first 16 years of the Space Age (between 1957 and 1972) for example, it published only 28 cover stories with 1962 and 1969 tying with 7.[47] With access, time, ingenuity and imagination, photojournalists

---

Apollo–Saturn 4 and 6 Missions" (Houston, TX, National Aeronautics and Space Administration Lyndon B. Johnson Space Center, December 1976) *http://eol.jsc.nasa.gov/sseop/metadata/Apollo-Saturn_4-6.html* (accessed March 18, 2008).

46. Norman Mailer, *Of a Fire on the Moon* (New York, NY: Signet Book/New American Library,1971), p. 136.

47. See Time, Inc., *Life, the First 50 Years: 1936–1986* (Boston, MA: Little, Brown and Company,1986).

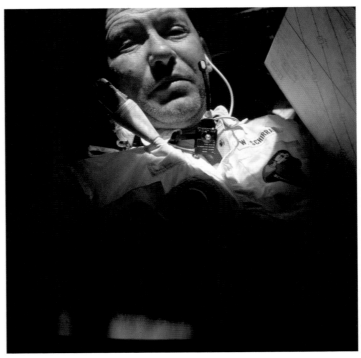

Snapshot portraits of the Apollo 7 crew in flight: October 1968. (left top) Donn Eisele, (left bottom) Wally Schirra (Cunningham/NASA), (above) Walt Cunningham. (Schirra/NASA)

like Ralph Morse, Otis Imboden and Jon Schneeberge revealed the early years of space exploration that often revealed the "manners" of the men of NASA.[48] These photographers provided their (stylistic) interpretations of astronaut training to the launchings of the Atlas, Titan and Saturn–Apollo. However, access to these photographers' complete work is challenging because it tends to be centralized within the news organization that they represented. Editorial photography made under contract by in-house photographers like Morse, Imboden, and Schneeberge is essentially owned and copyrighted by the magazine corporation, such as *Time-Life* and *National Geographic*.

Unlike access to NASA's public domain photographs, either in a flight center's archive or through its online Web portal, the decentralization of photographic archives among news media outlets creates a major challenge for

---

48. Other significant wire service photographers that covered the early years of the American space program were Jim Kerlin, Russ Yoder, Frank Beattie, and Hugo Wessels.

research.[49] Such is the case with Bruce Weaver, a former AP photographer currently with Agence France Press who has covered the American space program since the early 1980s. His 1986 photography of the *Challenger* launch (which are among those that revealed the subsequent puff-of-black smoke) is over shadowed by his clearly capturing the explosion 77 seconds later. In Weaver's case, the exposed roll of film was the property of AP, and subsequently, Weaver didn't have any say on the final edit nor did he have access to the film he had shot even though the photograph selected was among the most widely used in communicating the horror of the accident worldwide. It is impossible to track down Weaver's entire take from the AP archives, even today. And his "iconic" image of the explosion remains revenue intensive for the AP, but not for Weaver.

Contrary to this is the extensive and accessible photography by Bill Taub, the senior photographer at NASA headquarters in Washington, DC, from 1958 through 1975. During the early years of NASA, Taub was able to set the precedents in intimately capturing the nascent space agency. Taub's place was similar to the conditions that framed de Salignac's reign as sole photographer for the City of New York. However, de Salignac never saw his work published nor exhibited while Taub, on the other hand, was able to have his work credited in numerous stories in *National Geographic* during the early years of the space program. In addition, many of the public domain images in the NASA archive at Headquarters reflect his body of work.

Beyond those photographers already noted, there is a distinct group of photojournalists who have been covering the American space program since the 1980s for the AP, UPI, Reuters, *Florida Today*, and the *Orlando Sentinel*.[50] The scope of these photographers work is worth examining (although the restraints mentioned above will be formidable) given their continuous coverage and access to both human and robotic space missions from the Kennedy and Johnson Space Centers. There are, however, two photojournalists whose accessible work distinguishes itself. The first of these is Scott Andrews who has been photographing every shuttle and nearly every unmanned rocket/satellite launch since STS-1 in April 1981.

The quality of Andrews's work reflects a significant and stylistic approach in the reporting and documentation of the industrial landscape of space exploration photography. Within this landscape, Andrews portrays Mailers "elegance in the design" by constantly reexploring his subject matter. His subject matter is typically rockets, rocket launches, active and historically inactive launch pad complexes, and rocket assembly facilities.

---

49. Typically a photographer's space coverage is part of a magazine or media corporation's archive. In building its archives, a news group like AP selects only the most historically relevant or iconic to save and catalogue as a revenue bearing profit center for the organization. As a result, it may be difficult, even next to impossible, to examine a photographer's complete body of unedited work for possible alternative choices beyond the familiar or existing iconic.

50. Joe Skipper-Reuters; Bruce Weaver-Agence France Press; Pete Cosgroves-UPI; Phill Sandlin-AP; Mike Brown-*Florida Today*; Red Huber-*Orlando Sentinel*; and James Nielson-*Houston Chronicle*.

July 4, 2006: Shuttle *Discovery* Return to Flight. (©Scott Andrews)

Considered the guru of remote rocket launch photography by his peers, Andrews has been able to document both the magic and the elegance of the machines that launch into space through a combination of imagination and inventiveness.[51] With no two launches ever identical, he achieves his imagery by strategically placing multiple motor driven cameras—with varying focal length lenses—in and around a rocket's launch pad complex regardless of weather and time of day.[52] The scope of Andrew's work can be examined in context to

---

51. Andrews is a Washington, DC based photographer whose work has been published in magazines such as *Time, Newsweek, U.S. News and World Report, Smithsonian,* and *Discover.*

52. According to Andrews, the automated firing of the remote cameras around the launchpad, regardless of the time of day, is activated by the sound or vibration of the rocket engines at ignition. When the sound or vibration reaches a predetermined level, the camera trigger will fire the linked cameras.

April 26, 2003: Expedition 7 Soyuz Launch. (©Scott Andrews)

the photography of American railroads: specifically, the survey, building and operation of the American railway system by some of the pioneer late 19th century landscape photographers for the U.S. Geological Survey, and in the 20th century, the imaginative and stylistic (steam) railroad photography of O. Winston Link and Richard Steinheimer.[53] Their documentation of the fleeting era of steam locomotives set in context to the American landscape has been revered for detail. These photographers portrayed the scope and scale of the American railroad system with care and exactitude. This attention to detail of scope and scale exists in Andrews's work as well.

Andrews's photograph documenting the second return-to-flight of the Space Shuttle *Discovery* offers several references to the history of photography and space. Both the placement and framing of his remote Hasselblad camera is reminiscent of the remote video images of Apollo 17's LEM blasting off from the surface of the Moon. However, it is Andrews choice of black-and-white film and one of the dried out ponds that typically surround the launch pad complexes in dry weather that distinguishes this image. As a result, the foreground patterns of dried out clay playing against the ascending shuttle has aesthetics reminiscent of western American desert images captured by 19th century landscape photographers like T. H. O'Sullivan and J. K. Hillers. The other worldliness patterns of dried out clay can also be compared to some of the MRO (Mars Reconnaissance Orbiter) surface images of Martian polar cap regions. In contrast, Andrews remote close-up of a Russian Soyuz rocket just seconds in to its liftoff, offers an insightful document of the launch pad—the same one used to launch Sputnik nearly 40 years earlier. It also portrays the elegance of Russian rocketry, long hidden and secretive during the former Soviet era.

Next comes the photography of Bill Ingalls, currently the senior in-house photographer at NASA headquarters in Washington, DC. Ingalls has been the first NASA photographer to routinely document NASA's collaboration with the Russian space program. Ingalls tends to explore what Mailer describes as the "manners of the men" through unique access and time in which to photograph both the American and Russian human space flight programs. Since 1999 his continuous documentation has captured the cultural similarities and differences between both spaceflight programs. His documentation of the solemnity of the Russians in the training of their cosmonauts and launching of their rockets from the same launch complexes that supported Sputnik and Gagarin warrants examination. As of this writing, Ingalls remains the only photographer to have continuous access to every Russian launch involving

---

53. See O. Winston Link's *Steam, Steel and Stars* (Harry N. Abrams); *The Last Steam Railroad in America* (Harry N. Abrams); Richard Steinheimer's *A Passion for Trains* (W. H. Norton & Company) and Walker Evans photographic studies of railroad car insignias for a 1956 *Fortune* article "Before They Disappear."

Expedition 8 crew meets the media. (©Bill Ingalls/NASA)

Soyuz rocket being transported to launch pad. (©Bill Ingalls/NASA)

American astronauts.[54] For example, his photograph of the Expedition 8 crew (in a closed off, rather cage-like glass room) offers a distinct point of view of how the Russian's present a fully suited crew—only hours from their launch to the ISS—to space officials and the media. As a result, Ingalls has achieved continued access to a space exploration infrastructure, once remote and secretive, that harkens to de Salignac's continuous three decade documentation of New York City's urban infrastructure. By examining Ingalls's somber photograph of Russian security guards accompanying a Soyuz rocket being transported to its Baikanor launch pad, offers not only a contrast to shuttle launch preparations at the Kennedy Space Center in Florida. It provides an insightful observation on what the past history of Soviet era space exploration may have looked like in the desolateness of Kazakhstan.[55]

VI. The Past, Present and Future Photographic Documentation of NASA as a Labor Force

I saw doing space history as investigating what space flight efforts could reveal about a particular time and place.[56]

—Margaret A. Weitekamp

The examination of space exploration photography to the history of photography has demonstrated relationships to landscape photography, photographic documentation, and the evolution of the snapshot. Now it is time to consider another photographic genre, documentary portraiture and its relationships to space exploration photography, specifically the people that make going into space happen. In her essay, "Critical Theory as a Toolbox," Margaret Weitekamp considers historically examining NASA as a labor force:

> Many other aspects of NASA as a labor force remain unexamined . . . . Although the individual stories of astronauts, flight controllers and rocket scientists have been recorded,

---

54. See "Roads Less Traveled" by the photographer Jonas Bendiksen in *Aperture* 170, Spring 2003. Bendiksen documents the spent lower stages of Russian rockets that crash (and pollute) in the often populated and desolate areas of Kazakhstan and Siberia.

55. Since the return-to-flight of the Space Shuttle in 2005, Ingalls has also been able to frame all subsequent Shuttle launches from the perspective of the NASA Administrator in the firing room at Kennedy Space Center. Mission after mission, these photographs represent some of the most prolific unstaged documentation of a NASA Administrator and senior management during the moments of launch in the history of American space exploration.

56. Margaret A. Weitekamp, "Critical Theory as a Tool Box: Suggestions for Space History's Relationship to the History Subdisciplines" in *Critical Issues in the History of Spaceflight*, ed. Steve Dick and Roger D. Launius (Washington, DC: National Aeronautics and Space Administration, 2006), p. 562.

the collective stories of the thousands of people who made particular space projects work offer many opportunities for thinking about the space agency as a workplace.[57]

In considering the last 50 years of existing photographic coverage of NASA as a workplace, (often within a framework of industrial photography), inspires close examination and comparisons to the genre of documentary portrait photography and to photographers like August Sanders, Lewis Hine, Walker Evans, W. Eugene Smith, Irving Penn, Richard Avedon and Arnold Newman.[58] In fact the work of Lewis W. Hine and W. Eugene Smith, specifically their photography of American industrialization and its labor (before and after World War II), offer a significant reference and comparison to the wide range of NASA's photography of its own labor force.

In examining the photography of Lewis Hine, I am immediately drawn to two seminal images: the black-and-white photograph of a goggled welder working on the Empire State Building and the photograph of a T-shirted powerhouse mechanic using a massive wrench to tighten a steam valve. In the circa 1932 photograph, an unnamed goggled worker is pictured amid an elliptically shaped piece of steel with a cut circle rimmed with bolts as he holds (in gloved hands) an ignited welding torch. Leaning slightly into the steel, his dark-colored goggles are juxtaposed to the circular black hole in the steel piece. The convex shape of his soft tweed-like hat that covers his head juxtaposes off of the concave cut into the steel. The harmony of this laborer with the work before him can be

---

57. Ibid., p. 563.

58. The stylistic approach of 20th century portrait photographers like August Sanders, Irving Penn, and Richard Avedon yields distinct bodies of work that reveals the photographer's connection to his subject matter in the hope of capturing moments of vulnerability in body language and eye contact. Sanders's documentation of the physiognomy of Germans before World War II, for example, is significant in its breath and honesty. He often portrayed his subjects from all walks of life—pastry chef, musician, teacher, judge, lawyer, etc.—in their own work or personal environment. Sanders's influence is felt in the portraits of Irving Penn and of my own. Penn's black-and-white series on American and French working professionals and those of various Peruvian and African tribes (or even on American subcultures like Hell's Angels) are seminal in the cultural history of 20th century photography. Specifically, Penn's 1950s portraits of workers in Paris, London, and New York like bakers, butchers, waiters, charwomen, deep sea divers, and rag-and-bone men resulted in a seminal body of work called "The Small Trades." Each individual is photographed on a mottled gray background illuminated with natural daylight from windows or skylights. Richard Avedon believed that a photograph has a life of its own anchored in the era in which it was made. As a result, his approach was to both isolate and interpret his subjects without a definable location nor identifiable background. Avedon's portraits typically document an individual with a quality of flat lighting posed against a bath of pure white light. In some respects, this is not too different than the expected aesthetics of a passport photograph. By composing his subjects in different positions within the camera's frame, Avedon was able to capture the essence of an individual through the emotional expression of his or her body language and his or her direct or indirect eye contact.

compared to Paul Riedel's industrial image of a NASA TIG welder in the clean room of the Technical Services Building at the former Lewis Research Center.[59] Here a high tech welder in 1963 is similarly goggled however he is welding metallic objects by (safely) inserting his hands through protective gloves that securely enter a closed protective chamber. In both images each photographer discovers and frames a kind of unity between laborer and place of labor.

By comparing Hine's iconic 1920 staged portrait of a powerhouse mechanic to a 1964 NASA technician working on a "9 thruster ion engine array" furthers both the notion of unity and juxtaposition of laborer and place of labor.[60]

These four images convey a sense of aesthetics in which a discovered harmony exits in the very choreography of the subject matter. As a result, the visual harmony between worker, workplace, and the labor itself is a useful foundation for viewing and analyzing labor-related imagery from the archives of NASA's various space centers.

My photographic documentation of NASA's labor force has its roots in the writing of the author Studs Terkel. Terkel's *Working* (based on oral interviews) provides both a narrative and intellectual framework for examining NASA as a labor force.[61] To be sure, there are the engineers who, piece by piece, have pre-assembled every screw and electrical connection to all the parts of the ISS. Then there are the technicians who for the last 27 years have repaired and replaced tiles on shuttle orbiters. And there are the technicians who check, replace and check again the Space Shuttle's hydraulics and avionics. In any given Center on any given day, there exist countless photographic opportunities to document the labor force. Those men and women who work behind the scenes day-to-day. They are the laborers who make space exploration possible. Terkel understood the importance of the work that goes on behind-the-scenes in America. He celebrated that work and noted that it often goes unrecognized. In his book, *Working*, Terkel interviewed a steel worker, Mike Le Fevre:

> It's not just the work. Somebody built the pyramids. Somebody's going to build something. Pyramids. Empire State Building—these things just don't happen. There's hard work behind it. I would like to see a building—say the Empire State, I would like to see on one side of it a foot wide strip from top to bottom with the name of every brick layerer, the

59. See Glenn Research Center GRC Image Net C-1963-63814: "TIG welder located in the clean room of the technical services building TSB – The inert gas welding facility is used for welding refractory metals in connection with the Columbium Liquid Sodium Loop project."

60. See Glenn Research Center GRC Image Net C-1964-71003: "9 thruster ion engine array in tank 6 at the Electrical Propulsion Laboratory EPL."

61. Studs Terkel, *Working – People Talk About What They Do All Day and How They Feel About What They Do* (New York, NY: The New Press, 1972).

"Goggled Welder": c.1932. (Lewis Hine/Library of Congress)

name of every electrician, with all the names. So when a guy walked by, he could take his son and say "see, that's me over there on the 45th floor. I put the steel beam in" Picasso can point to a painting what can I point to? A writer can point to a book. Everybody should have something to point to.[62]

62. Ibid., p. xxxii.

"TIG Welder": 1963. (Paul Riedel/NASA)

My own interest in documenting some of the technicians who work on a shuttle orbiter was sparked by reading interviews like Le Fevre's.[63] My desire to

---

63. Since 2005, I have been photographically documenting NASA's New Horizons mission to Pluto and the Kuiper Belt at the Goddard Space Flight Center, Kennedy Space Flight Center, Johns Hopkins University Applied Physics Lab, and the Lowell Observatory. I expect to continue this documentation through New Horizons flyby of Pluto in the summer of 2015. My documentation

Powerhouse mechanic with wrench: 1920. (Lewis Hine/Library of Congress)

1964: Engineer at a 9 thruster ion engine array. (NASA)

visualize the relationship between a worker and their place of work is evident
in my black-and-white portrait of a United Space Alliance technician. In this
photograph made in one of the Space Shuttle's Orbiter Processing Facilities

and portraiture of the STS-125 crew and SM4 engineers, scientists, and technicians of the last
NASA service mission to the Hubble Space Telescope began in February 2007. Like New Horizons,
it is an ongoing exploration into the relationship between people and their place of work.

Technician at the engine base of shuttle orbiter *Discovery*. Kennedy Space Center: 2006. (©Michael Soluri)

(OPF) at the Kennedy Space Center, he is juxtaposed by one of the three main Shuttle engine insert positions. In scope and scale, I sought to explore the workspace of the people whose day job is working on a spaceship. In another exploration, I sought to document the engineers and technicians integrating the New Horizons spacecraft in the pre-clean room at the Goddard Space Flight Center. Working in a near sterile environment, these two electrical engineers are juxtaposed with a rolling cart of their tools and instruments. In both explorations, I sought the dignity of the space worker in context to his working environment.

While examining the photography of NASA's labor force, it is impossible to ignore the photography of the workplace itself. John Sexton has made Kennedy Space Center his industrial landscape. His highly crafted approach to black-and-white photography is informed by the history of landscape and documentary photography and his close working relationship with Ansel Adams.[64] Sexton's documentation of the Space Shuttle over a period of about

---

64. Sexton was Ansel Adams's technical and photographic assistant from 1979 until Adams's death in 1984.

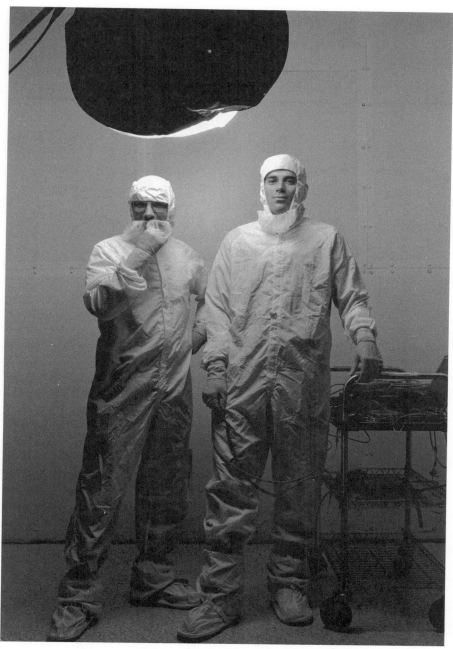

Two "New Horizons" electrical engineers and their cart of tools in the Goddard Space Flight Center pre-Clean Room: 2005. (©Michael Soluri)

Forward Reaction Control System Detail, *Columbia*. Kennedy Space Center, Florida.

Atlantis Vertical Vehicle Assembly Building. Kennedy Space Center, Florida. (©2000 John Sexton. All rights reserved)

eight years may very well be the most insightful and classic portrayal of an American spaceship to date.[65] It is unfortunate that Mercury, Gemini, nor

---

65. See John Sexton, *Places of Power: The Aesthetics of Technology* (Carmel Valley, CA: Ventana Editions, 2000).

Apollo spacecraft and their launch complexes were afforded such stylistic and personal interpretation.[66]

Having successfully navigated NASA's Public Affairs Office (PAO) between the post-*Challenger* era of 1990 and the events of 9/11, Sexton was able to document in large format black-and-white, both the elegance and dignity of the most complex machine ever created by humans. With the coveted combination of quality access and time, Sexton, over a period of 13 visits to the Kennedy Space Center, carefully portrayed various angles of the orbiter in context to the cathedral-like interior of the Vehicle Assembly Building (VAB) as it was being processed for spaceflight. He also systematically identified and isolated distinctive elements of the Space Shuttle orbiter like one of its three main rocket engines, landing gear, the flight deck's glass cockpit, the thermal protection system tiles, and the skeletal interior of a stripped down cargo bay.

The subtleness and monumentality to the scale of surroundings, and the distinct qualities of light that typically emerge from an Adams landscape photograph like "Moonrise over Hernandez, New Mexico" can be examined in many of Sexton's nature and industrial landscape images. In his many interpretations of the Space Shuttle orbiter—in this example, hanging vertically by its nose in preparation for mating to its fuel tank and two solid rocket boosters—he conveys a similar adherence to scale, monumentality, and qualities of light. Comparisons to Sexton's sense of organic shapes in his nature work can also be seen in his composition of one of the Space Shuttle's Forward Reaction Control Thrusters.

VII. RESOURCEFUL EDITING, AUTHORING, AND THE APPROPRIATION OF SPACE EXPLORATION PHOTOGRAPHY

> *It is time to remember that the camera lures. Then compels a man to create through seeing. It demands that he learn to make the realm of his responses to the world the raw material of his creative activity. Creative understanding is more camera-like than invention.*
>
> – Minor White[67]

In examining the context of different photographic genre to space exploration photography, it would be remiss to ignore fine arts photography. Although limited in number, a significant body of work by photographer-editors and fine art photographers has resulted in a range of significant, single-themed books and fine art photography that references the history of space exploration photography.

---

66. There are no known industrial landscape studies of the Apollo and Skylab configured launchpads, their crawler transports, VAB, and firing room configurations by fine art photographers.

67. Nathan Lyons, *Photographers on Photography* (Englewood Cliffs, NJ: Prentice-Hall, 1966), p. 164.

While there have been many high-end picture books that have published the iconic and familiar photography of space exploration, few have been edited to reflect an astronaut's insight while in space. One title that conveys the astronaut's perspective is *The Home Planet*.[68] The book remains one of the earliest "coffee table" books to cohesively portray an astronaut's photographic exploration of Earth and the Moon. The book brought together aesthetically compelling images (considering lighting, angle, composition) chosen from not just the American space program but also, for the first time since the space era began, from the then Soviet space exploration photography. The result is a book with visual continuity with images of Earth and Earth-related phenomena (weather, river and ocean patterns, and geologic formations). The photographs were mostly taken by astronauts in orbit around Earth and in orbit or on the surface of the Moon. What distinguishes this volume today is the quality of the editing. Complementing the minimalist design, layout and superb printing is the bilingual flow of first-person narrative in a variety of languages like, English, Russian, German, French, Hindu, and Bulgarian.

Hardly a decade later, the landscape photographer Michael Light published his book *Full Moon* in 1999. Light's book portrays the Moon as landscape and examines the sublime quality of light and detail from Project Apollo's space flight photography. Over a period of four years, Light researched and edited a substantial number of in-flight photographs made from Project's Gemini and Apollo (1965–1972). By gaining access to NASA's photo archive, he was able to make the first drum-scanned digital files from essentially second generation copies of the original flight films.[69] The resulting editing and juxtaposition of superbly reproduced full page black-and-white and color images created an aesthetic flow of a journey to the Moon and back via the historic timeline of manned spaceflight, beginning with the explosive fireball ignition of the Saturn Apollo's five rocket engines and ending with a view of the Pacific Ocean as seen through the window of a just landed Apollo module.

The subtext of Light's editing and editorial structure suggests themes of exploration and discovery. The quality of light and shadow in the photographs is significant. As a result, the actual quality of detail found on the printed page is stunning. This quality is impossible to appreciate in the familiar and iconic photography published by NASA and the print media. Even Light's handling of the iconic Apollo images discussed earlier (for example, the before and after photographs of Aldrin's boot print in the lunar dust) reveals a tonal range, and subtly of detail and texture that is impossible to draw out from the same images in familiar consumer print media.

Of particular interest to the author is the lighting captured by the Apollo astronauts. As explorers carrying cameras, the Apollo astronauts were similar to

68. See *The Home Planet*, ed. Kevin W. Kelly for the Association of Space Explorers, Reading, Massachusetts (Reading, MA: Addison-Wesley Publishing Company, 1988).

69. Michael Light, "The Skin of the Moon," *Full Moon* (New York, NY: Alfred A. Knopf, 1999).

the pioneer photographers of the 19th century who, with their cameras, were responding to the unknowns of the American West. However, unlike the physical conditions of the 19th century photographers, the Gemini and Apollo crews were responding to the new and unexplored by photographing their experiences inside their spacecraft and outside in the vacuum of space. They were able to capture qualities of light and shadow that were impossible to experience or capture on Earth. Commenting on the lunar landscape, Light puts it this way:

> Issues of big and small comprise but one half of the sub-lime landscape; for me, the other is the rule of light itself. Concerns about light always lurked behind all the others that drove my selection, and so to me the (NASA) archive's most important images will always remain the black-and-white ones, in part because their finer grain carries a higher visual acuity and renders more detail, but mostly because of the way they distill light in a world without air.[70]

As a result, not only does the handheld black-and-white photography from Apollo's 12 and 14 through 17 offer further aesthetic reexamination, the imagery from the automated metric and panoramic cameras (installed in the service module) also offer opportunities in visual exploration and discovery.

The writer, film maker, and photographer Michael Benson chose a different approach in conveying the sense of light, scale and landscape in space exploration photography. Benson researched and edited the still imagery that had been captured and beamed back to Earth by the robotic exploration of our solar system. Published in 2003, *Beyond, Visions of the Interplanetary Probes* often displays rediscovered photographs that have never been registered on a negative.[71] The subject matter is most of our solar system's inner and outer planets, their Moons, asteroids and the Sun. The images were taken by Earth and Moon orbiting, and interplanetary robotic space probes like OrbView, Terra, Aqua, Galileo, Lunar Orbiter, Magellan, Solar and Heliospheric Observatory (SOHO), Viking Orbiters and its landers, Near Earth Asteroid Rendezvous (NEAR), Voyager and the Hubble Space Telescope.

Benson addresses his experience researching and editing the images. The results of this became a kind of philosophical treatise on the history of the photographic imaging systems used in these robotic probes. Benson's ideas on the nature of art in these images become relevant here:

> I meditate on the fact that questions of authorship would tend to disqualify a space probe's pictures from serious

---

70. Michael Light, "The Skin of the Moon," *Full Moon* (New York, NY: Alfred A. Knopf, 1999).

71. Michael Benson, *Beyond-Visions of the Interplanetary Probes* (New York, NY: Harry N. Abrams, 2003) p. 295.

consideration as works of art—even though its scientific discoveries are undeniable, and attributed. Yet those same questions are very much present in the rarefied art-world air these days. Even Ansel Adams was only Ansel Adams part of the time. Like most photographers, he shot a lot of pictures and then selected those few that today constitute the work we connect with his name . . . What's left is choice —curatorship. And I would argue that these pictures   .fy for another reason: their mysterious, Leonardo-esque smile.[72]

One can look at a range of Benson's choices and make associations between a probe's photography and that of a photographic master. For example, Voyager's flyby of Jupiter takes on a modernist approach with its abstract colors and organic shapes. Next, there is Magellan's Minor White-like-exploration of Venus. The Synthetic Aperture Radar (SAR) imagery is all black-and-white, and White only worked in black-and-white. Also, Magellan captured unusual surface features, the quality of which are reminiscent to some of White's more interpretive work of objects and landscapes. And then there is the black-and-white exploration of Mars by the Viking Orbiters and Mars Global Surveyors that recalls the work of Ansel Adams, Edward Weston, and Aaron Siskind. Benson's choices also reinforce the notion that the photography from space exploration can be researched, edited and curated in a way that reflects the artistic proclivities of an author-curator.

While Benson and Light sought to convey various editorial approaches to the aesthetic possibilities of robotic and astronaut space flight photography, there are also artists who appropriate—take, borrow or are inspired by—space exploration in combination with photography, its related imaging technologies and even art. The work of the Spanish artist Joan Fontcuberta, for example, is distinctive for its humor, inventiveness and intellect. By experimenting with computer mapping software—used by cartographers to create realistic three-dimensional models and maps—Fontcuberta has created images of unimagined landscapes that look as though they were made by Earth orbiting satellites. Scanning examples of iconic paintings from artists like Rousseau, Turner, Cezanne, or Dali and feeding a digitized file from one of the paintings into the mapping software Fontcuberta achieved results in a fully rendered, realistic landscape of mountains, hills, valleys, rivers and lakes. The landscapes are, of course, visual fiction as experienced in his book fittingly entitled, *Landscapes without Memory*.[73] In context, Fontcuberta's fictional landscapes have an uncanny

---

72. Ibid., p. 304.

73. Joan Fontcuberta, *Landscapes Without Memory* (New York, NY: Aperture, 2005).

Henri Rousseau, "The Dream,": 1910. Oil on canvas. (The Museum of Modern Art, New York)

Derived photograph from "The Dream." (©Joan Fontcuberta)

SRTM image of Santa Monica Bay. ([PIA02779]/NASA)

resemblance to the imagery produced from the Shuttle Radar Topography Mission (SRTM) in 2000.[74]

When the digital scan of, say, Rousseau's 1910 painting *The Dream* is generated into the mapping software, the resulting fictional landscape image of a lake, sharply defined mountains, the sky, and clouds can be contrasted to a SRTM radar image of Santa Monica Bay to Mount-Baden Powell, California.[75] In the SRTM bay photograph, the distinct quality of the mountains, and the colors of water and sky have a three-dimensional quality such as that experienced in the mountain and lake image derived from *The Dream*.

In a turn from a series of fictional landscapes with their own imbedded story, Fontcuberta has also experimented with stories based on invented photographs and created ephemera. One such story is based on a fabricated organization, the Sputnik Foundation, and a fictitious cosmonaut, Ivan

74. In addition to the SRTM of Earth, some of the images in Fontcuberta's book, *Landscapes without Memory*, also have an image quality similar to the imagery received from the Magellan space probe's radar imaging of the surface of Venus.

75. See JPL's "Shuttle Radar Topography Mission" online at *http://www2.jpl.nasa.gov/srtm/mission/htm*.

"Ivan Stochnikov and Kloka in their historical EVA," from the Sputnik project: 1997.
(©Joan Fontcuberta)

Istochnikov. In 1998, Fontcuberta purports that the Sputnik Foundation sponsored an extensive, researched installation which examined the artifacts, details
and life of the Soviet cosmonaut, Istochnikov who had apparently disappeared
during the fight of Soyuz 2 in 1968. The reality, however, was that Fontcuberta
had created an entirely fictitious narrative that reflected his actual research of
the Soviet culture and its space program. Fontcuberta's portraits of Istochnikov
are, in fact, those of Fontcuberta himself. The artist also created a series of

Eugene Cernan's EVA from GT9: June 1966. (NASA)

convincingly staged photographs of Ivan and his dog Kloka in their historical EVA from the flight of Soyuz 2. The photographs are reminiscent of (mid-1960s) EVA's by, among others, the American astronaut Ed White on Gemini 4 and Eugene Cernan on Gemini 9.

Another example of the appropriation of space exploration photography by members of the fine arts community is the conceptual piece, *The Apollo Prophecies*.[76] On first blush, its title would suggest yet another assertion that the Apollo trips to the Moon were staged on vast sound sets (such as those used in the making of the HBO series "From the Earth to the Moon.") While that television series was staged to recreate some of the actual historical events of Apollo,

76. Nicholas Kahn and Richard Selesnick, *The Apollo Prophecies* (New York, NY: Aperture Foundation, 2006).

The Apollo Prophecies are, in fact, a series of staged, edited and photoshopped black-and-white photographs that document a purely fictional event. The work centers on the Apollo landing on the Moon and the subsequent discovery of a lost mission of Edwardian astronauts who colonized the Moon nearly three quarters of a century earlier.

The Apollo Prophecies is both humorous and well researched. Drawing from both the history of photography and the history of the Apollo missions, the two fine art photographers Nicholas Kahn and Richard Selesnick wrote, staged, acted, photographed, and edited two seamlessly woven, multi-page panoramas that portray a two-man crew, their launch, journey, landing, discovery, and return to Earth. In addition, the two photographers also wrote and created an accompanying booklet, "Apollo: A Prophecy" that conceptually chronicles the mission profiles from Apollo 1 to Apollo XXXI. The booklet contains portraits made to appear turn-of-the 20th century and hand-drawn. The illustrations detail the Edwardians and the artifacts from their lunar colonization. In the "Editor's Note" the authors offer insight into the philosophical subtext of their work:

> It is a little known fact that when the Apollo astronauts returned from the Moon, they brought back evidence of a previously unknown lunar expedition. This evidence comprised several cardboard canisters containing lunar breccia and, more significantly, a document written by the early explorer that prophesied the future arrival of the NASA astronauts themselves.
>
> Most saw the documents as a forgery, not least because the early explorers viewed the coming astronauts as cosmic deities. Whether the prophecy is authentic or not, its vision is hard to deny—if any man is to be transformed in to a god, what better candidate is there than the one who has ascended into the celestial sphere and stood alone on a distant world?[77]

On the surface the tone of the narrative could seem authentic but as readers and viewers, well aware of space history, we know it is not. In fact, we muse at Kahn and Selesnik's posturing. Calling the Apollo astronauts "cosmic deities" in the eyes of the early explorers is absurdly comical. Yet, there is also an irreverence that alludes to a belief on the part of some people that the Apollo landings were staged.

Within the panoramas of the book, though, a number of Gemini flight and Apollo surface photographs can be referenced to compare with some

---

77. Ibid., see accompanying pamphlet "Apollo: A Prophecy" (without page numbers), additional text by Erez Lieberman.

Panorama from *Apollo Prophecies*. (Kahn & Selesnick [Aperture, 2006]. All rights reserved)

"Apollo Bug," 1963. (Bill Bowles/NASA)

John Young, Commander of Apollo 16 on the lunar highland plains of Descartes April 1972. (Charles Duke/NASA)

of the staged scenes created by the photographers. As in Fontcuberta's EVA photograph of Ivan and Kloka, it would appear that Kahn and Selesnick appropriated the visual feeling of 1960s era EVA photography from Gemini missions like White's GT4, Cernan's GT9, and Aldrin's GT12. For example, on the outbound journey to the Moon, the crew of two astronauts emerge from their Gemini-inspired capsule on an EVA to study nearby asteroids. As the story continues, the crew lands and explores the lunar surface. The astronauts pick up rock samples and set up something that looks like the Far UV Camera/ Spectrograph.[78] The detail and aesthetics of rock samples and camera recall the

---

78. See further information on George Carruthers (of the Naval Research Lab), who was the PI and inventor of the Lunar Surface Ultraviolet Camera and its related imagery from Apollo 16 at *http://apod.nasa.gov/apod/ap960610.html.*

surface imagery from Apollo 16. It also recalls a staged black-and-white NASA Lewis Research Center photograph—in the construction spirit of a Bonestell—of an early model of the LEM on a faux lunar surface with stars in a black sky and a Buck Rogers style of astronaut.[79]

In the photograph that begins the second panorama of the book, the two astronauts have emerged from their lunar rover and have just discovered the Edwardian's camp. With the long-coated Edwardians' backs to the camera, the "Apollo" explorers survey the camp's infrastructure of housed rocks and space-suited pets, like a dog and elephant. The visual tone on the lunar surface can be compared to a black-and-white image of the Apollo 16 astronaut John Young.[80] In the photograph Young is just breaking into the right side of the camera frame and the lunar rover—its antenna aimed towards Earth—is parked in the background of the Descartes region landing site. The two images together create a contemplative, non-conspiratorial juxtaposition between historical fact and staged historical fiction.

VIII. CHOICES, TRANSITIONS, AND OPPORTUNITIES

> Who built the seven towers of Thebes?
> The books are filled with the names of kings.
> Was it kings who hauled the craggy blocks of stone…
> In the evening when the Chinese Wall was finished where did
> the masons go?
>
> —Bertolt Brecht[81]

The first 50 years of space exploration has been visualized largely through the publication of iconic photography, those few identifiable and often repeated images. As I have argued, however, there are other images that exist. For the most part, these photographs have been largely overlooked or even undiscovered, and yet they too can be placed alongside these iconic images and be considered within the context of the history of photography. I have also been discussing the aesthetics of space exploration photography in terms of landscape, documentary, and snapshot photography as a means to a visual literacy. How then can *some* of these photographs of space exploration be defined as artistic? The noted landscape photographer Robert Adams offers some thought on art and the making of photographs:

---

79. See Glenn Research Center GRC Image Net: C-1963-65465, "Model of Apollo Bug to Simulate Lunar Landing" by Bill Bowles.

80. See Lunar and Planetary Institute *Apollo Image Atlas – 70mm Hasselblad Image catalog* for Apollo 16 http://www.lpi.usra.edu/resources/apollo/catalog/70mm/mission/?16.

81. Studs Terkel, *Working – People Talk About What They Do All Day and How They Feel About What They Do* (New York, NY: The New Press, 1972), p. xxxi.

It seems to me that what art has historically, traditionally focused on are these moments of recognition and insight. By looking closely at specifics in life, you discover a wider view. And although we can't speak with much assurance about how this is conveyed, it does seem to me that among the most important ways it's conveyed by artists is through attention to form.

The notable thing, it seems to me, about great pictures is that everything fits. There is nothing extraneous. There is nothing too much, too little, and everything within that frame relates. Nothing is isolated. The reason that becomes so moving is that the artist finally says that the form that he or she has found in that frame is analogous to form in life. The coherence within that frame points to a wider coherence in life as a whole.[82]

It is my observation that a significant quality of space exploration photography is unique and idiosyncratic in form. The form is revealed in landscape and documentation as it spans through a whole host of photographic imagery (in color and black-and-white) not just produced and created on earth, but created from robotic spacecraft, human spaceflight, and Earth-based telescopes. Furthering the discussion on the aesthetics of art photography, it is worth noting the historical relevance that black-and-white space exploration photography has had on contemporary fine art photographers like Michael Light, Michael Benson, John Sexton, and even myself. Charlotte Cotton, a curator and writer on photography, offers some thought on the reascendancy of black-and-white in a recent online essay titled "The New Color: The Return of Black-and-white":

I am sure I'm not alone in beginning to think that the more complex, messy, unfashionable, and broad territory of black-and-white photography is where we are going to find some of the grist to the mill in photography's substantive and longer-term positioning within art.[83]

Indeed, the rich tonalities of black-and-white photography—from velvety blacks to grays to pure whites—have defined the aesthetics of landscape and documentary photography since the late 19th century. The recent documentary and landscape work by contemporary photographers such as Sebastian Salgado

---

82. See more on Robert Adams "Photography, Life and Beauty" on PBS's *Art in the Twenty First Century*, *http://www.pbs.org/art21/artists/adams/clip2.html* (accessed February 13, 2008).

83. Charlotte Cotton, "The New Color: The Return of Black-and-white." Contribution to The Tip of the Tongue forum, March 2007, *http://www.thetipofthetongue.com* (accessed March 2007, now defunct).

(*Genesis*), Bernd and Hilla Becher (*Water Towers*), and John Sexton (*Places of Power*) re-affirms the significance of black-and-white as a means of exploration and discovery. Even the recent exploration of the daguerreotype process by the artist Chuck Close has resulted in finely crafted black-and-white portraits of contemporary artists and writers. The consideration of black-and-white as the "new color" is significant. To Cotton, it may even move photography to a new plane:

> Contemporary black-and-white photography . . . has moved my thinking about the present state of photography onto a much more optimistic platform. Through these contemporary manifestations, the true, maverick character of photography, of our medium's history, is far from lost. Indeed, these threads of the past are given new and meaningful effect.[84]

One could conclude that the enthusiastic response of a few contemporary photographers to the abundance of black-and-white photography that NASA produced during the first 50 years of space exploration is no surprise.[85] It is compelling that Apollo 12 and 14 through 17 produced a significant amount of black-and-white photography, and it is the scope of that photography which tends to be referenced and appropriated by fine art photographers. This appropriation is significant because the inherent nature of black-and-white imaging foreshadows what art photography could contribute to the present and evolving history of space exploration photography, particularly in the documentation of people and place as human spaceflight transitions from the Space Shuttle and the ISS into the Constellation program.[86]

---

84. Ibid.

85. It should be noted, however, that the institutional choice (by photographic engineers) at NASA to use color film on human spaceflight missions (since Project Mercury) was not arbitrary. The decision was pragmatic because it reflected the practical needs of the engineering and scientific communities. A technical philosophy unchanged as human spaceflight activities evolved from Project Apollo and Skylab, to the Space Shuttle and ISS programs. Although during the first 20 years of NASA, the PAO (Public Affairs Office) typically relied on black-and-white photography to record the day-to-day and staged media events given the medium's immediacy and the historical nature of its use in the print media. However, by the late 1970s and with the emerging Space Shuttle program, black-and-white was eased out in favor of the immediacy (and cultural preference) that color coverage could provide. A choice unchanged as the second 50 years of space exploration begins, except to note that color films were gradually eased out in the late 1990s in favor of color digital technology.

86. The reemergent relevance of black-and-white, however, does not diminish the significance that color continues to have. For example, note the works by photographers like William Eggleston, Stephen Shore, Joel Meyerowitz, Robert Adams, and John Pfahl. It also must be considered that over the last several decades artist photographers, curators and teachers have emerged from MFA programs like those found at the Art Institutes of Chicago, San Francisco, Yale, and the Rochester Institute of Technology.

The quality of this transition to the Constellation program, one that has not been experienced since the close of Apollo and the emergence of the Space Shuttle, reinforces Weitekamp's suggestion that the history of NASA be examined in terms of its labor force. Weitekamp's proposition fits squarely within the landscape and documentary aesthetics found in the history of 20th century photography. This thinking can be considered as a catalyst in leading conversations that examine the existing photography of the labor force and its various work sites. Examining the first 50 years can help prepare and plan for a systematic and managed documentation of the next 50 years of space exploration.

In essence, it may be well worth evaluating the necessity for the researching and cataloguing NASA's photographic archives and the collections of its contractors as a means for creating the criteria for the next 50-year cycle. This would include the photography from Kennedy, Marshall, Michoud, Stennis, Dryden, Goddard, Glenn, Langley, Johnson, Ames, Vandenberg, and JPL. The results of such research may unveil collections of insightful work like high-speed engineering (Schlieren photography), industrial, portraiture, and the day-to-day workings of the labor force by one or more unrecognized photographers. For example, there are, in fact, decades of remarkably sophisticated work by the industrial photographers Bill Bowles, Paul Riedel, and Martin Brown at NASA's Glenn Research Center (formerly Lewis Research Center). I contend that research like this may yield historically significant discoveries (not unlike Michael Lorenzini's curatorial research, discovery, and exhibition of Eugene de Salignac's engineering and infrastructure photography). Perhaps, in conjunction with the NASA History Division and a flight center's archives, a combination of graduate students and doctoral candidates with affinities towards the realm of the curatorial would be likely resources in implementing such a long-term undertaking.

As the second 50 years of space exploration begins, so does its resulting photography. Rather than solely depending on NASA flight center photography, which is largely reactive to the moments at hand, I suggest taking a more proactive approach. Why not engage fine art photographers and photojournalists who may offer historians, curators, policymakers, and, ultimately, the public a more perceptive understanding of the current and future American space programs before they vanish forever. In many respects, why wait until an American space exploration program like the Space Shuttle is completed, scrapped, and rusted like remnants of Project Apollo's infrastructure.

Take for examples Scott Andrews prescient documentation of discarded and "abandoned in place" launch pads at the Kennedy Space Center. Well beyond the dusk of the Apollo years, Andrews documented (over a number of visits beginning in the late 1990s) all that remains of Launch Complex 34.[87] Andrews

---

87. See chapter two, Charles D. Benson and William Barnaby Faherty, *Moonport: A History of Apollo Launch Facilities and Operations* (NASA Special Publication-4204 in the NASA History Series, 1978), *http://history.nasa.gov/SP-4204/contents.html* (accessed March 27, 2008).

Saturn Apollo Launch Complex 34, Cape Canaveral Air Force Station: October 1999.
(©Scott Andrews)

captured in black-and-white the discarded isolation of an elevated launch pad resting on four rectangular legs of steel and eroding concrete. The skeletal launch pad looms over cracked slabs of concrete like the "shattered visage" from Shelley's "Ozymandias."[88] When Andrews's prosaic image is placed in context to the nostalgic years of Apollo, certain events become more understandable: the early testing of the Saturn rocket, the death of the first Apollo astronauts (trapped in fire within their Apollo capsule), the subsequent test launches of Saturn and the first manned mission of Apollo 7, leading to the quest to land on the Moon. What studied documentation of people and place that might have been photographed

88. Percy Bysshe Shelley, "Ozymandias," Alexander W. Wilson, et al., *The Norton Anthology of Poetry* (New York, NY: W. W. Norton & Co., 1983), p. 619.

of Apollo, Skylab, and the initial emergence of the Space Shuttle program may never be known. What small in-roads of photographic documentation that have been made during the shuttle era may be considered a foreshadowing of what may be documented in the coming 50 years of American space exploration.

It is tantalizing to consider that if Project Apollo had had a combination of landscape and documentary photographers interpreting the scope and scale of this program from 1962 through 1975 (much like the photography that was directed and managed by the government supported Farm Securities Administration or FSA in the 1930s), the aesthetic range would have been invaluable to space historians and the public alike. Perhaps, the creation and implementation of a modestly funded artist and writers program (likened to some extent to the National Science Foundation's "Antarctic Writers and Artists" program) can be considered. Such a program could be an extension of NASA's educational outreach, its history office and even the Smithsonian's National Air and Space Museum (NASM). In combination with the research and curatorial needs of both NASA and the NASM, such a program could contribute to the direction of how space exploration photography is curated and documented. These efforts could also result in a foundation of relevant photographic documentation, and at the same time, identify iconic imagery from other space fairing nations like ESA, Russia, China, Japan, and India. All of which, of course, would contribute to an understanding of space exploration and its history on a worldwide level.

The necessity for planning and implementing a methodology of research and archiving becomes evident. It can result in an emerging visual literacy that is in sync with the proactive photographic documentation of the American space program over the next 50 years. Such a methodology would need to memorialize not just the intended scientific, technical and day-to-day record-keeping, but an aesthetic that embraces the essential labor force responsible for actualizing the next 50 years of human and robotic space exploration. From this a more salient visual literacy emerges which broadens and deepens the understanding of human exploration.

IX. Epilogue

As Brecht wondered about the Chinese masons and the Great Wall, I wonder about the photographic documentation of what remains of the Space Shuttle and ISS programs, and the visual evidence that both historians and curators will be able to examine, publish, and exhibit in the decades to come. I embrace a photographic approach whose framework encompasses the discovery of the past and the documentation and interpretation of the present in context to the evolving history of both photography and space exploration. I liken this approach to my poetic journey when walking through Richard Serra's *Sequence*—a vast sculpture consisting of a series of connected 13-foot high torqued, curving, rust-colored steel plates. From an elevated position, *Sequence* resemble two gigantic violin scrolls standing back to back.[89] As I journey inward within one of the spirals, I experience an enclosed elliptical space where shafts of light cascade down the sides of the steel walls. I notice how the sunlight plays upon the scale of the space within the ellipse. When I move back into the wall's shadows, I discover a remarkable angled slit, an apparent exit. I exit and proceed through a narrow, disorienting corridor of curved steel, until I discover another interior space, similar yet different than the first. I retrace my steps back to the first only to realize that this time the path has lead me to yet another possible exit. Moving within these one of a kind structures, I could not have anticipated my journey or my exit. As with the commitment to an interprative photographic documentation of people and place during the next 50 years in space exploration, we cannot begin to know where our efforts might lead.

APPENDIX

In all the history of mankind, there will be only one generation that will be the first to explore the solar system, one generation for which, in childhood, the planets are distant and indistinct discs moving through the night sky, and for which, in old age, the planets are places, diverse new worlds in the course of exploration.

—Carl Sagan[90]

As I contemplate the next 50 years of space exploration photography, I find myself thinking about the serendipitous relevance between two rather unremarkable images from the photographic annals of Project Apollo. The very first image, recorded from the first launch of a Saturn V, came from Apollo 4's robotically controlled onboard camera. It is the image of a waxing crescent planet Earth in November 1967.[91] Some five years later in December 1972, the very last image made by one of the Apollo 17 crew (before stowing the camera away for reentry) was of a waxing crescent planet Earth.[92] Perhaps, these two images foreshadowed the next nearly 40 years of human spaceflight. The images

---

89. Tom Christie and Holly Meyers, "Steeling Beauty, Richard Serra's Advance Party," *LA Weekly*, August 15, 2007, *http://www.laweekly.com/art+books/art/steeling-beauty/17007/* (accessed February 16, 2008).

90. Dava Sobel, *The Planets* (New York, NY: Viking, the Penguin Group, 2005), pre-contents page (not numbered).

91. See Apollo 4 images at the Lunar and Planetary Institute *Apollo Image Atlas – 70mm Hasselblad Image catalog, http://www.lpi.usra.edu/resources/apollo/catalog/70mm/magazine/?01.*

92. See Apollo 17 images at the Lunar and Planetary Institute *Apollo Image Atlas – 70mm Hasselblad Image catalog, http://www.lpi.usra.edu/resources/apollo/catalog/70mm/magazine/?152.*

foreshadowed space flight limited to only the low Earth orbiting ventures of Skylab, the Space Shuttle and the International Space Station.

Now at the outset of the next 50 years of space exploration, I contemplate the extraordinary imagery beaming down from the Hubble Space Telescope: distant nebulas and galaxies (and the confirmation of the existence of an organic compound in the atmosphere of a planet in a near-by star system), Cassini's exploration of Saturn's moons Titan and Enceladus for water, the Mars Reconnaissance Orbiter's discovery of water deposited clay in a dry lake bed, and Messenger's first flyby of Mercury since Mariner 10.

As a result, I am drawn to the visual possibilities that will originate from both robotic spacecraft and human spaceflight. So it is reasonable for me to postulate what still photographic images may be reasonable candidates for "iconic" during the next 50 year cycle, among them:

- first discernable image of a water planet—with evidence of oceans, clouds, continents—in another solar system

- first image of alien life forms either alive or in fossil form

- first image capturing the earliest light of the universe just after the "Big Bang"

- Jupiter and some of its moons as seen from the surface of Europa

- first panoramic image from the surface of Europa illuminated by the reflected light of Jupiter, not the Sun

- Saturn and its rings as (possibly) seen from the surface of Titan

- first detailed image of the surface of Pluto in the foreground with Charon and/or other Plutonian moons in some crescent phase in the background

- defining color image of Earth and the Moon from the surface of Mars

- clear discernable image of Earth's "pale-blue-dot" taken from the outer fringes of our solar system

- first sequential or montage image of a Kuiper Belt Object

- image of the first group of civilian "tourists" to orbit Earth in a spaceship ll type of spacecraft

- during SM4, the last human mission to the Hubble Space Telescope, a 180 degree montage of overlapping images of HST, the orbiter, Earth, and space taken from various vantage points in the Space Shuttle's cargo bay and from its robotic arm

- first detailed image of the Apollo 11 landing site by a robotic spacecraft (ideally in low angled sunlight) showing the LEM lander, American flag, ALSEP, discarded artifacts (camera bodies, etc.), and boot prints

- Earth and the Moon rising above the irregular horizon line of the first asteroid to be explored by humans

- astronauts (not waving) by their lander (in low angled or backlit sunlight) on the surface of the first asteroid to be explored by humans

- self-portrait of the first 21st century astronauts—with their lander amid lunar hills, mountains, boot and wire-rim tire tracks—to have landed on the surface of the Moon since Apollo 17's December 1972 exploration

- time exposed image from the lunar surface—during the two-week lunar night—of lunar surface geography and the stars above

- defining photograph from a northwest position—in low angled sunlight from the east— by the first astronauts to visit (but not enter) Apollo 11's Tranquility Base landing site

- first Chinese astronauts, by their flag and lander, to land on the Moon

- a series of black-and-white images—in film—made with a space hardened Leica by the first crew to inhabit the first lunar outpost created on the surface of the Moon

- a series of available light "self" portraits (without flash) of the first crew heading to Mars: one set made just as they pass the Moon; the second made half way in their journey; the third made 24 hours or less before they set out to first land on the surface; the fourth made during the first EVA on the surface of Mars by humans

- astronauts on the surface of Mars by just-discovered evidence of actual or recent water flow

- first astronauts (on the surface of Mars) at the location of the first discovered evidence of either life forms or fossilized life forms

- the first en route interstellar probe's imagery of its intended destination of a nearby star system and some of its planets

CHAPTER 16

ROBERT A. HEINLEIN'S INFLUENCE ON SPACEFLIGHT

Robert G. Kennedy, III

Robert Heinlein is one of the most influential science fiction authors of all time. His writings not only inspired numerous people to enter the sciences and engineering in general—and the field of spaceflight in particular—but also shaped the way that people thought about spaceflight. Thus, even though Sputnik was a strategic surprise for the United States, there were legions of young Americans predisposed to step up and get to work on the challenging task of winning the space race. Heinlein's influence can currently be seen in the activities of numerous private spaceflight entrepreneurs.

LOOKING BACKWARD

Science fiction has changed history. We know this happened at least once in a very direct and far-reaching way by the documented influence of the science fiction writer H. G. Wells upon the yet-to-be Manhattan Project physicist Leo Szilard—one of the seven so-called "Men from Mars,"[1]—when crossing a London street in 1933. As Richard Rhodes relates this story in his Pulitzer Prize-winning *The Making of the Atomic Bomb*:

> On February 27, 1932 . . . physicist James Chadwick of the Cavendish Laboratory at Cambridge University . . . announced the possible existence of a neutron . . . The neutron . . . had no electric charge, which meant it could pass through the surrounding electrical barrier and enter into the nucleus. The neutron would open the atomic nucleus to examination. It might even be a way to force the nucleus

---

1. The seven famous Hungarian Jewish physicists who emigrated to America before World War II were all products of the famous Minta Gimnasium in Budapest. Two of them would go on to win Nobel Prizes. They were in birth order: Theodor von Karman, George de Hevesy, Michael Polanyi, Leo Szilard, Eugene Wigner, John von Neumann, and Edward Teller. The joke among their American colleagues was that they were actually from Mars and not Hungary as they claimed because they possessed unearthly brilliance, spoke English with an impenetrable Central European accent, and nobody knew what a Hungarian accent really sounded like anyway.

to give up some of its enormous energy. Just then, in 1932, Szilard found or took up for the first time that appealing orphan among H. G. Wells's books that he had failed to discover before: *The World Set Free*. . . . It was a prophetic novel, published in 1914, before the beginning of the Great War [World War I]. As Szilard recalled, Wells described

> The liberation of atomic energy on a large scale for industrial purposes, the development of atomic bombs, and a world war which was apparently fought by an alliance of England, France, and perhaps including America, against Germany and Austria, the powers located in the central part of Europe. He places this war in the year 1956, and in this war the major cities of the world are all destroyed by atomic bombs.[2]

It is difficult to read this story, even at nearly a century's remove, without chills running down one's spine in much the same way that a first reading of "Future Prospects of the United States" in *Democracy in America* by Alexis de Tocqueville produced during the depths of the Cold War.[3] Such prescience and perspicacity is almost inhuman. According to Rhodes:

> In London . . . across from the British Museum in Bloomsbury, Leo Szilard waited irritably one gray Depression morning for the stoplight to change . . . Tuesday, September 12, 1933

---

2. Richard Rhodes, *The Making of the Atomic Bomb* (New York, NY: Touchstone/Simon & Schuster, 1986), pp. 23-24.

3. de Tocqueville, Comte Alexis, Democracy in America (1835), chapter 21. "On the Future Prospects of the United States. There are at the present time two great nations in the world, which started from different points, but seem to tend towards the same end. I allude to the Russians and the Americans. Both of them have grown up unnoticed; and whilst the attention of mankind was directed elsewhere, they have suddenly placed themselves in the front rank among the nations, and the world learned their existence and their greatness at almost the same time." All other nations seem to have nearly reached their natural limits, and they have only to maintain their power; but these are still in the act of growth. All the others have stopped, or continue to advance with extreme difficulty; these alone are proceeding with ease and celerity along a path to which no limits can be perceived. The American struggles against the obstacles which nature opposes to him; the adversaries of the Russian are men. The former combats the wilderness and savage life; the latter, civilization with all its arms. The conquests of the American are therefore gained by the ploughshare; those of the Russian by the sword. The Anglo-American relies upon personal interest to accomplish his ends, and gives free scope to the unguided strength and common sense of the people; the Russian centres all the authority of society in a single arm. The principal instrument of the former is freedom; of the latter, servitude. Their starting-out point is different, and their courses are not the same; yet each of them seems marked out by the will of Heaven to sway the destinies of half the globe."

. . . Szilard stepped off the curb. As he crossed the street time cracked open before him and he saw a way to the future, death into the world and all our woe, the shape of things to come . . . Without question, Szilard read *The Times* of September 12, with its provocative sequence of headlines:

THE BRITISH ASSOCIATION

BREAKING DOWN THE ATOM

TRANSFORMATION OF ELEMENTS

Szilard was not the first to realize that the neutron might slip past the positive electrical barrier of the nucleus . . . but he was the first to imagine a mechanism whereby more energy might be released in the neutron's bombardment of the nucleus than the neutron itself supplied . . . As the light changed to green and I crossed the street," Szilard recalls, "it . . . suddenly occurred to me that if we could find an element which is split by neutrons and which would emit *two* neutrons when it absorbs *one* neutron, such an element, if assembled in sufficiently large mass, could sustain a nuclear chain reaction . . . In certain circumstances, it might be possible to . . . liberate energy on an industrial scale, and construct atomic bombs."[4]

The accidental discovery of x rays and radioactivity in 1895-96 upset everyone's notion of the immutable atom and the eternal clockwork universe, opening up grand new vistas of disturbing change. There at the turn of the century to interpret these mysterious new findings and extrapolate their potential meaning was one Herbert George Wells, a consumptive who in the fine tradition of impoverished tubercular writers before (Robert Louis Stephenson) and after (Robert A. Heinlein) was unable to do any heavier work than writing for a living. In 1899, he had already produced what Heinlein would call "the greatest speculative novel ever written," *When the Sleeper Wakes*.[5] In this single novel, just one among many, Wells conceived:
  a) heavier-than-air engine-powered warplanes, including their major types (fighter, bomber, and large transport), as well as thought-out doctrine for

---

4. Rhodes (ref. 2), pp. 13, 27-28.

5. A good subtitle might have been The Miracle of Compound Interest.

their application in air battles (note that this was four years before the Wright brothers flew 300 feet in their contraption at Kitty Hawk)

b) a variety of so-called "Babble Boxes"—audio media machines to appeal to every demographic segment that anticipated narrowcasting, blogs, and the World Wide Web—and "televisors" that resembled the information-retrieval capabilities of the Internet

c) mass-transit systems such as slidewalks, automatic high-speed surface freight, and airports

In addition to predicting the Bomb (as well as related concepts that we would recognize as decapitating first strike, strategic atomic exchange by air, and mutual assured destruction) in *The World Set Free* (1914), in other novels, Wells forecast suburbia and many other political and social developments that would accompany these innovations and, like de Tocqueville, the superpower status of America.[6] All during his own life, Robert A. Heinlein described H. G. Wells as his single greatest literary and intellectual influence.[7]

HEINLEIN'S INFLUENCE

So what about Heinlein himself? He was more technically prolific than even the incredible Wells, but his influence was regrettably less direct than the example above. This may simply be a characteristic of how things go in a naïve versus mature ecosystem, in which 80 percent of the significance is determined within the first 20 percent of the timespan.

Direct Effects on Society and Spaceflight via Technological Innovation

The Web site *http://www.technovelgy.com* attributes 120 (so far) inventions, novel devices (e.g., the waterbed), and neologisms (e.g., "free fall" and "grok") to Robert A. Heinlein. An incomplete list of just some of his space-related ideas includes: various electromagnetically-levitated transport systems also known as "mass-drivers," a hands-free helmet, the "parking" orbit, a Space Shuttle, stealth, and the gravity slingshot maneuver. This polymath's skill at innovating was not limited to science, technology, and engineering either, which handicapped most of the writing in what came to be known as "the pulp era." Heinlein brought originality to his craft, pioneering the literary technique of "Future History" used by many top writers of the genre since (implicitly or

---

6. Paul Crabtree, "The Remarkable Forecasts of H. G. Wells," *The Futurist* 41, no. 5 (Sept./Oct. 2007): 40-46.

7. Michael Hunter, "First Look: the Influences of H. G. Wells on Robert A. Heinlein's For Us, the Living," *The Heinlein Journal*, no. 14 (January 2004): 15-18.

explicitly), and refining Wells trick of "domesticating the impossible" (canonical instance: "the door dilated").[8] Heinlein eventually grew impatient with what he called his Procrustean bed. According to Elwood Teague, a contemporary of his, Heinlein, who read as widely as Wells, was obsessed with "the coming of the Bomb" even in the late 1930s. This surely must have been the Wellsian influence. Heinlein did, in fact, manage to meet Wells in Los Angeles about that time, and would have seen the groundbreaking epic motion picture, *Things to Come*, which was based on Wells's work, both dystopian and utopian. He was in frequent correspondence with scientists such as the physicist Robert Cornog, as well as engineers who would go on to the Manhattan Project, informing and being informed, and using the new discoveries to lend the essential Heinleinesque verisimilitude to his art. In keeping with his deep sense of discretion and military honor, this speculative phase ended instantly when his editor at *Astounding Science Fiction*, John W. Campbell, Jr., told him in December 1940 that discussion of uranium-235 had "gone black" in the technical literature. (His salient novella, *Solution Unsatisfactory* was already in press by then. The story is remarkable even today for the essential political truths it captured.) Being the Renaissance Man of the world he was, Heinlein knew exactly what this blackout portended.[9] He maintained this self-imposed censorship throughout the war years, though it is obvious he never stopped thinking about it.[10] Others were not so discreet. For example, Cleve Cartmill, his fellow habitue of the Manana Literary Society (MLS) that met in the Heinleins' living room in prewar Los Angeles, published a short story called "Deadfall" in the March 1944 issue of *Astounding* that was so technically accurate, it resulted in a visit to Campbell's editorial office by the naturally irate FBI. Heinlein's contact with the community of "rocket science" (meaning rocketry, nuclear weapons, and strategic matters) resumed after Hiroshima and continued for the rest of his life. One group of atomic scientists eventually became the Federation of American Scientists, principally interested in disarmament and arms control. Another later group became the Citizen's Action Committee for Space, the first proponent of what came to be called the Strategic Defense Initiative. Heinlein was apparently never troubled by the hobgoblin of consistency.

Though an engineer by training and inclination, Heinlein did not promote engineering per se. In his frequent lectures on the value of a liberal education, he would only say that the stool of knowledge has three legs: mathematics, (foreign) language, and history. Three legs are all that are necessary to stand:

---

8. Bill Patterson, "A Study of 'If This Goes On . . . '," *The Heinlein Journal*, no. 7 (July 2000): 29–42.

9. Heinlein also predicted the time, mode, and method of the Japanese attack on Pearl Harbor a week before the event, based on his own experience as a Navy gunnery officer aboard an aircraft carrier participating in a simulated attack exercise nine years before.

10. Robert A. Heinlein to J. S. Kean, "Tentative Proposal for Projects to be carried on at NAMC," August 14, 1945, Heinlein Archives, *http://heinleinarchives.net/* (accessed August 19, 2007).

neither engineering, technology, nor science are mentioned. The first leg exists because mathematics is the universal language of science. The second leg exists because one will never really understand one's own language, so one cannot know the true shape of one's mind until one has seen it from the outside through a foreign language (there's probably a connection to Godel's Incompleteness Theorem). The third leg exists because one will never be prepared for the future until one has first learned to see the present in the light of the past.

Direct Effects on Spaceflight via People

After the Japanese attack on December 7, 1941, Robert Heinlein tried to rejoin the Navy. After being turned down, he used his Annapolis connections to get an engineering job as a civilian in an aeronautical factory at the Philadelphia Navy Yard, relocating there from Los Angeles with his second wife, Leslyn.[11] His Navy classmates, who were well aware of Heinlein's gifts, at first had him spotting engineering talent before giving him a materials research position at the Navy Air Materials Center (NAMC). (It was here he met the woman who would become his third wife—and so important to his later work—a Navy WAVE lieutenant (j.g.) named Virginia Gerstenfeld, forever known to history as "Ginny.") Turning to the science-fiction community, Heinlein recruited his fellow writers L. Sprague deCamp and Isaac Asimov to work at the Navy Yard in aeronautical engineering as well. De Camp took a Navy commission and, under Heinlein's guidance, eventually turned to work on high-altitude pressure suits at NAMC. Towards the end of the war, combat aircraft—particularly long-range heavy bombers—were flying so high that mere warm clothing and oxygen masks could not protect the crews from the elements. Heinlein's troubles with tuberculosis, which had invalided him into early retirement from the Navy in 1934, precluded his direct participation in the altitude chamber and other experiments. But it is certain that the science fiction background of all three men—namely in regard to what would be called "spacesuits"—informed the work. The 1940 short story "Misfit" contains an accurate description of what a space suit should be.

One of Heinlein's new (and less famous) hires was Edward L. "Ted" Hays, a mechanical engineer like Heinlein himself. Hays worked as a flight test engineer on problems associated with carrier operations. (Heinlein's first billet after graduating from Annapolis was on the most advanced warship of her day, the carrier USS *Lexington*.) Hays went on to safety and survival equipment, became deeply involved in the development of Navy pressure suits, moved to NASA in 1961 after Project Mercury was underway, and eventually ending

---

11. Robert James, "Regarding Leslyn," *The Heinlein Journal*, no. 9 (July 2001): 17-36.

up on the Apollo program as chief engineer of life support systems where he specialized in, of course, space suits![12]

Indirect Effects on Society and Spaceflight via Literature

Because his literary genius was recognized so early by his readers and fellow writers in the late 1930s, Heinlein left an indelible imprint on the entire genre of science fiction, which might not have happened in a later more fractured and competitive age.[13] His prewar influence on the other writers (e.g., L. Sprague deCamp, Frederick Pohl, Isaac Asimov) of what came to be known as the postwar "Golden Age of science fiction" was simply profound. It is no exaggeration to say that writing in the field experienced a quantum leap in quality compared to its pulp roots. Postwar, Heinlein even managed to bring his chosen genre out of the ghetto into the respectable "slicks" (glossy, large-format color weeklies such as the *Saturday Evening Post* and *Collier's*, bygone media of a bygone age), where he would continue to be published. A full generation later, he was still mentoring and guiding major new writers (e.g., Larry Niven and Jerry Pournelle for their seminal "First Contact" novel, *The Mote in God's Eye* (1973).[14] Heinlein not only transmitted literary technique to his colleagues—his values of service and sacrifice, an individualistic outlook, and ethos of competence also came through and were propagated to millions of these authors' readers in turn.

Indirect Effects on Society and Spaceflight via Politics

Though his postwar writing was certainly more polished and sophisticated, Heinlein's prewar thinking was more original and imaginative in some ways. His earliest work contained themes that were politically revolutionary even by today's jaded standards. *Revolt in 2100* comprised the novella *If This Goes On . . .* (1939), and two short stories "Misfit" (1939), and "Coventry" (1940)—graced with the best science fiction cover art ever—are in this genre.[15]

The first of these stories is perhaps the purest example of what MLS-member Henry Kuttner called "the innocent eye"—no surprise that it's among the earliest work before professionalism sets in. It is set in a world in which the United States has turned its back on interplanetary exploration and science

---

12. Bill Higgins interview, Kansas City, MO, July 7, 2007, (unpublished article forthcoming).

13. Again, early players are generally more significant in a field than later ones, notwithstanding their absolute level of skill. This is a fundamental property of evolution.

14. Robert A. Heinlein to Larry Niven and Jerry Pournelle, "Motelight," June 20, 1973, (copy in U.C. Santa Cruz Heinlein Archives, UCSC Library, Santa Cruz, CA); Robert A. Heinlein to Larry Niven and Jerry Pournelle, "The Mote in God's Eye," August 1973, (copy in Heinlein Archives).

15. Robert A. Heinlein, *If This Goes On . . .*, "Coventry," and "Misfit" in *Revolt in 2100* (New York, NY: Signet Books, 1953).

Robert Heinlein's book, *Revolt in 2100*, was made up of the novella *If This Goes On . . .* (1939) and two short stories, "Misfit" (1939) and "Coventry" (1940).

after a period of "Crazy Years," falling into a theocratic police state. One does not have to imagine how Depression-era readers received these words, either: we have their letters describing how awestruck they were and their immediate realization at what a talent they had in their mailboxes. (Regrettably, the background story of the novella *If This Goes On . . .* seems less outlandish now than it did 70 years ago.)

After Hiroshima, Heinlein wrote a remarkable (yet usual for him) valedictory memorandum to his superiors notifying them that the Bomb would put them out of business. Then he promptly resigned and returned to Los Angeles with his wife, Leslyn. He endured several lean years of hardship, during which his second marriage broke up, before returning to writing fiction. This was the period of 1945-47 when he engaged in what he disingenuously called "world saving"—articles for the general public about the significance of the new Atomic Age. Why distinguish his activity as disingenuous? Because, despite his protestations that he wrote just to keep the wolf from his door and his frequent declarations equating the value of his writing with the reader's beer money, Heinlein was in fact deeply interested in politics—he ran for the California State Assembly in 1938—and educating his fellow human beings. These pungent articles were mostly ignored by the mainstream and never saw print except for one major exception. An early postwar collaboration with his Annapolis classmate Captain Caleb Laning called "Flight into the Future" appeared in the August 30, 1947 issue of *Collier's*, which described a nightmarish vision of an atomic arms race in space. (The concept eventually became the core of the second juvenile novel, *Space Cadet*.) The article (which was mostly Heinlein's work) did attract a lot of attention but ultimately led nowhere.

Why didn't "Flight" succeed? Why were its prescriptions and prognostications ignored by the military establishment and the policymakers? One must recall that the USAF was once called United States Army Air Force (USAAF) before being split off from the Army by President Truman in 1947 in the same Act that created the Department of Defense (DOD) and the Central Intelligence Agency (CIA). Aviation's roots in this country are in the Army, not the Navy.[16] Recall also that the Manhattan Project, and the related Operation Paper Clip and Project ALSOS (netted the German nuclear scientists as well as the rocket scientists including Wernher von Braun), were primarily Army operations. Likewise, Project RAND—an R&D department spun off from Douglas Aircraft that drafted the seminal "Design of an Experimental World-Circling Spaceship" (1946)—was supported by the USAAF. Rocket research was chronically underfunded by Navy until the submarine-launched ballistic

---

16. Analagous to the situation in Russia, the Russian strategic missile forces and Russian rocketry in general have their roots in artillery (a classic army mission), not aerospace as is usually the case elsewhere, which has led to some interesting differences in design philosophy, doctrine, and operating procedures compared to the West.

missile (SLBM) program started in the mid-1950s.[17] Caleb Laning was as original a thinker as Heinlein, but despite starting earlier, strategic weapons and, by extension, their platforms, were always the Army's rice bowl. Perhaps the Navy's early expression of interest stimulated the nascent USAF to actively take over satellite portfolio.

Indirect Effects on Spaceflight via Pop Culture

After his breakup with Leslyn towards the end of his hard times in 1947, Robert hooked up with Ginny and hit the road. The first of his juvenile novels, *Rocket Ship Galileo*, appeared, which would become the basis for the movie *Destination Moon*. Robert and Ginny worked out the *modus vivendi* that would guide the rest of their lives together. She became his first reader and indispensable partner. A long string of juveniles alternating with adult novels followed during an amazingly prolific decade.

In 1949, some of Heinlein's connections from his prewar Hollywood days led to a collaboration with the producer George Pal as technical advisor on the Oscar-winning science-fiction motion picture *Destination Moon* (1950). Heinlein enjoyed an unusually close (by Hollywood standards) productive relationship with the film's director, Irving Pichel, who took most of Heinlein's advice. Thus this film still looks remarkably good by today's standards and raised the bar for science fiction on the silver screen.[18] *Destination Moon* led to the trio of great science fiction films by George Pal: *When Worlds Collide* (1951), *War of the Worlds* (1953), and *The Time Machine* (1960). These classic films with their high production values certainly had at least indirect effects on pop culture. It is interesting that Pal, starting at Heinlein, came around to Wells.

It is surely no accident that, a decade after *Rocket Ship Galileo* (1947) and the whole series of juvenile novels that inspired millions of people who were teenaged in 1947-1959, legions of young professionals were ready to answer the challenge of Sputnik and to choose technical careers, entering the workforce just as the space race began.[19] One (current) NASA Administrator is the apparent exception, declaring that an interest in spaceflight led him to science fiction, not the usual way around.[20] The predominantly libertarian people who work in the free space movement almost universally cite Heinlein as their principal inspira-

---

17. R. Cargill Hall, "Earth Satellites, A First Look at the United States Navy," *Proceedings of the Fourth History Symposium of the International Academy of Astronautics*, Konstanz, German Federal Republic (October 1970): 253-277.

18. Like most of the crew, Heinlein did not enjoy the film's financial rewards, again standard for Hollywood. Also, and regrettably, it did not lower the bar on a lot of bad science fiction flicks yet to come, but one must remember that Sturgeon's Law is always in effect.

19. Bill Patterson, "A Study of 'Misfit'," *The Heinlein Journal*, no. 3 (July 1998): 24-32.

20. Michael Griffin, "The Future of NASA," speech delivered at the Robert A. Heinlein Centennial Conference 1907-2007, Kansas City, MO, July 6, 2007. This was apparently the first time in history that a serving NASA Administrator addressed a science-fiction audience.

tion, including the most recent winner of the Ansari X-Prize.[21] Oddly enough, Heinlein's values lap over into pop culture by another unexpected route—namely the computing/cyberpunk community, which has a high degree of congruence with the sets of libertarians, space enthusiasts, and science fiction fans.

CONCLUSION

Heinlein's diluted meta-gift of values—independence and liberty, technical competence and self-sacrifice, a paradoxically well-informed innocent eye— passed down and paid forward, may well turn out to be his greatest contribution to spaceflight.

We'll see.

Acknowledgments

I extend thanks and sincere appreciation to Dr. Dwayne A. Day, Mr. Timothy B. Kyger, Mr. Bill Patterson, and Dr. Bill Higgins

Cassutt, Michael, "More Conversation: On the Meaning of Periods," *The Heinlein Journal*, no. 20 (January 2007): 3.

Doherty, Brian, "Robert Heinlein at 100," *Reason* 39, no. 4 (August/ September 2007): 48-54.

Gifford, James, *Robert A. Heinlein: A Reader's Companion* (Sacramento, CA: Nitrosyncretic Press, 2000.

Laning, Caleb, and Robert A. Heinlein, "Flight Into The Future," *Collier's* 60, no.34 (August 30, 1947): 19, 36-37.

McCurdy, Howard E. "Fiction and Imagination: How They Affect Public Administration," *Public Administration Review*, 55, no. 6 (November–December, 1995): 499-506.

Patterson, Bill, "Robert A Heinlein: A Biographical Sketch," *The Heinlein Journal*, no. 5 (July 1999): 7-36.

Patterson, Bill, "Review of Leon Stover's annotated H.G. Wells, *When the Sleeper Wakes: A Critical Text of the 1899 New York and London First Edition*," *The Heinlein Journal*, no. 7 (July 2000): 43-46.

Patterson, Bill, ed., "Critique of Stover, Leon, *Science Fiction from Wells to Heinlein*," *The Heinlein Journal*, no. 10 (January 2002): 48.

Patterson, Bill, "Virginia Heinlein: A Biographical Sketch," *The Heinlein Journal*, no. 13 (July 2003): 2-26.

Patterson, Bill, "H. G. Wells and the Country of the Future," *The Heinlein Journal*, no. 15 (July 2004): 4.

---

21. Peter Diamandis, "Private Human Spaceflight," paper presented at the Robert A. Heinlein Centennial Conference 1907-2007, Kansas City, MO, July 6, 2007.

www.technovelgy.com (accessed October 21, 2007).

Wells, H. G. *When The Sleeper Wakes* (London: McMillan & Co., Ltd., 1899).

Wells, H. G., *The World Set Free: A Story of Mankind* (London: McMillan & Co., Ltd., 1914).

CHAPTER 17

AMERICAN SPACEFLIGHT HISTORY'S MASTER NARRATIVE AND THE MEANING OF MEMORY

Roger D. Launius

INTRODUCTION

The term master narrative typically refers to a set of sociocultural interpretations of events agreed upon by most of the interpreters of the event or age, and these are abundantly apparent when considering the history of the Space Age. They offer what might best be considered secure knowledge formed to delineate the trajectory of the historical event and center it in its appropriate cultural place. Master narratives are ubiquitous in American history. They serve important purposes in helping to create a useable past for the nation and its peoples. Historians, perhaps unmindfully, accept the master narrative about whatever subject they are examining with relative ease most of the time and facilitate its creation and maintenance as bulwarks upon which the national, or other, story rests. In this instance they support a group identity, whether it be a subgroup or a nation-state, exhibiting varying degrees of commitment to, as well as detachment from, the concepts of the groups that they serve. They move between these two poles to construct historical perspectives that will be of value to the group. Rarely do historians create from whole cloth a master narrative, instead usually reinforcing the dominant perceptions, or master narrative, already held by the group.[1]

It may be argued that there are four narratives that have emerged concerning the U.S. space program, one that is a master narrative and three minor variations. These include: 1) the overwhelmingly dominant narrative of American triumph, exceptionalism, and success; 2) the counter narrative of criticism of the space program from the left, wasting funds on a worthless expense that yielded little when so many Americans could have benefited from spending on social programs; 3) a more recent narrative of criticism of spaceflight from the right of the political spectrum focusing on the program as a representation of liberal taxing and spending strategies; and 4) a fringe narrative that sees in the U.S.

---

1. I have explored this issue an another context in Roger D. Launius, "Mormon Memory, Mormon Myth, and Mormon History," *Journal of Mormon History* 21 (Spring 1995): 1-24.

NASA's original Mercury 7 astronauts posing with a U.S. Air Force F-106B jet aircraft in 1959. These astronauts epitomized the perceived "American exceptionalism" that was considered to be such an intrinsic part of the national character. From left to right: M. Scott Carpenter, L. Gordon Cooper, John H. Glenn, Jr., Virgil I. "Gus" Grissom, Jr., Walter M. "Wally" Schirra, Jr., Alan B. Shepard, Jr., and Donald K. "Deke" Slayton. (NASA)

space program a close tie to all manner of nefarious activities. This last narrative emphasizes conspiracy theories—of extraterrestrial visitation, abduction, and government complicity, of denials of the Apollo Moon landings in favor of a deep-seated conspiracy, as part of a larger militarization scheme aimed at world domination, and a host of strange and bewildering conspiracies affecting the lives of normal Americans in negative ways. Each of these narratives has a place in the American consciousness as it remembers the Space Age. This essay will seek to discuss these four narratives and how they have interrelated over the 50 years of the Space Age.

CIVIL SPACEFLIGHT AS AMERICAN TRIUMPH AND EXCEPTIONALISM

The history of American spaceflight has rested for some 50 years on the master narrative of an initial shock to the system, surprise, and ultimate recovery with success after success following across a broad spectrum of activities. It is a classic story of American history in which a vision of progress, of moving from nothing to something, dominates the story. That master narrative offers comfort to the American public as a whole, but most especially to the governing class who take solace in how the nation responded to crisis.

For example, the surprising Soviet success with Sputnik, so the master narrative relates, created a furor and led the United States to "catch up" to the Soviet Union in space technology. This crisis forced the Eisenhower administration to move quickly to restore confidence at home and prestige aboard. With mounting pressure, the Eisenhower response became typical of earlier crises within the United States; politicians locked arms and appropriated money to tackle the perceived problem. In this effort, both the civilian and military space efforts benefited, one openly and the other in secrecy. The Department of Defense approved additional funds for an Army effort, featuring Wernher von Braun and his German rocket team, to launch an American satellite. The Army's Explorer project had been shelved earlier in favor of concentrating on Vanguard as the first American scientific satellite, but drastic times called for drastic measures and suddenly the atmosphere in Washington had changed. The Army was told to orbit the first satellite by February 1, 1958, only four months after the first Sputnik. Von Braun and his team went to work on a crash program with a modified Jupiter C ballistic missile. The first launch took place on January 31, 1958, placing Explorer 1 in orbit. On this satellite was an experiment by James A. Van Allen, a physicist at the University of Iowa, documenting the existence of radiation zones encircling Earth. Shaped by Earth's magnetic field, what came to be called the Van Allen Radiation Belts partially dictates the electrical charges in the atmosphere and the solar radiation that reaches Earth.[2]

Following this, Congress passed and Eisenhower signed the National Aeronautics and Space Act of 1958. This legislation established NASA with a broad mandate to explore and use space for "peaceful purposes for the benefit of all mankind."[3] The core of NASA came from the earlier National

---

2. Robert A. Divine, *The Sputnik Challenge: Eisenhower's Response to the Soviet Satellite* (New York, NY: Oxford University Press, 1993), pp. 93–96; Roger D. Launius, *NASA: History of the U.S. Civil Space Program* (Malabar, FL: Krieger Publishing Co., 1994), pp. 26–27; James A. Van Allen, *Origins of Magnetospheric Physics* (Washington, DC: Smithsonian Institution Press, 1983).

3. "National Aeronautics and Space Act of 1958," Public Law #85-568, 72 Stat., 426. Signed by the president on July 29, 1958, Record Group 255, National Archives and Records Administration,

Advisory Committee for Aeronautics with its 8,000 employees, an annual budget of $100 million, and its research laboratories. It quickly incorporated other organizations into the new Agency, notably the space science group of the Naval Research Laboratory in Maryland, the Jet Propulsion Laboratory managed by the California Institute of Technology for the Army, and portions of the Army Ballistic Missile Agency in Huntsville, Alabama.[4] This set in train the necessary capabilities for the achievement of considerable success in space exploration during the 1960s.

According to this master narrative, the experience from Sputnik through the Apollo Moon landings have represented an epochal event that signaled the opening of a new frontier in which a grand visionary future for Americans might be realized. It represented, most Americans have consistently believed, what set the United States apart from the rest of the nations of the world. American exceptionalism reigned in this context, and Apollo is often depicted as the critical event in the United States' spaceflight narrative, one that must be revered because it shows how successful Americans could be when they try. At a basic level, Apollo served as a trope of America's grand vision for the future. This exceptionalist perspective has also dominated the public characterizations of spaceflight in general, and Apollo in particular, regardless of the form of those characterizations.[5] For example, expressing this central perspective on Americanism, not long after the first lunar landing in July 1969 Richard

---

Washington, DC; Alison Griffith, *The National Aeronautics and Space Act: A Study of the Development of Public Policy* (Washington, DC: Public Affairs Press, 1962), pp. 27–43.

4. Launius, *NASA*, pp. 29–41.

5. Several years ago I prepared "A Baker's Dozen of Books on Project Apollo," and I have updated it periodically since. These are singularly worthwhile books, but all support the dominant trope in the historiography. The titles include: Donald A. Beattie, *Taking Science to the Moon: Lunar Experiments and the Apollo Program* (Baltimore, Maryland: Johns Hopkins University Press, 2001); Roger E. Bilstein, *Stages to Saturn: A Technological History of the Apollo/Saturn Launch Vehicles* (Washington, DC: National Aeronautics and Space Administration SP-4206, 1980); Courtney G. Brooks, James M. Grimwood, and Loyd S. Swenson, Jr., *Chariots for Apollo: A History of Manned Lunar Spacecraft* (Washington, DC: National Aeronautics and Space Administration SP-4205, 1979); Andrew Chaikin, *A Man on the Moon: The Voyages of the Apollo Astronauts* (New York, NY: Viking, 1994); Michael Collins, *Carrying the Fire: An Astronaut's Journeys* (New York, NY: Farrar, Straus and Giroux, 1974); Edgar M. Cortright, ed., *Apollo Expeditions to the Moon* (Washington, DC: National Aeronautics and Space Administration SP-350, 1975); David M. Harland, *Exploring the Moon: The Apollo Expeditions* (Chicester, England: Wiley-Praxis, 1999); Stephen B. Johnson, *The Secret of Apollo: Systems Management in American and European Space Programs* (Washington, DC: Johns Hopkins University Press, 2002); W. Henry Lambright, *Powering Apollo: James E. Webb of NASA* (Baltimore, MD: Johns Hopkins University Press, 1995); John M. Logsdon, *The Decision to Go to the Moon: Project Apollo and the National Interest* (Cambridge, MA: The MIT Press, 1970); Walter A. McDougall, . . . *the Heavens and the Earth: A Political History of the Space Age* (New York, NY: Basic Books, 1985); Charles A. Murray and Catherine Bly Cox, *Apollo, the Race to the Moon* (New York, NY: Simon and Schuster, 1989); David West Reynolds, *Apollo: The Epic Journey to the Moon* (New York, NY: Harcourt, Brace, 2002).

Nixon told an assembled audience that the flight of Apollo 11 represented the most significant week in the history of Earth since the creation.[6] Clearly, the President viewed the endeavor as both path breaking and permanent, a legacy of accomplishment that future generations would reflect on as they plied intergalactic space and colonized planets throughout the galaxy. Perhaps the Americans were responsible for the second most important week in the history of the cosmos, placing an essentially godlike cast upon the nation.

Spaceflight has persistently represented a feel-good triumph for the nation and its people. It conjured images of the best in the national spirit and served, in the words of journalist Greg Easterbrook, as "a metaphor of national inspiration: majestic, technologically advanced, produced at dear cost and entrusted with precious cargo, rising above the constraints of the earth."[7] Certainly Apollo represented this in the imagery that became iconic in the public consciousness— an astronaut on the Moon saluting the American flag served well as a patriotic symbol of what the nation had accomplished—but so have the later human missions of the Space Shuttle, the International Space Station, and the robotic probes and observatories that have pulled back the curtain to reveal a wondrous universe. This self-image of the United States as a successful nation has been repeatedly affirmed in the spaceflight master narrative since 1957.[8]

It might be argued that spaceflight represented an expression of national power in the context of the "positive liberal state" offered the world by the United States. In essence, this position celebrates the use of state power for "public good." Human exploration of the solar system was always viewed as reasonable and forward-looking and led to "good" results for all concerned, or so adherents of this master narrative believed. Without perhaps seeking to do so, Apollo offered an important perspective on a debate that has raged over the proper place of state power since the beginning of the republic. As one historian remarked about this philosophy of government, the state would actively "promote the general welfare, raise the level of opportunity for all men, and aid all individuals to develop their full potentialities." It would assert active control in this process, seeking improvements to society "both economic and moral, and they did not believe in leaving others alone."[9]

The Democrats of the 1960s believed in activist government, and examples on the part of the Kennedy and Johnson administrations abound. This translated into an ever increasing commitment to the use of the government to

---

6. *10:56:20 PM, EDT, 7/20/69* (New York: CBS News, 1969), p. 159.

7. Greg Easterbrook, "The Space Shuttle Must Be Stopped, *Time*, February 2, 2003, available online at *http://www.mercola.com/2003/feb/8/space_shuttle.htm* (accessed February 24, 2006).

8. I made this argument in relation to Apollo in Roger D. Launius, "Perceptions of Apollo: Myth, Nostalgia, Memory or all of the Above?" *Space Policy* 21 (May 2005): 129–139.

9. Daniel Walker Howe, *The Political Culture of the American Whigs* (Chicago, IL: University of Chicago Press, 1979), p. 20.

achieve "good ends"—the war on poverty, the Peace Corps, support for civil rights, numerous Great Society programs, space exploration, and a host of other initiatives are examples. These all represented a broadening of governmental power for what most at the time perceived as positive purposes.

Such statements of triumph and exceptionalism have permeated the narrative of spaceflight from the beginning. For only one example among many that might be discussed, Andrew Chaikin's 1994 *A Man on the Moon* oozes the narrative of triumph in the context of Apollo.[10] Alex Roland captured the importance of this book best when he proposed that Chaikin offered a retelling of a specific myth and in that retelling it performed a specific purpose. It is not so much history as it is "tribal rituals, meant to comfort the old and indoctrinate the young." He added:

> All the exhilarating stories are here: the brave, visionary young President who set America on a course to the moon and immortality; the 400,000 workers across the nation who built the Apollo spacecraft; the swashbuckling astronauts who exuded the right stuff; the preliminary flights of Mercury and Gemini—from Alan Shepard's suborbital arc into space, through John Glenn's first tentative orbits, through the rendezvous and spacewalks of Gemini that rehearsed the techniques necessary for Apollo. There is the 1967 fire that killed three astronauts and charred ineradicably the Apollo record and the Apollo memory; the circumlunar flight of Christmas 1968 that introduced the world to Earth-rise over the lunar landscape; the climax of Apollo 11 and Neil Armstrong's heroic piloting and modest words, "that's one small step for a man, one giant leap for mankind"; the even greater drama of Apollo 13, rocked by an explosion on the way to the moon and converted to a lifeboat that returned its crew safely to Earth thanks to the true heroics of the engineers in Houston; and, finally, the anticlimax of the last Apollo missions.

Roland finds that Chaikin had to struggle to maintain a triumphal narrative of Apollo, however, for the missions became a deadend rather than a new beginning and no amount of heroic prose could overcome that ironic plot twist.[11]

American exceptionalism has dominated the vast majority of the national discussion of spaceflight, represented perhaps best in popular culture. As only one example among many, the late comedian Sam Kinison once ranted to other

---

10. Chaikin, *A Man on the Moon.*

11. Alex Roland, "How We Won the Moon," *New York Times Book Review,* July 17, 1994, pp. 1, 25.

nations seeking to replicate the "greatness" of America: "You really want to impress us! Bring back our Flag!"[12] This statement of American exceptionalism and triumph in relation to Apollo, while it has a form of jingoism at its center, expresses a core truth about how these efforts have been embraced and celebrated in the United States. While Kinison's challenge symbolized for many Americans national superiority—at the same time signaling the inferiority of all others—NASA situated its spaceflight aspirations within the arena of international prestige. If anything this trope of national exceptionalism and triumph has only intensified over time, clearly dominating discussion of America's spaceflight efforts.[13]

At sum, Americans have usually viewed space exploration as a result of a grand visionary concept for human exploration that may be directly traced to the European voyages of discovery beginning in the 15th century.[14] Given this observation, these endeavors have been celebrated as an investment in technology, science, and knowledge that would enable humanity—or at least Americans—to do more than just dip their toes in the cosmic ocean, to become a truly spacefaring people. Accordingly, Americans have taken as a measure of the majesty of this vision the length of time, complexity, and expense of the program, and the linkage of the length of time, complexity, and expense of its space exploration activities to earlier explorations. The Spanish exploration of the Americas proved time consuming, complex, and expensive. So did the efforts of other European powers in the sweepstakes of exploration and imperialism that took place over long periods made possible by these explorations. The exploration of space was much the same only more so, and this made it special and grand and visionary.

---

12. See "Bush to announce goal of returning to the moon," online at *http://forums.pcper.com/printthread.php?threadid=277513* (accessed April 19, 2004).

13. Representative works include, Frutkin, *International Cooperation in Space*; Roger Handberg and Joan Johnson-Freese, *The Prestige Trap: A Comparative Study of the U.S., European, and Japanese Space Programs* (Dubuque, IA: Kendall/Hunt Publishing Co., 1994); Brian Harvey, *The New Russian Space Programme: From Competition to Collaboration* (Chichester, England: Wiley—Praxis, 1996); Dodd L. Harvey and Linda C. Ciccoritti, *U.S.-Soviet Cooperation in Space* (Miami, FL: Monographs in International Affairs, Center for Advance International Studies at the University of Miami, 1974); Joan Johnson-Freese, *Changing Patterns of International Cooperation in Space* (Malabar, FL: Orbit Books, 1990); Joan Johnson-Freese, "Canceling the U.S. Solar-Polar Spacecraft: Implications for International Cooperation in Space," *Space Policy 3* (February 1987): 24-37; Joan Johnson-Freese, "A Model for Multinational Space Cooperation: The Inter-Agency Consultative Group," *Space Policy 5* (November 1989): 288-300; John M. Logsdon, "U.S.-Japanese Space Relations at a Crossroads," *Science* 255 (January 17, 1992): 294-300.

14. The best example of this is Stephen J. Pyne, "Space: A Third Great Age of Discovery," *Space Policy* 4 (August 1988): 187-199.

Celebrants of spaceflight have long argued that returns on investment in this age of exploration changed Americans' lives.[15] As President Lyndon B. Johnson remarked at the time of the third Gemini flight in August 1965, "Somehow the problems which yesterday seemed large and ominous and insoluble today appear much less foreboding." Why should Americans fear problems on Earth, he believed, when they had accomplished so much in space?[16] In this triumphalist narrative, the reality of spaceflight demonstrated that anything we set our minds to we could accomplish. "If we can put a man on the Moon, why can't we . . ." entered the public consciousness as a statement of unlimited potential.[17] Spaceflight, of course, remains a powerful trope of American exceptionalism to the present.

CRITICISM OF SPACEFLIGHT FROM THE POLITICAL LEFT AS WASTEFUL GOVERNMENT SPENDING

A counter narrative to the master account of American triumph, exceptionalism, and success also emerged in the 1950s and argued that a large space exploration program deserved criticism from the left as a waste of expenditures of federal funds that could have been much more effectively used to feed the poor, help the elderly, care for the sick, or otherwise carry out critical social programs.[18] Left-leaning critics argued that NASA's efforts

---

15. Stephen J. Pyne, *The Ice: A Journey to Antarctica* (Iowa City, IA: University of Iowa Press, 1986); Nathan Reingold, ed., *The Sciences in the American Context: New Perspectives* (Washington, DC: Smithsonian Institution Press, 1979); Norman Cousins, et al., *Why Man Explores* (Washington, DC: NASA Educational Publication-125, 1976); Sarah L. Gall and Joseph T. Pramberger, *NASA Spinoffs: 30 Year Commemorative Edition* (Washington, DC: National Aeronautics and Space Administration, 1992).

16. Lyndon B. Johnson, "President's News Conference at the LBJ Ranch," *Public Papers of the Presidents*, August 29, 1965, p. 944-45. See also Lyndon B. Johnson, "Michoud Assembly Facility, Louisiana," *Weekly Compilation of Presidential Documents*, December 12, 1967, p. 1967.

17. To determine how widespread this question is, in 2001 I undertook a search of the DowJones database, which includes full text of more than 6,000 newspapers, magazines, newswires, and transcripts. Some of the publications go back to the 1980s but most have data only from the 1990s. Except for perhaps Lexis-Nexis, DowJones is the largest full-text database available. There are more than 6,901 articles using this phrase, or a variation of it, in the database. Among them was a statement by former White House Chief of Staff, Mack McClarty concerning Mexico on National Public Radio's "All Things Considered," entitled, "Analysis: President Bush to visit Mexico and its President." Maria Elena Salinas, co-anchor at Miami-based Spanish-language cable network Univision, used this phrase when discussing her decision to list the Apollo Moon landings as first in the top 100 news events of the 20th century. Levinson A. Atomic bombing of Hiroshima tops journalists' list of century's news. Associated Press. February 24, 1999.

18. Among those criticisms, see Hugo Young, Bryan Silcock, and Peter Dunn, *Journey to Tranquillity: The History of Man's Assault on the Moon* (Garden City, NY: Doubleday, 1970); Erlend A. Kennan and Edmund H. Harvey, Jr., *Mission to the Moon: A Critical Examination of NASA and the Space Program* (New York, NY: William Morrow and Co., 1969); John V. Moeser, *The Space Program and the Urban Problem: Case Studies of the Components on National Consensus* (Washington, DC: Program of Policy Studies in Science and Technology, George

were, in the words of aerospace historian Roger E. Bilstein, "a cynical mix of public relations and profit-seeking, a massive drain of tax funds away from serious domestic ills of the decade, or a technological high card in international tensions during the Cold War."[19] Some of those attacks were sophisticated and involved, others were simplistic and without appeal to all but those with the predilection to believe them.

For example, Vannevar Bush, a leading and well-respected scientist who appreciated the marshaling of the power of the federal government in the furtherance of national objectives, questioned large expenditures for spaceflight. He wrote to NASA Administrator James E. Webb in April 1963 voicing his concerns about the cost, versus the benefits, of the human space exploration program. He asserted "that the [Apollo] program, as it has been built up, is not sound." He expressed concern that it would prove "more expensive than the country can now afford," adding that "its results, while interesting, are secondary to our national welfare."[20]

Sociologist Amitai Etzioni was even more critical. In a reasoned, full-length critique of the Moon landing program in 1964, he deplored the "huge pile of resources" spent on space, "not only in dollars and cents, but the best scientific minds—the best engineering minds were dedicated to the space project." Could not those resources have been better spent on improving the lives people in modern America?[21] Etzioni bemoaned the nation's penchant for embracing both high technology and unsustainable materialism: "we seek to uphold humanist concerns and a quest for a nobler life under the mounting swell of commercial, mechanical, and mass-media pressure."[22]

As Etzioni remarked in a 1962 article that also expressed his concern about spaceflight: "If private foundations or some university professors wish to continue to satisfy their own and the common human desire to know about outer space, fine. But can the public spend 30 billion dollars—the amount required to send one man to the moon—to answer some questions about the shape of the moon? Are we that curious, when the same amount of money would serve to develop . . . India?" Furthermore, he noted, "As emotional as it might sound, this is truely [sic] a question of investment in feeding starving children

Washington University, 1969); Edwin Diamond, *The Rise and Fall of the Space Age* (Garden City, NY: Doubleday and Co., 1964).

19. Roger E. Bilstein, *Testing Aircraft, Exploring Space: An Illustrated History of NACA and NASA* (Baltimore, MD: Johns Hopkins University Press, 2003), p. 200.

20. Vannevar Bush to James E. Webb, Administrator, NASA, April 11, 1963, p. 2, Presidential Papers, John F. Kennedy Library, Boston, MA.

21. Amitai Etzioni, *The Moon-Doggle: Domestic and International Implications of the Space Race* (New York, NY: Doubleday, 1964), p. 70. See Alton Frye, "Politics—The First Dimension of Space," *Journal of Conflict Resolution* 10 (March 1966): 103-12, for a review of *Moon-Doggle*.

22. Etzioni, *Moon-Doggle*, p. 195.

as against improving the maps of Van Allen belts, of suppressing ignorance and disease on earth as against finding new moons in the skies."[23]

Other critics were even more impulsive in their censure. If spaceflight was truly about demonstrating to the world American capabilities, cautioned nuclear physicist Leo Szilard, "we are making the wrong choice." Americans could demonstrate this in other more positive ways. "To race the Russians to the moon and let our old people live on almost nothing is immoral," he remarked specifically about the Apollo lunar program. "The moon is not science—not bread. It is circus. The astronauts are the gladiators. It's lunacy, I say."[24] As time passed, for Szilard and a minority of other Americans, space exploration seemed like an increasingly embarrassing national self-indulgence.[25]

Several of the leaders in the U.S., especially those within the Democratic Party, found that support for NASA's space exploration agenda clashed with supporting funds for social programs enacted through "Great Society" legislation. They disparaged Apollo both as too closely linked to the military-industrial complex and defense spending and too far removed from the ideals of racial, social, and economic justice at the heart of the positive liberal state the Democrats envisioned. Liberal senators such as J. William Fulbright, Walter Mondale, and William Proxmire challenged the Johnson administration every year over funding for NASA that they believed could be more effectively used for social programs. Accordingly, Bureau of the Budget Director Charles Schultze worked throughout the middle part of the 1960s to shift funds from NASA to such programs as the war on poverty. Johnson even tried to defend NASA as a part of his "Great Society" initiatives, arguing that it helped poor southern communities with an infusion of federal investment in high technology. Nonetheless, this proved a difficult sell and the NASA budget declined precipitously throughout the latter half of the 1960s.[26]

Indicative of this concern, even as Apollo 11 was being prepared for launch from the Kennedy Space Center in Florida on July 16, 1969, Rev. Ralph Abernathy led a protest at the gates of the Center for 150 protesters and 4

23. Amatai Etzioni, "International Prestige, Competition and Cooperative Existence," *Archives of Europeenees de Sociologie* 3, no. 1 (1962): 21–41, quotes from pp. 38–39.

24. Quoted in Oscar H. Rechtschaffen, ed., *Reflections on Space: Its Implications for Domestic and International Affairs* (Colorado Springs, CO: USAF Academy, 1964), p. 118, available from Defense Technical Information Center, accession no. AD0602915.

25. W. Henry Lambright, *Powering Apollo: James E. Webb of NASA* (Baltimore, MD: Johns Hopkins University Press, 1995), pp. 140–141.

26. Robert Dallek, "Johnson, Project Apollo, and the Politics of Space Program Planning," in Roger D. Launius and Howard E. McCurdy, eds., *Spaceflight and the Myth of Presidential Leadership* (Urbana, IL: University of Illinois Press, 1997), pp. 75–88.

mules. His aim was to call attention to the plight of the poor even as the U.S. government spent lavishly on flights to the Moon.[27]

In contrast to the triumphalist, exceptionalist narrative that celebrates space exploration, this narrative views the endeavor as a waste, a missed opportunity to further important and necessary goals in America. Indeed, the triumphalist narrative of spaceflight has been so powerful a memory that most people in the United States reflecting on it believe that NASA enjoyed enthusiastic support during the 1960s and that somehow the Agency lost its compass thereafter.[28] Contrarily, at only one point prior to the Apollo 11 mission, October 1965, did more than half of the public favor the lunar landing program. Americans have consistently ranked spaceflight near the top of those programs to be cut in the federal budget. Such a position is reflected in public opinion polls taken throughout the Space Age when the majority of Americans ranked NASA as the government initiative most deserving of reduction, and its funding redistributed to Social Security, Medicare, and numerous other programs. While most Americans did not oppose space exploration per se, they certainly questioned spending on it when social problems appeared more pressing.[29] At some level it was like the characterization of the overlanders en route westward on the Oregon Trail who opined that the Platte River that they followed was a mile wide and an inch deep. Support for space exploration was broad but not deep and almost always lost in comparison to other federal initiatives.

Since the heyday of Apollo, little has changed in this support for NASA and its space exploration agenda. Many on the left view spaceflight, usually characterized exclusively as the human space program, as a waste of resources that might be more effectively deployed to support other good ends. Many find themselves nodding in agreement when Josh Lyman, the White House Assistant Chief of Staff in the fictional *West Wing* television series told NASA officials that his one priority for the space Agency was that it stay out of the newspapers with tales of mismanagement and woe. He added that his agenda included using precious federal funds here on Earth to help people rather than to conquer space.[30]

27. Bernard Weinraub, "Some Applaud as Rocket Lifts, but Rest Just Stare," *New York Times*, July 17, 1969, p. 1.

28. James L. Kauffman, *Selling Outer Space: Kennedy, the Media, and Funding for Project Apollo, 1961-1963* (Tuscaloosa, AL: University of Alabama Press, 1994); Mark E. Byrnes, *Politics and Space: Image Making by NASA* (New York, NY: Praeger, 1994); Neil de Grasse Tyson, "Expanding the Frontiers of Knowledge," in Stephen J. Garber, ed., *Looking Backward. Looking Forward: Forty Years of U.S. Human Spaceflight Symposium* (Washington, DC: NASA SP-2002-4107, 2002), pp. 127-136.

29. Roger D. Launius, "Public Opinion Polls and Perceptions of U.S. Human Spaceflight," *Space Policy* 19 (August 2003): 163-175.

30. "The Warfare of Genghis Khan," Episode #513, *West Wing*, broadcast: February 11, 2004.

Of course, criticism of the space exploration initiatives of NASA have taken myriad turns within the space community itself, as losers in the debate question the course taken. Because NASA pursued the Space Shuttle program in the aftermath of Apollo it was unable to undertake other projects that might have been more fruitful, the argument goes. There is no question that this is true, and usually critiques along these lines take one of three forms. The first is a criticism that NASA spent the last 30 years in Earth orbit and it could have—indeed should have—used the same funding that it received for the Space Shuttle and Space Station to return to the Moon or to explore Mars.

Robert Zubrin, a persistent advocate for a mission to Mars, made this case in testimony before the U.S. Senate in 2003. He said:

> In today's dollars, NASA['s] average budget from 1961-1973 was about $17 billion per year. This is only 10% more than NASA's current budget. To assess the comparative productivity of the Apollo Mode with the Shuttle Mode, it is therefore useful to compare NASA's accomplishments between 1961-1973 and 1990-2003, as the space agency's total expenditures over these two periods were equal.

He concluded: "Comparing these two records, it is difficult to avoid the conclusion that NASA's productivity in *both* missions accomplished *and* technology development during its Apollo Mode was at least ten times greater than under the current Shuttle Mode."[31]

A second criticism of "paths not taken" comes from representatives of the scientific community and usually involves questioning the role of humans in space at the expense of science missions. University of Iowa astrophysicist and discoverer of the radiation belts surrounding Earth that bears his name, James A. Van Allen, never believed that human spaceflight was worth the expense. In 2004 he remarked, "Risk is high, cost is enormous, science is insignificant. Does anyone have a good rationale for sending humans into space?"[32] Undoubtedly, large numbers of scientific missions could have been developed had funding used for human missions been used instead to fund other types of scientific efforts. But it is not a zero-sum-game, and there is little reason to believe that

---

31. Testimony of Robert Zubrin to the Senate Commerce Committee, October 29, 2003, p. 2, available online *at http://www.marssociety.org/content/Zubrin102903.PDF* (accessed February 26, 2006).

32. James A. Van Allen, "Is Human Spaceflight Obsolete?" *Issues in Science and Technology,* vol. 20 (Summer 2004), available online at *http://www.issues.org/20.4/p_van_allen.html* (accessed August 3, 2004).

reducing funding for human spaceflight would translate into greater funding for robotic missions.[33]

Most recently, the NASA Administrator, Mike Griffin, questioned the human space exploration agenda of NASA since Apollo, calling the Space Shuttle program the result of a "policy failure" that was relentlessly pursued by NASA for more than a generation. "It is now commonly accepted that was not the right path," Griffin told *USA Today* in an interview that appeared as a page one story on September 28, 2005. "We are now trying to change the path while doing as little damage as we can." When asked pointedly if the shuttle had been a mistake the NASA Administrator responded, "My opinion is that it was . . . It was a design which was extremely aggressive and just barely possible."[34]

A subtext in all of this is that conservative political decisions, especially Richard Nixon's decision to approve only the Space Shuttle in the aftermath of Apollo, set course down a wasteful, useless road when it might have been possible to reach other decisions and pursue much more productive paths. All of these criticisms about the place of space exploration in modern America have become part of a larger counter to the master narrative that questions the dominant story of American exceptionalism. At some level, as political scientist Howard E. McCurdy remarked, space exploration "was to America what the pyramids were to Egypt. It's one of our great accomplishments . . . . But when you go back and look, there were people, at the time who are expressing public misgivings. And in private—where you can get those kinds of conversations—[they] are pulling their hair out about this program."[35]

This theme has been played out repeatedly in the American left since the beginning of the Space Age. For example, NASA came under congressional fire even as it tried to pursue new space exploration initiatives at the beginning of the 1970s. Faced with domestic unrest, urban problems, and escalating military spending in Vietnam, Congress was eager to cut whatever programs it could, and NASA presented an appealing target. As New York Congressman Ed Koch mused, "I just for the life of me can't see voting for monies to find out whether or not there is some microbe on Mars, when in fact I know there are rats in the Harlem apartments." Even some pro-space legislators questioned the necessity of further space exploration after Apollo and wondered if NASA had fully considered its options. Congressman Joseph Karth led the opposition because of what he considered NASA's hubris. "NASA must consider the members of

---

33. See Daniel S. Goldin, speech at California Institute of Technology, December 4, 1992, NASA Historical Reference Collection, NASA History Division, Washington, DC.

34. Traci Watson, "NASA Administrator says Space Shuttle was a Mistake," *USA Today*, September 28, 2005, p. 1A.

35. "Transcript: Washington Goes to the Moon, Part 1: *Washington, We Have A Problem*," aired May 25, 2001, WAMU FM, transcript available online at *http://wamu.org/d/programs/special/moon/opp_show.txt* (accessed October 17, 2007).

the Congress a bunch of stupid idiots," he complained. "Worse yet, they may believe their own estimates—and then we are really in bad shape."[36]

A persistent drumbeat of criticism from the left for NASA's efforts in human space exploration has sometimes reached crescendo proportions. Critics have long condemned NASA for "overselling" the space exploration agenda and then failing to deliver on that promise. Many liberal Americans have agreed with Leo McGarry, the White House Chief of Staff in the fictional *West Wing* television series when asked about NASA's overreach: "Where's my jet pack, my colonies on the Moon? Just a waste."[37] More recent ventures in space exploration, and especially their failure, have wrought even more energetic criticisms.[38]

CRITICISM OF SPACEFLIGHT FROM THE POLITICAL RIGHT

From the beginning of the Space Age, some figures on the right of the American political spectrum have also criticized NASA's exploration agenda as an excess of federal power, another counter to the master narrative. In their view, the federal government should not do much of anything, offering a persistently libertarian position that emphasized individual prerogative and personal freedom over state action. As an example, for this reason Eisenhower believed that empowering NASA to accomplish the Apollo Moon landings of the 1960s was a mistake. He remarked in a 1962 article: "Why the great hurry to get to the moon and the planets? We have already demonstrated that in everything except the power of our booster rockets we are leading the world in scientific space exploration. From here on, I think we should proceed in an orderly, scientific way, building one accomplishment on another."[39] He later cautioned that the Moon race "has diverted a disproportionate share of our brain-power and research facilities from equally significant problems, including education and automation."[40] Likewise, in the 1964 presidential election, Republican candidate Senator Barry Goldwater urged a reduction of the Apollo commitment to pay for national security initiatives.

With the coming of the successful Moon landings, however, the American right largely retreated from any high profile criticism of Apollo. That position dominated until the 1980s when a full-scale assault on the "Great Society" efforts

---

36. Quoted in Ken Hechler, T*oward the Endless Frontier: History of the Committee on Science and Technology, 1959-79* (Washington, DC: Government Printing Office, 1980), p. 274.

37. "The Warfare of Genghis Khan," Episode #513, *West Wing*, broadcast: February 11, 2004.

38. Greg Easterbrook, "The Case Against NASA," *New Republic,* July 8, 1991, pp. 18-24; Alex Roland, "Priorities in Space for the USA," *Space Policy* 3 (May 1987): 104-114; Alex Roland, "The Shuttle's Uncertain Future," *Final Frontier,* April 1988, pp. 24-27.

39. Dwight D. Eisenhower, "Are We Headed in the Wrong Direction?" *Saturday Evening Post,* August 11-18, 1962, p. 24.

40. Dwight D. Eisenhower, "Why I Am a Republican," *Saturday Evening Post,* April 11, 1964, p. 19.

of the Democrats in the 1960s emerged in the public realm. A questioning of the Apollo program became part of a conservative strain in American political discourse that increasingly found expression during the Reagan era of the 1980s. Percolating for many years, it emerged full-blown during the era to reconsider the history and policy of liberal ideology in the United States. In the process, reappraisals have castigated the social upheaval of the 1960s, defeat in Vietnam, and Great Society programs as failures of American politics.[41] There was also a conservative space history, as well as a conservative space policy, that emerged during the same era. Some have even hinted that criticism of Apollo was appropriate as part of a larger assault on the "products of the maniacal 1960s."[42]

No one has been more successful in offering a conservative critique of the early efforts to explore space than Walter A. McDougall, who published a Pulitzer Prize-winning "political history of the Space Age."[43] His situation of the history of Apollo in the context of the United States' well-documented political "right turn" may well represent the central thrust of space history and policy since the 1980s, for many have followed in his footsteps.[44] This critique has emphasized a derogation of government programs as wasteful and inefficient, a celebration of private sector space initiatives, a relaxation of the regulatory environment,

---

41. The reinterpretation of America in the 1960s has been a major cottage industry in recent years, and the reassessment has as often as not been negative. Anyone wishing to pursue study of the reorientation of American society in the 1960s should read Milton Viorst, *Fire in the Streets: America in the 1960s* (New York, NY: Simon and Schuster, 1979); Allen J. Matusow, *The Unraveling of America: A History of Liberalism in the 1960s* (New York, NY: Harper and Row, 1984); William L. O'Neill, *Coming Apart* (Chicago, IL: Quadrangle Books, 1971); Godfrey Hodgen, *America in Our Time: From World War II to Nixon, What Happened and Why* (Garden City, NY: Doubleday and Co., 1976); Morris Dickstein, *Gates of Eden: American Culture in the Sixties* (New York, NY: Basic Books, 1977). For works that question the "Great Society" and the social upheaval of the 1960s, see Myron Magnet, *The Dream and the Nightmare: The Sixties Legacy to the Underclass* (New York, NY: William Morrow and Company, 1993); Thomas C. Reeves, *The Empty Church: The Suicide of Liberal Christianity* (New York, NY: Free Press, 1996); Charles Murray, *Loosing Ground: American Social Policy, 1950-1980* (New York, NY: Basic Books, 1984); Irwin Unger, *The Best of Intentions: The Triumph and Failure of the Great Society Under Kennedy, Johnson and Nixon* (Naugatuck, CT: Brandywine Press, 1995); Gareth Davies, *From Opportunity to Entitlement: The Transformation and Decline of Great Society Liberalism* (Lawrence, KS: University Press of Kansas, 1997); Arthur Benavie, *Social Security Under the Gun* (New York, NY: Palgrave Macmillan, 2003); Ellen Schrecker, ed., *Cold War Triumphalism: The Misuse of History After the Fall of Communism* (New York, NY: New Press, 2004).

42. Walter A. McDougall, "Technocracy and Statecraft in the Space Age: Toward the History of a Saltation," *American Historical Review* 87 (October 1982): 1010-1040, quote from p. 1025.

43. Walter A. McDougall, . . . *the Heavens and the Earth: A Political History of the Space Age*, in *Journal of American History* (New York: Basic Books, 1985).

44. Darryl L. Roberts, "Space and International Politics: Models of Growth and Constraint in Militarization," *Journal of Peace Research* 23 (September 1986): 291-298.

and a redistribution of federal research and development funds from traditional sources to organizations less tied to Democratic administrations.[45]

Nothing expresses this "right turn" better than the rehabilitation of Dwight D. Eisenhower as President. He has emerged as the hero of the Space Age, seeking to hold down expenditures, refusing to race the Soviet Union into space, and working to maintain traditional balances in policy, economics, and security. As Alex Roland noted concerning Water McDougall's study of the subject, Eisenhower stands "alone against the post–Sputnik stampede, unwilling to hock the crown jewels in a race to the moon, confident that America's security could be guaranteed without a raid on the Treasury, and concerned lest a space race with the Russians jeopardize America's values and freedoms and drag us down to the level of the enemy." Conversely, the Democrats—especially Kennedy and Johnson—emerge as villains in this drama, ever seeking to enhance the power of big government to reshape the landscape of the United States as a means of facilitating their schemes of social revolution. Indeed, as NASA Administrator James E. Webb asked, if we can accomplish Apollo "why can't we do something for grandma with Medicare?"[46] The linkage of space policy and social policy may seem tenuous at first, but in this critique the power of the federal government and the state system to "intrude" in individual lives required denunciation.

Critiques from the right also noted that the mandate to complete Apollo on President John F. Kennedy's schedule prompted the space program to become identified almost exclusively with high-profile, expensive, human spaceflight projects. This was because Apollo became a race against the Soviet Union for recognition as the world leader in science and technology and, by extension, in other fields. For example, McDougall juxtaposed the American effort with the Soviet space program and the dreams of such designers as Sergei P. Korolev. While he recognized the American effort as a significant engineering achievement, he concluded that it was also enormously costly both in terms of resources and the direction to be taken in state support of science and technology. In the end, NASA had to stress engineering over science, competition over cooperation, and international prestige over practical applications.

---

45. This subject has been discussed in Andrew J. Butrica, *Single Stage to Orbit: Politics, Space Technology, and the Quest for Reusable Rocketry* (Baltimore, MD: Johns Hopkins University Press, 2003); W. D. Kay, "Space Policy Redefined: The Reagan Administration and the Commercialization of Space," *Business and Economic History* 27 (Fall 1998): 237-247. An element of manipulation science data has also surfaced. For instance, this may be found in such works as Mark Bowen, *Thin Ice: Unlocking the Secrets of Climate in the World's Highest Mountains* (New York, NY: Henry Holt and Co., 2005), which talks at length about NASA and censorship concerning global climate change.

46. Alex Roland, "How Sputnik Changed Us," *New York Times*, April 7, 1985, pp. 1, 6, quote from p. 6.

Most importantly, McDougall argued that the Space Age gave birth to a state of "perpetual technological revolution" because of the technocracy that arose to support this incredibly complex set of machines and activities. In essence, driven to respond to the Soviet challenge the United States recreated the same type of command technocracy that the Soviets had instituted. McDougall concluded that the space race led to nothing less than "the institutionalization of technological change for state purposes, that is, the state-funded and managed R&D explosion of our time."[47] As McDougall wrote:

> [I]n these years the fundamental relationship between the government and new technology changed as never before in history. No longer did state and society react to new tools and methods, adjusting, regulating, or encouraging their spontaneous development. Rather, states took upon themselves the primary responsibility for generating new technology. This has meant that to the extent revolutionary technologies have profound second order consequences in the domestic life of societies, by forcing new technologies, all governments have become revolutionary, whatever their reasons or ideological pretensions.[48]

And once institutionalized, technocracy has not gone away. McDougall concluded that it was enormously costly to the nation, and not just in public treasure. Emphasizing the effect of the space race upon American society, this critique focused on the role of the state as a powerful promoter of technological progress—to the detriment of the nation as a whole.

The spaceflight critique from the right bemoaned fundamentally what one observer called so much nostalgia for "the lost world of Thomas Jefferson and Adam Smith, its seeming faith in the untrammeled operation of the marketplace, its occasionally strident anticommunism, or its neo-orthodox assertions about humanity's sinful nature."[49] Whether or not such a world ever actually existed was problematic, but in reality the debate over spaceflight from the right revolved around how much activity by the federal government is appropriate. Conservatives question an activist government and spaceflight clearly demonstrated activism in a most significant manner. While most Americans accepted at face value the benign nature of this power, conservatives tended to challenge its legitimacy.[50]

---

47. McDougall, . . . the Heavens and the Earth, p. 5.

48. Ibid., pp. 6–7.

49. Robert Griffith, "Roots of Technocracy," Science 230 (December 6, 1985): 1154.

50. Ralph E. Lapp, The New Priesthood: The Scientific Elite and the Uses of Power (New York, NY: Harper & Row, 1965), pp. 227-228. Similar cautions, but aimed at the use of science and

Though distinctive in many respects, critics from the right believed the power accrued by NASA corrupted it, making it exploitative of others and engendering in them cynicism toward those they dominated. They may have tried to conceal that fact by laying claim to the dominant myths and symbols of the American frontier, invoking heroes from American folklore, positivist images of "manifest destiny," and happy visions of white-topped wagon trains traveling across the prairies, but conservative critics declared that only a ruse. Through space exploration the federal government enhanced its power and while many Americans celebrated this use of federal power, conservatives bemoaned its intrusion into their vision of individual liberty for the future. That concern has enjoyed a persistent presence in the American spaceflight community since the 1980s.

Of course, absent the power sharing relations present on Earth—state to state, local to national, philosophy to philosophy—the regime above Earth's atmosphere must be ruled by concentrated state power, much of it U.S. power, often hidden behind beguiling masks. They have been reminded by conservative critics of the subtle nature of strenuous and sometimes capricious governmental power in this experience. The region has, of course, been the scene of intense struggles over power and hierarchy, not only between nations but also between classes, genders, and other groups. The outcome of those struggles has a few distinctive features found nowhere else in America, especially power elites that are not much like those in other areas, particularly those elites located at intersections between the federal agencies, corporations, and interest groups. At sum, these concerns suggest an uneasy relationship to the bureaucracy that made possible the advance of space exploration.

This is seen in at least one criticism of space exploration from the political right in the mid-1990s when then Speaker of the House Newt Gingrich (R–GA) criticized NASA as having too much power and becoming muscle-bound. He said that while he generally favored science and technology investment by the federal government he always believed that NASA should have been dismantled after Apollo. In the aftermath of the Moon landings, Gingrich said, NASA had become a bureaucracy in the worst sense of the term. "If you keep people there," he contended, "they become obsolescent."[51] That was a metaphor for the whole of NASA as it moved beyond its glory of the Moon landings.

In an irony too great to ignore, criticism of space exploration—especially the Apollo program—from the right has largely been juxtaposed with support for NASA from conservative politicians in the years since the Moon landings. Whereas the Apollo program, expensive and large and successful, had been the

technology to dupe Americans, may be found in Robert L. Park, *Voodoo Science: The Road from Foolishness to Fraud* (New York, NY: Oxford University Press, 2000); Amitai Etzioni, *The Limits of Privacy* (New York, NY: Basic Books, 2000).

51. "Gingrich Says NASA Should Have Folded," *New York Times*, February 5, 1995, p. 24.

initiative of a Democratic President; the period since has been dominated by a Republican political consensus that has become increasingly conservative. That criticism then took on the added flavor of enthusiasm for private sector space activities instead of large government efforts. Core questions plaguing space policy since the 1950s have revolved around the role of the government versus the private sector in facilitating space exploration. Should all activities be undertaken by the federal government? Should there be some type of public/private partnership put into place to accomplish these tasks? Should the government leave these activities entirely to private companies, involving itself only insofar as required to assure safety of its citizens? Should some entirely different model be employed to ensure space exploration? If the macroeconomic studies sponsored by NASA were an indication, the returns on investment in space research and development were astounding. The Midwestern Research Institute (MRI) study of 1971 determined that NASA R&D provided an overall 7:1 return. Essentially, for every dollar spent on R&D, seven dollars were returned to the GDP. MRI refined its study in 1988, calculating this time an even higher 9:1 return on investment. Chase Econometrics performed a more sophisticated study in 1975 that reported a whopping 14:1 return on investment.[52] If this was true, said the conservative critique of spaceflight, should not the private sector pursue this objective free from government interference.

The Reagan administration of the 1980s certainly thought so and proceeded to privatize spaceflight. At Reagan's behest Congress passed the Commercial Space Launch Act of 1984 that ensconced in law the desire to open space access to private sector providers.[53] There followed a series of moves intended to create a commercial space capability while reducing government funding for space exploration.[54] Increasingly since the Reagan era spaceflight has become increasingly private, in no small measure the result of efforts to reduce the role of the federal government.

Many examples exist. Beginning in the mid-1990s, several start-up companies were organized to undertake new space initiatives. Indeed, 1996 marked something of a milestone in the history of spaceflight as worldwide

---

52. "Economic Impact of Stimulated Technological Activity," Final Report, Midwest Research Institute, October 15, 1971, Contract No. NASW-2030; Michael K. Evans, "The Economic Impact of NASA R&D Spending," Chase Econometric Associates, Inc., Bala Cynwyd, PA, April 1976; "Economic Impact and Technological Progress of NASA Research and Development Expenditures," Midwest Research Institute, Kansas City, MO, for the National Academy of Public Administration, September 20, 1988; BDM, "Economic Return on Technology Investments Study: Final Report," September 30, 1994.

53. "Commercial Space Launch Act of 1984, Public Law 98-575," in John M. Logsdon, gen. ed., *Exploring the Unknown: Selected Documents in the History of the U.S. Civil Space Program, Volume IV, Accessing Space* (Washington, DC: NASA SP-4407, 1999), pp. 431-440.

54. Space Launch Policy Working Group, "Report on Commercialization of U.S. Expendable Launch Vehicles," April 13, 1983, p. 3, NASA Historical Reference Collection.

commercial revenues in space for the first time surpassed all governmental spending on space, totaling some $77 billion. This growth continued in 1997, with 75 commercial payloads lofted into orbit, and with approximately 75 more military and scientific satellites launched. This represented a threefold increase over the number the year before. Market surveys for the period thereafter suggested that commercial launches would multiply for the next several years at least.[55] In that context many spaceflight advocates believed that the market had matured sufficiently that government control was no longer necessary. Instead, they asked that the federal government simply "get out of the way" and allow the private sector to pursue their efforts in space free from bureaucratic controls.[56]

This critique has also found expression in the first decade of the 21st century. Even as NASA was given a new responsibility to return to the Moon, conservative policymakers refused to appropriate the federal funds necessary to accomplish the task. By 2007, accordingly, it had become highly uncertain that the initiative could be realized. It appeared increasingly that this proposal would follow the path of the aborted Space Exploration Initiative (SEI) announced with great fanfare in 1989 but derailed in the early 1990s.[57] Indeed, one candidate for the presidency during the 2008 election, Senator Hillary Rodham Clinton (D–NY), has already stated her opposition to continuing George W. Bush's Vision for Space Exploration should she become President. As reported in the *New York Times*, "Travel to the Moon or Mars 'excites people,' she said, 'but I am more focused on nearer-term goals I think are achievable'."[58] It seems that critics of human space exploration on the left were intent on ending this large space initiative because they viewed it as taking funds away from more pressing social needs while critics on the right were unwilling to put much funding into it and emphasized greater private sector involvement.

55. Tim Beardsley, "The Way to Go in Space," *Scientific American*, March 1999, special issue on "The future of Space Exploration."

56. Craig R. Reed, "Factors Affecting U.S. Commercial Space Launch Industry Competitiveness," Business and Economic History 27 (Fall 1998): 222-236; Andrew J. Butrica, "Commercial Spaceports: Hitching Your Wagon to a VentureStar," *Space Times: Magazine of the American Astronautical Society* 37 (September/October 2000): 5-10.

57. Frank Sietzen, Jr. and Keith L. Cowing, *New Moon Rising: The Making of the Bush Space Vision* (Burlington, Ontario: Apogee Books, 2004); Craig Cornelius, "Science in the National Vision for Space Exploration: Objectives and Constituencies of the 'Discovery-Driven' Paradigm," *Space Policy* 21 (February 2005): 41-48; Wendell Mendell, "The Vision for Human Spaceflight," *Space Policy* 21 (February 2005): 7-10.

58. Patrick Healy and Cornelia Dean, "Clinton Says She Would Shield Science From Politics," *New York Times*, October 5, 2007.

SPACE EXPLORATION AND THE CULT OF CONSPIRACY

Americans, certainly, and perhaps all the cultures of the world, love the idea of conspiracy as an explanation of how and why many events have happened. Certainly this is the case in one of the counter narratives of spaceflight. These conspiracy theories play to the innermost human fears and hostilities that there is a well-organized, well-financed, and Machiavellian design being executed by some malevolent group, the dehumanized "them," which seek to rob "us" of something we hold dear. Usually the "something" being robbed is one of the constitutionally defined rights of all Americans: life, liberty, or property.

Conspiracy theories abound in American history. Oliver Stone's film, *J.F.K.*, while presenting a truly warped picture of recent American history, shows how receptive Americans are to believing that Kennedy was killed as a result of a massive conspiracy variously involving Cuban strongman Fidel Castro, American senior intelligence and law enforcement officers, high communist leaders in the Soviet Union, union organizers, organized crime, and perhaps even the Vice President, Lyndon B. Johnson. Stone's film only brought the assassination conspiracy to a broad American public. For years amateur and not-so-amateur researchers have been churning out books and articles about the Kennedy assassination conspiracy. It has been one of the really significant growth industries in American history during the last 40-some years.[59]

Conspiracy theories, of course, have been advanced to explain many other historical events in the United States. A favorite is the "backdoor to war" conspiracy thesis of U.S. entry into World War II. As stated, President Franklin D. Roosevelt had intelligence information about the Japanese attack on Pearl Harbor hours beforehand and with the help of other highly placed national leaders withheld that information from the Navy's Pacific Fleet so that it would be destroyed—all so he could get the American people behind a war with Germany.[60] Another conspiracy argues that there has been a grand intrigue in the 20th century "to control the foreign and domestic policies of the United States, subvert the Constitution, and establish a totalitarian society."[61]

---

59. David R. Wrone and DeLloyd J. Guth, *The Assassination of John F. Kennedy* (Westport, CT: Greenwood Press, 1980), listed more than 5,000 publications dealing with the subject. The number has grown substantially since that bibliography was published.

60. Charles A. Beard, *President Roosevelt and the Coming of the War, 1941: A Study in Appearances and Realities* (New Haven, CT: Yale University Press, 1948) makes the case for conspiracy. Countervailing positions are argued in Roberta Wolstetter, *Pearl Harbor: Warning and Decision* (Palo Alto, CA: Stanford University Press, 1962) and Gordon A. Prang, *"At Dawn We Slept":The Untold Story of Pearl Harbor* (New York, NY: McGraw-Hill, 1981). A superb discussion of the memory of the Pearl Harbor attack may be found in Emily Rosenberg, *A Date Which will Live: Pearl Harbor in American Memory* (Durham, NC: Duke University Press, 2003).

61. Chesly Manly, *The Twenty-Year Revolution: From Roosevelt to Eisenhower* (Chicago, IL: n.p., 1954), p. 179, as cited in Richard Hofstadter, *The Paranoid Style in American Politics and Other Essays* (New

What are the general attributes of these historical conspiracy theories writ large? A central point revolves around how to define a "conspiracy." At its most innocuous a conspiracy is simply the planning and execution of some activity by a group of people. All actions of any consequence require some planning with others and could be considered conspiracies in that sense. The dictionary definition of conspiracy, however, is "a joining secretly with others for an evil purpose," a connotation first acquired during the politically charged 1960s, and most planning efforts, therefore, do not qualify.[62] One could argue that conspiracies do indeed exist, even when using the dictionary definition. Even so, much rides on what defines an "evil purpose," for very often that is a matter of perspective. From the American perspective, whether or not Roosevelt was involved matters not, in the strictest sense of the term Pearl Harbor was attacked as a result of a conspiracy, for the Japanese high command struck an evil plot against the United States. Even so, from a Japanese perspective it was not so much a conspiracy as good strategic planning. The definition of a conspiracy, therefore, is subjective.

At least in the minds of conspiracy theorists, however, there is always a belief that there is or has been a vast and well-organized plot to carry out some sinister goal, often the very destruction of a way of life. At its extreme form the theorist might consider the conspiracy the vast and prime mover of history. Thus, Americans on the political right have interpreted many of the world's events in the 20th century as a "communist conspiracy" against which the "free world" had always to react.[63] As a result, opponents fighting a perceived conspiracy see themselves as the last bastion of what is good and just and true in the world. There is an especially powerful apocalyptic vision that motivates those who accept such conspiracy ideologies. These opponents have often had an almost messianic belief in the rightness of their cause and that the time remaining to salvage whatever is at stake is running out. At a fundamental level, conspiracy theories serve as a "particular narrative form of scapegoating that frames demonized enemies as part of a vast insidious plot against the common good, while it valorizes the scapegoater as a hero for sounding the alarm."[64]

Additionally, those who truly believe that a conspiracy has been afoot do not have any interest in talking over differences. They are at war with a malicious, sinister, powerful, ubiquitous personification of evil. That evil is responsible for most of the negative events that happen. It makes crises; starts

York, NY: Vintage books, 1965), p. 25; Kevin Phillips, *American Theocracy: The Perils and Politics of Radical Religion, Oil, and Borrowed Money in the 21st Century* (New York, NY: Viking, 2006).

62. *The New Lexicon Webster's Dictionary of the English Language* (New York, NY: Lexicon International-Publishers Guild Group, 1989), p. 208.

63. See Daniel Bell, ed., *The Radical Right* (New York, NY: Vintage Books, 1963).

64. Chip Berlet and Matthew N. Lyons, *Right-Wing Populism in America: Too Close for Comfort* (New York, NY: The Guilford Press, 2000), p. 9.

economic depressions, wars, and disasters; and enjoys the misery foisted upon the culture under attack. Advocates of conspiracy assign demonic omnipresence to whatever and whomsoever they have decided are a part of the conspiracy. They possess a special source of power which is used malevolently against others, especially those who have learned about the conspiracy and are seeking to combat it. Any suggestion from non-believers that a presumed conspiracy might be just as easily and accurately explained by some less diabolical method is met with a sharp rebuke that the non-believer is either a willing participant or a dupe being used by the conspirators.[65]

Almost from the point of the first spaceflight missions, a small group of Americans began to spin conspiracy theories. These range from fantastical theories of extraterrestrial visitation, abduction, and government complicity, to the development of secret technologies such as the Aurora ultra-secret spaceplane, to elaborate collusions between great powers to subvert human liberties. By far the most important of these were those that emerged to question the Moon landings undertaken by NASA during Project Apollo, and it is this conspiracy theory that I intend to discuss in this short essay. The Moon landing had, they argued, been faked in Hollywood by the federal government for purposes ranging—depending on the particular Apollo landing denier—from embezzlement of the public treasury to complex motivations involving international intrigue and murderous criminality. For example, Andrew Chaikin commented in his massive history of the Apollo Moon expeditions that at the time of the Apollo 8 circumlunar flight in December 1968 some people thought it was not real; instead it was "all a hoax perpetrated by the government." Bill Anders, an astronaut on the mission, thought live television would help convince skeptics since watching "three men floating inside a spaceship was as close to proof as they might get."[66] He could not have been more wrong.

Some of those skeptical of the Apollo flights made their cases based on naïve and poorly constructed knowledge, but imagery from space did not seem to help. For example, my paternal grandfather, Jeffrey Hilliard Launius, was a 75-year-old farmer from southern Illinois at the time of the first Moon landing in 1969. A Democrat since the Great Depression of the 1930s—because, as he said, Roosevelt gave him a job with the Work Projects Administration (WPA) when he could not feed his family and was on the verge of losing everything—his denial of the Moon landing was based essentially on lack of knowledge and

---

65. This conspiracy motif has been an important part of the "political correctness" debate currently raging in which everyone who does not accept at face value the arguments about minority oppression and the means of ending it, even though those committed to the goal do not themselves agree on the proper means, are charged with racism, chauvinism, or prejudice. See Aaron Wildovsky, *The Rise of Radical Egalitarianism* (Washington, DC: American University Press, 1991).

66. Chaikin, *A Man in the Moon*, p. 100.

naïveté. In his estimation such a technological feat was simply not possible. Caught up in the excitement of Apollo 11 in the summer of 1969, I could not understand my grandfather's denial of what appeared obvious to me. He did not assign any conspiratorial motives to the government, especially the Democrats; after all, it was a party he had trusted implicitly for more than 35 years. Even now I still cannot fully fathom his conflicting position of trust of the Democrats in government and unwillingness to believe what they said about the Moon landing. In his insular world change came grudgingly, however, and a Moon landing was certainly a major change. As a measure of his unwillingness to embrace change, my grandfather farmed his entire life with horses rather than adopting the tractor because in his estimation tractors were "a passing fad." Jeff Launius still did not believe that America had landed on the Moon at the time of his death in 1984.

President Bill Clinton recalled in his 2004 autobiography a similar story of a carpenter he worked with not long after the Apollo 11 landing. As he wrote about him in August 1969:

> Just a month before, Apollo 11 astronauts Buzz Aldrin and Neil Armstrong had left their colleague, Michael Collins, aboard spaceship Columbia and walked on the Moon, beating by five months President Kennedy's goal of putting a man on the Moon before the decade was out. The old carpenter asked me if I really believed it happened. I said sure, I saw it on television. He disagreed; he said that he didn't believe it for a minute, that "them television fellers" could make things look real that weren't.

Clinton thought him a crank at the time and since, a homespun skeptic. He then allowed that a healthy criticism of everything was not necessarily a bad idea.[67]

How widespread were the skeptics about the Moon landings in the 1960s? That is almost impossible to say. For example, the *New York Times* science reporter John Noble Wilford remarked in December 1969 that "A few stool-warmers in Chicago bars are on record as suggesting that the Apollo 11 moon walk last July was actually staged by Hollywood on a Nevada desert."[68] More important, the *Atlanta Constitution* led a story on June 15, 1970, with: "Many skeptics feel moon explorer Neil Armstrong took his 'giant leap for mankind' somewhere in Arizona." It based its conclusion that an unspecified "many" questioned the Apollo 11 and 12 landings, and presumably the April 1970 accident aboard Apollo 13, on an admittedly unscientific poll conducted by the Knight

---

67. Bill Clinton, *My Life* (New York, NY: Alfred A. Knopf, 2004), p. 244.

68. John Noble Wilford, "A Moon Landing? What Moon Landing?" the *New York Times*, December 18, 1969, p. 30.

The launch of Apollo 11 on July 16, 1969. On the essential character of this experience, Ray Bradbury wrote: "When the blast of a rocket launch slams you against the wall and all the rust is shaken off your body, you will hear the great shout of the universe and the joyful crying of people who have been changed by what they've seen." (NASA)

Newspapers of 1,721 U.S. citizens in "Miami, Philadelphia, Akron, Ohio, Detroit, Washington, Macon, Ga., and several rural communities in North and South Carolina." Those polled were asked, "Do you really, completely believe that the United States has actually landed men on the moon and returned them to earth again?" While numbers questioning the Moon landing in Detroit, Miami, and Akron averaged less than five percent, among African Americans in such places as Washington, DC, a whopping 54 percent "doubted the moon voyage had taken place." That perhaps said more about the disconnectedness of minority communities from the Apollo effort and the nation's overarching racism than anything else. As the story reported, "A woman in Macon said she knows she couldn't watch a telecast from the moon because her set wouldn't even pick up New York stations."[69]

Not everyone who denied the Moon landings at the time were so naïve in their assessments. Some spun conspiracy theories of complex structure and shocking intent. As Howard McCurdy opined, "To some, the thrill of space can't hold a candle to the thrill of conspiracy."[70] Over the years many conspiracy scenarios have been concocted, and it sometimes appears that the various theorists are even more cantankerous toward rival theories than they are toward NASA and the Apollo program. An early and persistent theme has been that as a cold war measure the U.S. could not afford to lose the race to the Moon, but when failure loomed NASA faked the landing to save face and national prestige. It used the massive funds dedicated to the effort to "pay off" those who might be persuaded to tell the truth; it also used threats and in some instances criminal actions to stop those who might blow the whistle.[71] One of the most common assertions has been that in the latter 1960s the U.S. government was in disarray because of the debacle of the Vietnam War, the racial crisis in the cities, and social upheaval. The Apollo program proved an ideal positive distraction from this strife, a convenient conspiracy designed to obscure other issues. One story published in 1970 stated this belief as expressed by an African American preacher: "It's all a deliberate effort to mask problems at home," Newsweek reported, "the people are unhappy—and this takes their minds off their problems."[72]

Other conspiracy motifs were more absurd. For example, William Brian asserted that perhaps Americans did go to the Moon, but they did so through the

---

69. "Many Doubt Man's Landing on Moon," Atlanta Constitution, June 15, 1970.

70. Howard A. McCurdy, "Moonstruck," Air & Space/Smithsonian, October/November 1998, p. 24.

71. All of these arguments, as well as variations on them, are offered in Bill Kaysing and Randy Reid, We Never Went to the Moon: America's Thirty Billion Dollar Swindle (N.P., 1974). This pamphlet has been reissued several times, notably in Pomeroy, OR: Health Research, 1976, and again in 2002.

72. Newsweek, July 20, 1970, quoted in Rogier van Bakel, "The Wrong Stuff," Wired 2 (September 1994): 108–113, 155.

means of some extraterrestrial technology. In his estimation NASA employed captured—or perhaps given—technology from beings beyond Earth to reach the Moon. This forced the Agency to create a cover story for more sinister purposes. "You can't let one bit of information out without blowing the whole thing," he noted. "They'd have to explain the propulsion technique that got them there, so they'd have to divulge their UFO research. And if they could tap this energy, that would imply the oil cartels are at risk, and the very structure of our world economy could collapse. They didn't want to run that risk." Likewise, others suggested that astronauts found evidence of alien civilization on the Moon, à la the 1968 feature film *2001: A Space Odyssey*, and had to fake imagery on the Moon to cover up that fact.[73]

The first conspiracy theorist to make a sustained case for denying that the U.S. landed on the Moon was Bill Kaysing, a journalist who had been employed for a few years in the public relations office at Rocketdyne, Inc., a NASA contractor, in the early 1960s. His 1974 pamphlet, *We Never Went to the Moon*, laid out many of the major arguments that have been followed by other conspiracy theorists since that time. His rationale for questioning the Apollo Moon landings offered poorly developed logic, sloppily analyzed data, and sophomorically argued assertions. Kaysing believed that the failure to land on the Moon all sprang from the fact that NASA lacked the technical expertise to accomplish the task, requiring the creation of a massive coverup to hide that fact. He cited as evidence optical anomalies in some imagery from the Apollo program, questioned the physical features of certain objects in the photographs (such as a lack of a star field in the background of lunar surface imagery and a presumed waving of the U.S. flag in an airless environment), and challenged the possibility of NASA astronauts surviving a trip to the Moon because of radiation exposure.[74]

Throughout the latter third of the 20th century and into the 21st, with confidence in the U.S. government by the American public declining—because of Vietnam, Watergate, and other scandals and malfeasance—it became somewhat easier for people to believe the worst about such a cover-up. For example, responding to a public opinion survey in 1964, 76 percent of the Americans polled expressed confidence in the ability of their national government "to do what is right" most or all of the time. This was an all-time high in the history of polling, and this goodwill helped lay the foundation for all manner of large initiatives during the 1960s, including all types of reforms. This consensus collapsed in the post-Vietnam and post-Watergate era of the 1970s, to a low of

---

73. van Bakel, "The Wrong Stuff," *Wired*, p. 112.

74. Kaysing and Reid, *We Never Went to the Moon*.

less than 25 percent of Americans believing that the government would seek to do right all or even a majority of the time by the early 1990s.[75]

Additionally, as time passed and more people were born and grew to maturity since the last of the Moon missions had been completed in 1972, youngsters became increasingly skeptical since they had no firsthand recollection of Apollo. Evidence of that issue was found in a 2004 poll about attitudes toward spaceflight among Americans. While polls had consistently shown that only about six percent of the public as a whole questioned the Moon landings, and a whopping 89 percent firmly believed in their reality, among Americans between 18 and 24 years old "27% expressed doubts that NASA went to the Moon," according to pollster Mary Lynne Dittmar in a 2004 study. Doubt is different from denial, but it was a trend that seemed to be growing over time.[76]

Major media sources, especially, fueled doubts. For example, folklorist Linda Degh asserted that the 1978 fictional feature film *Capricorn One*, in which NASA supposedly faked a landing on Mars, may have fostered greater acceptance of the denials of the Moon landings. No question, the February 2001 airing of the Fox special, *Conspiracy Theory: Did We Land on the Moon?*, changed the nature of the debate. In this instance a major network presented a conspiracy scenario without any serious rebuttal that might have been offered.[77] As *USA Today* reported in the aftermath of the "news special":

> According to Fox and its respectfully interviewed "experts"— a constellation of ludicrously marginal and utterly uncreden- tialed "investigative journalists"—the United States grew so eager to defeat the Soviets in the intensely competitive 1960s space race that it faked all six Apollo missions that purport- edly landed on the moon. Instead of exploring the lunar sur- face, the American astronauts only tromped around a crude movie set that was created by the plotters in the legendary Area 51 of the Nevada desert.[78]

75. Paul R. Abramson, *Political Attitudes in America* (San Francisco, CA: W. H. Freeman, 1983), p. 12. See also Seymour Martin Lipset and William Schneider, *The Confidence Gap* (New York, NY: Free Press, 1983).

76. Mary Lynne Dittmar, "Building Constituencies for Project Constellation: Updates to The Market Study of the Space Exploration Program," presentation at Building and Maintaining the Constituency for Long-Term Space Exploration workshop, George Mason University, Fairfax, VA, July 31-August 3, 2006; The Gallop Poll, "Did Men Really Land on the Moon?" February 15, 2001, available online at *http://www.galluppoll.com/content/?ci=1993&pg=1* (accessed June 26, 2007).

77. For a discussion of the claims made in this "documentary," as well as rebuttal to it, see Phil Plait, "Fox TV and the Apollo Moon Hoax," February 13, 2001, available online at *http://www.badastronomy.com* (accessed October 14, 2002).

78. "Faking a Hoax," *USA Today*, April 9, 2001.

While the program claimed to "Let the viewer decide" about the validity of the claims for denial of the Moon landings, it made no attempt whatsoever to offer point and counterpoint, thereby giving the viewers a seriously biased view of the issue and skewed evidence in favor of a hoax.

The Fox television show exposed the arguments of the Moon landing deniers to a much broader public than ever before. As Linda Degh noted, "The mass media catapult these half-truths into a kind of twilight zone where people can make their guesses sound as truths. Mass media have a terrible impact on people who lack guidance."[79] Without a proper rebuttal available from NASA—the Agency had taken an official position before of not responding to what it considered absurd claims—many young people publicly began to question the Apollo landings. Several astronauts stepped forward to affirm the legitimacy of the program, but others thought the charges too silly to warrant response. Many debated the issues in the emerging world of the Internet. Indeed, the Internet became a haven for conspiracy theorists of all stripes, and with the barrier for publication online so low anyone could put up any page they wished with any assertions they wished to make. But it also became a haven for counters to the conspiracy theorists and a healthy debate has resulted.[80]

At the same time, the twin features of modern society—a youth movement and post-modernism—helped to raise questions about the Moon landings. More than half the world's population had been born since the last of the Moon landings had taken place in December 1972. Consequently, they had not lived through the excitement of the experience. This raises the specter of how individuals view time and history. Mostly without even realizing it, individuals tend to divide time into three general, inconsistent, and individualistic spheres or cones of memory. The first is a sphere of personal experience. Events that individuals participated in personally or that had salience to their individual lives are the first and most immediate sphere. These differ from person to person, and include not only activities that the individual experienced firsthand but events of great importance that took place in their memory. For instance, there are colossal events that mark the time of our lives, and they hold great resonance for those participating in them. Virtually all Americans know where they were and what they were doing when they learned of the 9/11 attacks in New York and Washington. The same is true for other dramatic incidents in individual lives such as the Moon landings for those who remember them. It is this memory of our individual and immediate experiences that govern most people's perspective on the past. Roy Rosenzweig and David Thelen in their study of popular uses of history in American life noted that far from Americans

---

79. van Bakel, "The Wrong Stuff," *Wired*, p. 113.

80. A search on the term, "Moon hoax," will yield no fewer than 5,000 sites containing information of one type or another relating to this subject.

being disengaged from history, as has been routinely thought because of their detachment from national themes, most people have supplanted interest in these broader themes to the history of family and locale. Indeed, Rosenzweig and Thelen insist that Americans "pursue the past actively and make it part of everyday life."[81] They found that no more than 24 percent of their sample answered that the history of the United States was the past they felt was "most important" to them, as opposed to the 50-60 percent who identified a more intimate past as central to their lives.[82]

Less immediate but still resonating with Americans is a sphere of history that is not intimate to the individual but related by members of the family, by close friends, and by mentors. While the person may have no individual sense of history about World War II, for instance, they have heard stories about it and its effects on families and loved ones. It has a reverberation of meaning because of this connection. There are dark areas in this sphere of historical understanding that may be further illuminated through public presentations of the past, in whatever those forms of presentation might take, but they will never enjoy the salience reserved for personal experience in most people's minds.

The third sphere encompassing all humans is the past that has no special connection through loved ones or personal experience. In that context events, epochs, themes, and the like discussed throughout the broad expanse of history have essentially an equal importance. The Crusades, the Ming Dynasty, the English/French/American/Russian/or other revolutions all essentially stand at the same level for most of those who have no intimate connection to them. Difficulties in creating resonance with those events of the past abound, and always perspectives are obscure as this past is digested. It also has considerably more dark spaces than more immediate past events. An important challenge for all historians is how to breech that truly lost and forgotten past and offer its meaning to most people. This is done through many processes, especially rituals, public representations, reenactments, museums and historic sites, and a range of other possibilities for constructing and reinforcing meaning. There are numerous examples of this basic fact across a broad spectrum of American life, as master narratives of American history are reinforced rather than reinterpreted.[83]

For the younger members of society, the recollection of Apollo is distant to begin with and receding into the background quickly as time progresses.

---

81. Roy Rosenzweig and David Thelen, *The Presence of the Past: Popular Uses of History in American Life* (New York, NY: Columbia University Press, 1998), pp. 11-13, quote from p. 18.

82. Ibid., p. 237.

83. Jane Adams, "Melting Pot, Stew Pot, or Salad," available online at *http://mccoy.lib.siu.edu/~jadams/introduction_text.html* (accessed October 28, 2005). Sociologist Robert Bellah calls these "communities of memory." See Robert N. Bellah, et al., *Habits of the Heart: Individualism and Commitment in American Life* (New York, NY: Harper and Row, 1985).

Commemoration and ritual help to preserve these events for society as a whole, but if they are not taking place is the case for Apollo, then events dim.

Indeed, post–modernism suggests that reality is more a suggestion of meaning rather than an absolute. It blurs the line between fact and fiction, between realism and poetry, between the unrecoverable past and our memory of it.[84] This raising of the inexact character of historical "truth," as well as its relationship to myth and memory and the reality of the dim and unrecoverable past, has foreshadowed deep fissures in the landscape of identity and what it means to be American. Truth, it seems, has differed from time to time and place to place with reckless abandon and enormous variety. Choice between them is present everywhere both in the past and the present; my truth dissolves into your myth and your truth into my myth almost as soon as it is articulated. We see this reinforced everywhere about us today, and mostly we shake our heads and misunderstand the versions of truth espoused by various groups about themselves and about those excluded from their fellowship. They have given and continue to give meaning and value to individual human lives and to create a focal point for explaining the sufferings and triumphs of the group.

At some level there is no absolute; instead everything is constructed. If so, what might be the case of the Moon landings? Might this be, in essence, an issue of agreeing that something was true but could also be agreed that it never happened. If enough doubt could be cast on some particular narrative might it be overcome and obliterated? This has happened in history repeatedly, as versions of the past have replaced earlier versions that seemed so true. For more than a half–century, for example, the Frontier Thesis as enunciated by Frederick Jackson Turner reigned supreme as a critical explanation offered for the manner in which the U.S. character emerged. It was dismantled and destroyed and all but forgotten in the last quarter of the 20th century.[85]

The denials of the Moon landings excite the response of crank and crackpot from most who hear them. Indeed, those conspiracy ideas deserve disdain. But so to, do many other conspiracy theories that are now major elements of the

---

84. See the fascinating discussion of myth and history in Hayden White, *Metahistory: The Historical Imagination in Nineteenth-Century Europe* (Baltimore, MD: Johns Hopkins University Press, 1973); and Roland Barthes, "The Discourse of History," trans. Stephen Bann, *Comparative Criticism: A Yearbook* 3 (1981): 3–20; Dominick LaCapra, *Rethinking Intellectual History* (Ithaca, NY: Cornell University Press, 1983); Brook Thomas, *The New Historicism: And Other Old-Fashioned Topics* (Princeton, NJ: Princeton University Press, 1991).

85. Frederick Jackson Turner, "The Significance of the Frontier in American History," *The Frontier in American History* (New York, NY: Holt, Rinehart, and Winston, 1920), pp. 1–38; Richard Slotkin, *Gunfighter Nation: The Myth of the Frontier in Twentieth-Century America* (New York, NY: Atheneum, 1992); John Mack Faragher, *Rereading Frederick Jackson Turner: The Significance of the Frontier in American History, and Other Essays* (New York, NY: Henry Holt, 1994); Allan G. Bogue, *Frederick Jackson Turner: Strange Roads Going Down* (Norman, OK: University of Oklahoma Press, 1998); Ray Allen Billington, *America's Frontier Heritage* (Albuquerque, NM: University of New Mexico Press, 1974).

memory of the nation. For example, how many Americans believe that John F. Kennedy was assassinated by means of a massive conspiracy that involved the national security establishment? More than 45 years of a persistent churning over the data, near data, and wishful thinking has forced massive fissures in the conclusions of the Warren Commission. Might this happen in the future in relation to the Moon landings?

Conclusion

Finally, who has the right—not to mention the power—to interpret the past? It seems obvious that the fierceness of the discourse over the possible narratives of the past has arisen from the desire to secure a national identity of one nation, one people, coupled with a concern that the bulwarks of appropriate conceptions may be crumbling. Viewing history as largely a lesson in civics and a means of instilling in the nation's citizenry a sense of awe and reverence for the nation state and its system of governance ensures that this debate over narratives will be vicious and longstanding. The dominant master narrative of spaceflight fits beautifully into this approach to seeing the past. It is one of an initial shock to the system, surprise, and ultimately recovery with success after success following across a broad spectrum of activities. It offers general comfort to the American public as a whole and an exceptionalistic, nationalistic, and triumphant model for understanding the nation's past.[86] Small wonder that this story of spaceflight emerged as the narrative so dominant from the earliest days of the space program. It offered a subtle, usable past for the nation as a whole.

But that master narrative of both spaceflight and the larger American history began to break down with the rise of the new social history of the 1960s.[87] By the 1980s the consensus, exceptionalistic perspective on the American past had crumbled throughout academia, but it had not done so among the broader public and in the cultural institutions that sought to speak to the public.[88] In this setting it would seem that the alternative spaceflight

---

86. On American exceptionalism see Seymour Martin Lipset, *American Exceptionalism: A Double-Edged Sword* (New York, NY: W. W. Norton & Company, 1997); Charles Lockhart, *The Roots of American Exceptionalism: Institutions, Culture and Policies* (New York, NY: Palgrave Macmillan, 2003); Deborah L. Madsen, *American Exceptionalism* (Oxford, MS: State University of Mississippi Press, 1998); David W. Noble, *Death of a Nation: American Culture and the End of Exceptionalism* (Minneapolis, MN: University of Minnesota Press, 2002).

87. Peter Charles Hoffer, *Past Imperfect: Facts, Fictions, Fraud—American History from Bancroft and Parkman to Ambrose, Bellesiles, Ellis, and Goodwin* (New York, NY: Public Affairs, 2004), p. 63.

88. Frances Fitzgerald, *America Revised* (Boston, MA: Little, Brown, 1979), pp. 53-58; Michael Kammen, *In the Past Lane: Historical Perspectives on American Culture* (New York, NY: Oxford University Press, 1997), pp. 64-68; Neil Jumonville, *Henry Steele Commager: Midcentury Liberalism and the History of the Present* (Chapel Hill, NC: University of North Carolina Press, 1999), pp. 232-235.

narratives could emerge to challenge the master narrative, creating for their individual and individualistic followings a uniquely boutique but satisfactory interpretation of space exploration's history.

In the context of spaceflight, the duels between these four narratives have represented a battle for control of the national memory concerning this one area of the "lifeworld" of Americans. Would it be one that is unified—one people, one nation—or one that was fragmented and personal? This is an important issue and fully worthy of consideration by all in the marketplace of ideas. By taking action to fashion and champion alternative narratives, individuals reasserted a fundamental direction over meaning whether for good or ill. Political scientist Jürgen Habermas has suggested that when the "instrumental rationality" of the state intrudes too precipitously into the lifeworld of its citizenry, they rise up in some form to correct its course or to cast it off altogether. The lifeworld is evident in the ways in which language creates the contexts of interpretations of everyday circumstances, decisions, and actions. He argues that the lifeworld is "represented by a culturally transmitted and linguistically organized stock of interpretive patterns."[89] For a not inconsequential proportion of Americans the interpretation of space exploration that dominated the discourse has intruded into their lifeworld, as their alternative narratives certainly suggest. Accordingly, they have taken direct action to alter this perspective. Over time, their alternative narratives have come to challenge the master perspective invoked routinely.

This leads back to the question posed above, who has the authority to decide what the history says? An old baseball joke is apropos here. Three umpires were discussing how they call balls and strikes behind the plate. The first said, "I call them as they are," a pre-modern, absolutist position. The second said, "I call them as I see them," a position reflecting rationality and modernity. The third opined in a fit of post-modern existential angst, "They ain't nothin' til I call them." It seems that this last perspective is the critical element in considering these various narratives about the history of the Space Age. Perhaps the reality of what happened does not matter all that much; the only thing that is truly important is the decision about its meaning. That may well be an intensely personal decision predicated on many idiosyncrasies and perspectives. When will historians begin to explore the process whereby this has taken place and seek to document and understand its evolution?

---

89. Jürgen Habermas, *The Theory of Communicative Action, Volume 2: Lifeworld and System, A Critique of Functionalist Reason* (Boston, MA: Beacon Press, 1987), p. 124.

PART III.

REFLECTIONS ON THE SPACE AGE

CHAPTER 18

A MELANCHOLIC SPACE AGE ANNIVERSARY

Walter A. McDougall

My sincere thanks to Steven Dick, Roger Launius, and the entire space history and space policy communities for inviting an old dilettante like myself to this event. Some of you good people I've not seen since we commemorated the 40th anniversary of Sputnik, and some of you doubtless I shall not have occasion to meet again. That alone makes this a somewhat melancholy affair for me. But I also have a sense that the 50th anniversary of the birth of the Space Age is draped with a certain melancholy. Do you sense a mood of disappointment, frustration, impatience over the failure of the human race to achieve much more than the minimum extrapolations made back in the 1950s, and considerably less than the buoyant expectations expressed as late as the 1970s? After all, one modest prediction went like this: "There are few today who do not look forward with feelings of confidence that spaceflight will some day be accomplished. All that we require is to make rocket motors somewhat larger than those already in existence . . . the pooling of skills already available, and a good deal of money . . . . We may reasonably suppose that a satellite vehicle is entirely practicable now and that travel to the moon is attainable in the next fifty years."[1] That was Dr. Hugh Dryden in 1953, on the occasion of the 50th anniversary of the Wright brothers' flight. (Indeed, if all of us interviewed by the media this month have accomplished anything I think we have at last disabused journalists of the notion that the Eisenhower administration was "surprised" by the first satellite launch.) But what that means is that all the satellites, space probes, and human missions launched over 50 years amount pretty much to what Dryden took for granted would happen. Moreover, the fact that the Moon landing was achieved just 16 years after he wrote this only compounds the disappointment that it proved to be a dead end.

That disappointment is also evident, I think, in the false expectation I expressed this past spring in an essay written for the Foreign Policy Research Institute. I began it like this:

---

1. Hugh L. Dryden, "The Next Fifty Years," *Aero Digest* (July 1953).

It has gone down in history as 'the other world series': a championship match even more shocking than the Milwaukee Braves' upset victory over the New York Yankees in baseball's 1957 Fall Classic. That shot literally 'heard 'round the world' was Sputnik I, the first artificial Earth satellite that gave birth to the Space Age, and its 50th anniversary this October 4th is sure to inspire worldwide attention. By contrast, another anniversary of equal importance was all but ignored this past March. The birth certificate of that *other* age born 50 years ago was the Treaty of Rome which founded the European Community. Its charter members numbered just six and pledged only to coordinate some economic policies. But 50 years later Europe is a Union, not just a Community, counts 27 members, and has so deepened and broadened its purview that Europe today has become a veritable state of mind.[2]

In retrospect it has indeed been European integration—a boring, bureaucratic enterprise for the most part—that worked a metamorphosis across a whole continent over 50 years, whereas any global consciousness or Spaceship Earth mentality inspired by astronautics has worked no metamorphosis in national or international affairs. So perhaps it is fitting that the Sputnik anniversary passed without the great global eclat I predicted. For if Space Age technology had enabled a great portion of the human race to imagine itself a family sharing a fragile planet and cosmic destiny, then one might have expected a global celebration on the scale of that staged for Y2K. Instead, we got World Space Week sponsored by the United Nations Office for Outer Space Affairs. But the U.N. does Space Week every year between October 4 and October 10, the day the Outer Space Treaty was signed in 1967. And since the U.N.'s special attraction this year was Valentina Tereshkova, the first female cosmonaut, it reduced our species' first escape from its planet to a human interest story.

In the classroom October 4, I asked my 120 students if they knew the significance of the date. A few senior-citizen auditors and exactly one undergraduate knew the answer. My survey of Web sites was also deflating. *SearchEngineLand.com* reported that Google temporarily altered its logo in honor of Sputnik (and perhaps to hype its Lunar X Prize of $30 million to a private

---

2.   Walter A. McDougall, "Will Europe Survive the 21st Century? A Meditation on the Fiftieth Anniversary of the European Community," 2 parts, Part I: "The Other Age Born in 1957," *Foreign Policy Research Institute E-Note*, *http://www.fpri.org* (August 3, 2007).

inventor of a Moonship).[3] But then, Google also alters its logo in honor of St. Patrick's Day and Halloween. Other Internet portals treated the anniversary, if at all, like any other feature story. Nor did Web surfers display much interest outside of techie and trekkie blogs. *InformationWeek.com* invited discussion of its brief story on Sputnik and received exactly zero posts. The anniversary page on *Makezine.com* received just four posts, one of which was this forlorn message: "I was happy to see a Sputnik post on this historic day. Thanks." Another site reported the European Space Agency's plan to launch 50 miniature "nanosats" in honor of the anniversary, but complained, "the event has not been widely covered. I found only very short pieces of information, such as a press release from Arianespace."

The *New York Times* essay on the anniversary was elegant, insightful, and graceful because John Noble Wilford wrote it.[4] But his tone was nostalgic, and he closed with decidedly downbeat judgments from Gerald Griffin, John Logsdon, and Alex Roland, plus Neil Armstrong's lament over "external factors or forces which we can't control." Indeed, if the commentary of space experts has had any unified theme it is that politics and economics—both foreign and domestic—have always dictated the scale and trajectory of space programs, rather than a revolutionary technology transforming politics and economics. In short, there has been no paradigm shift but instead international behavior as usual. To be sure, one could point to the Outer Space Treaty, international conventions on geosynchronous satellites, telecommunications, remote sensing, scientific cooperation, and so forth. But those achievements are simply comparable to what the otherwise rival nation states of the 19th and 20th centuries did when they established regimes to govern telegraphy, undersea cables, postal service, maritime law, standard time zones, air travel, radio, and rules for global commons such as the seabed and Antarctica.

Another noteworthy tribute (noted by John Krige as well) ran in the *USA Today* science supplement on September 25. After making the conventional point that turning civilian spaceflight into a race undercut its appeal after Apollo, the author quoted Roger Launius to the effect that support for human spaceflight has always been "a mile wide and an inch deep."[5] That apt remark reminded me of the chapter in *Critical Issues in the History of Spaceflight* in which Launius listed five rationales for space technology (noted also by Asif Siddiqi): 1) human destiny and perhaps the survival of our species; 2) geopolitics and

3. "Google Logo Celebrates Sputnik" (accessed October 16, 2007), *http://searchengineland. com/071004-111609php*; "50 'Nanosats' for Sputnik's Fiftieth Anniversary" (accessed October 16, 2007), *http://www.primidi.com/2004/10/13.html*.

4. John Noble Wilford, "With Fear and Wonder in Its Wake, Sputnik Lifted Us Into the Future," the *New York Times* (September 25, 2007).

5. Traci Watson, "Sputnik's Anniversary Raises Questions About Future of Space Exploration," *USA Today* (September 25, 2007).

national prestige; 3) military defense; 4) applications and economics; and 5) science and discovery. (Another whimsical way of listing those rationales is to say human beings do five things in space: work, play, fight, boast, and worship.) It seems in retrospect that what happened between 1955, when the IGY satellite program was announced, and 1961, when Yuri Gagarin orbited, was the elevation of prestige to an inordinate, artificial primacy in that mix of rationales. That spawned a crash program that space enthusiasts believed was, or should be, the norm when in fact it was a grotesque aberration made even worse by the 1970s decision to throw the baby (Apollo/Saturn hardware) out with the bathwater.

Where we stand today with respect to global vs. national identities and rationales for spaceflight can be deduced by recalling two wise sayings from the otherwise not-always-wise Robert S. McNamara. First, he said space is not a mission or a cause; it is just a place. Second, he said the budget *is* the strategy. So let us look at humanity's budget. Let us indeed "follow the money." According to The Space Foundation's latest estimates the world's allocations for activities in the place called outer space totaled $74.5 billion in 2006.[6] By coincidence, that is almost identical to the supplemental appropriations the White House requests every year for Iraq. (Hence, space advocates need no longer rely on the quip that U.S. consumers spend more on tobacco or cosmetic surgery than the space program because they need only observe that the U.S. government spends more existing tax revenue on one dubious exercise in overseas state-building than the whole world does on space exploration.)

Equally significant is the fact that just under $60 billion, or about 80 percent of global investment in space, is America's share, so ipso facto the priorities of the human race are really the priorities of one nation state. I understand where Neil DeGrasse Tyson and Jim Garvin are coming from when they say that America has been standing still, that China, Japan, and India may spark the next space race, and that a manned mission to Mars will likely plant "a whole sheaf of flags" in the ruddy dust. But apart from such high-profile human endeavors as the ISS or planetary exploration, space technology remains overwhelmingly a national activity overwhelmingly dominated by the United States.

ESA contributes $3.5 billion or just below 5 percent, and all other national programs (led by Japan and China) about $11.4 billion or 15 percent. The motives of ESA derive largely from science and applications. The motives of national programs such as those of Japan, China, India, and France run mostly to defense, economics, and prestige. Needless to say, no one spends a euro or a yen on "human destiny and the survival of the species."

---

6. Data on space spending is Space Foundation, Colorado Springs, CO, "Government Budgets—The Space Report 2007 Update (October 11, 2007), *http://www.spacefoundation.org/news/story.php?id=419*.

The breakdown of American spending, precisely because of its scale, is even more telling. The biggest chunk—$22.5 billion—goes to the Pentagon, with another $20.5 billion going to black programs such as those of the National Reconnaissance Office and Geo-spatial Intelligence Agency. Thus, about $43 billion, or 58 percent of humanity's space budget, is spent on the defense of the U.S. and its allies. Perhaps that is necessary. It is a fundamental tenet of the national strategy that the United States maintain hegemony in the aerospace theater, and most other nations would much rather have America police that global commons than to see it contested or dominated by some other nation. But in the context of rationales and priorities, those budget numbers are the most telling evidence that defense outweighs all other spaceflight put together, several times over. By contrast, NASA, which is responsible for the human spaceflight program, science and exploration, satellite applications, new launch technologies, test-bed technologies, and even the "human destiny and survival" rationale if we count astrobiology and asteroid research, receives $16.6 billion. That amounts to 28 percent of U.S. space spending and 22 percent of global space spending.

To put it another way, if we add NASA's budget to that of the ESA and estimate that a third of the various national budgets are devoted to civilian pursuits, we arrive at a sum of about $24 billion or 32 percent of the Space Foundation's global figure. That means 68 percent—more than two-thirds—of planet Earth's space effort serves national defense and prestige. And that means the answer to today's question—"Has the Space Age fostered a new global identity?—is "No."

Has the Space Age at least fostered—especially among young people—a sense of awe, wonder, curiosity, and impatience to know, an urge to explore and a rekindled faith in progress, the future, and human nature, or perhaps even a postmodern, gnostic religious vision conflating transhuman evolution, biological or post-biological immortality, space colonization, and contact with extraterrestials? Those have been stock themes of science fiction authors like Isaac Asimov, Ray Bradbury, and Arthur C. Clarke, none of whom could be considered a crackpot.[7] Indeed, it was Captain Jacques Cousteau, not exactly a cult leader, who took the occasion of NASA's 1976 conference on "Why Man Explores" to echo Konstantin Tsiolkovskii's conviction that, in conquering gravity, humanity would conquer death. Perhaps the Space Age will alter the consciousness of a critical mass of people. Perhaps, as William Sims Bainbridge eloquently contends, such a quasi-religious consciousness may give rise to a new social movement transforming the scale and priorities of the human presence in space.

---

7. For the late Sir Arthur C. Clarke's wise reflections, see *Spectrum*, "Remembering Sputnik: Sir Arthur C. Clarke," *http://www.spectrum.ieee.org/print/5584*.

Perhaps, but not yet. Twice this year I myself was thrilled to experience anew the awe and wonder so many felt at the dawn of the Space Age. The first experience was a stroll on the surface of Mars! I luckily visited NASA headquarters on May 17, the very day Dr. Alfred McEwen of the University of Arizona revealed "Mars As You've Never Seen Before," courtesy of the Mars Orbiter and Phoenix rovers Spirit and Opportunity. The second experience a few weeks later occurred while I was on a VIP tour of JPL courtesy of Blaine Baggett, who is producing a documentary for the 50th anniversary of Explorer 1, America's first satellite. *Pace* Howard McCurdy (whose brilliant analysis of robots consigns them to the dying industrial age of human culture), I marveled at the magical robotic spacecraft designed and assembled in the hills above Pasadena. It is they who have made what Carl Sagan called the Golden Age of planetary exploration; and it is they who bear witness to what Samuel Florman called "the existential pleasures of engineering." Yet I also watched troop after troop of children on school field trips to JPL and could not help but wonder whether it made any impression on them. Can youth today feel the tingle that Homer Hickam felt the night Sputnik passed over West Virginia? Or have today's kids been so jaded by the far more spectacular virtual reality of Nintendo and Dreamworks that NASA cannot compete? Or will the excitement of virtual reality instead render brilliant young people impatient to accelerate the human thrust into space?

On young people—and the future—I have no authority to speak. But as an historian with some authority to pronounce on the past 50 years, I would suggest that the trajectory spaceflight has taken reflects the fact that the nation that drove the enterprise, the United States, has been perversely burdened by its responsibilities as defender of most of the world and is perversely ill-suited to what spaceflight requires. Not as ill-suited as that fraudulent technocracy, the Soviet Union, but ill-suited nonetheless. Given the costs, lead-times, and distances involved, the pioneering of space requires a coherent, sustainable, long-term approach, predictably financed and supported by a patient people willing to sacrifice and delay gratification even over a generation or more. Americans do not fit that description. Likewise (and I defer here to political scientists such as John Logsdon) the U.S. government does not exactly fit the description of a streamlined technocracy, given its checks and balances, contesting parties, rival bureaucracies, frequent elections and personnel turnovers, mixed public and private sectors, gigantic distractions both foreign and domestic, and reliance in all cases on a meandering, manipulable public opinion. Indeed, given those handicaps and the mistakes and false starts bound to occur in a venture of such scope and novelty, perhaps Sir Arthur C. Clarke was correct when he recently said, all disappointment aside, that a great deal has been accomplished in the first 50 years of the Space Age. Not least, I would stress, the cosmic advances in space science which, so far at least, have been strangely ignored in our proceedings.

Will the United States continue to dominate humanity's agenda in space? Or will we pass the baton to others, such as several countries in Asia? Or will some new, genuinely cheap and safe launch technology emerge to permit rapid expansion of the human footprint in space without any government having to lead? When and if that occurs, then private and corporate activity may indeed become an independent variable capable of transforming geopolitics and geoeconomics. When and if that occurs, a new generation of the sort McCurdy awaits may indeed hearken to Siddiqi's plea that we cease fearing our own imaginations. When and if that occurs, a tired old baby-boomer such as I will eagerly take Charles Murray's advice "to get a grand mission . . . give it to a new generation, and get the hell out of the way."[8]

---

8. Charles Murray, and Catherine Bly Cox, *Apollo: Race to the Moon* (New York, NY: Simon & Schuster, 1989).

CHAPTER 19

HAS SPACE DEVELOPMENT
MADE A DIFFERENCE?

John M. Logsdon

In his paper in this volume, J. R. McNeill writes that "It is in fact too soon to tell what the real significance of the Space Age may be. At the moment, space exploration, space flight, space research, all seem at most secondary next to the dominant trends of contemporary history. . . . The big things would probably be much the same, for better or for worse." He adds "space programs changed the history of our times, but not (yet) in any fundamental ways." Walter McDougall in his paper adds that he senses "that the fiftieth anniversary of the birth of the Space Age is draped with a certain melancholy. Do you sense a mood of disappointment, frustration, impatience over the failure of the human race to achieve much more than the minimum extrapolations made back in the 1950s, and considerably less than the buoyant expectations expressed as late as the 1970s?"

I beg to disagree, at least in part. The assignment for this paper was to discuss this question: "Has the Space Age fostered a new global identity, or has it reinforced distinct national identities? How does space history connect with national histories and with the histories of transnational or global phenomena . . . ?" It is an interesting mental exercise to imagine what today's world would be like, at least in the urbanized Northern hemisphere, if all space systems were shut down for 24 hours. I believe that we would quickly realize that those systems have become deeply integrated into the infrastructure of the modern world, and that neither the modern nation state nor the global economy could operate effectively without them. If the overall history of most of the past 50 years has not been fundamentally affected by the development of space capabilities, it is my view that the history being made today and in the recent past is in meaningful ways a product of how nation states and the private sector have incorporated the possibilities made available through space technology into their everyday operations.[1] In this sense, the ability to operate in outer space is part of history, not an independent variable shaping it.

---

1. Most of the papers in Steven J. Dick and Roger D. Launius, eds., *Societal Impact of Spaceflight* NASA SP2007-4801 (Washington, DC: Government Printing Office, 2007) provide evidence and analysis in support of this assertion.

The Impacts of Space Development

That reality may be part of the problem in identifying the impact of space development during its first half-century. As various capabilities have become operational, they have been subsumed into the larger pattern of human activity and not usually thought of separately as "space." McNeill suggests that "Some things would have been a bit different without spy satellites, communications satellites, weather satellites, earth-observation satellites, and so forth," but, in his view, not dramatically different. He asks whether "the current surge of globalization has derived some of its momentum from an enhanced awareness that we are all in the same boat, all stuck on the same small blue dot spinning through the darkness? Or could it owe something to instantaneous communications via satellites?" His view is that "the best answer is: yes, but not much. If no one had ever seen photos of the earth from space, and if information from India and Indonesia still arrived by telegraph and took a day or two to reach other continents instead of a second or two, would globalization be substantially different?"

For at least the latter of his two questions, my answer would be "yes." It is really difficult to imagine today's world absent instantaneous information flow, and space systems are a crucial part of the global information transmission network that makes such flow possible. Whether the view of Earth from cosmic distances—Earthrise over the barren lunar surface or the "pale blue dot" most recently glimpsed by the Cassini spacecraft as it orbits Saturn—has created a global consciousness is more debatable. Certainly, the Earthrise image became the icon of the environmental movement in the 1970s and references to "Spaceship Earth" still appear in admonitions of the Green movement. But, as McDougall comments, "any global consciousness or Spaceship-Earth mentality inspired by astronautics has worked no metamorphosis in national or international affairs."

Somewhat the same can be said for the other space capabilities that McNeill cites. For nations with global or regional security interests—during the Cold War, the United States and the Soviet Union, and today an additional small number of other nation states—the ability to obtain near-real-time information on potential security threats is a stabilizing element in international security affairs. But space-derived intelligence information is merged with intelligence from other sources, and it is not possible to measure its independent contribution to avoiding or ameliorating (or abetting) conflict. Information regarding the variables determining short- and longer-term weather patterns obtained from meteorological satellites is integrated with other information; there are many projections of the billions of dollars and hundreds of lives not lost due to better weather forecasts.[2]

---

2. See, for example, the discussion in Henry R. Hertzfeld and Ray A. Williamson, "The Social and Economic Impact of Earth Observing Satellites" in Launius and Dick, *The Societal Impact of Spaceflight,* pp. 237-263.

McNeill does not discuss the impact of satellites delivering positioning, navigation, and timing services. But such satellites, most notably to date the U.S. GPS system, have become the basis for a global utility with multiple applications from guiding precision weapons to their targets to providing the timing information that makes the Internet possible. Again, one does not often think of the space-based source of these capabilities; what matters is the application, not the means that enables it.

Though not the focus of this and the other papers in this volume, it would be remiss to avoid discussing the impact of space capabilities on warfighting in an assessment of the importance of the last 50 years of space development. So far, only the United States has made its approach to power projection and fighting wars strongly dependent on the use of space systems. It is well beyond the scope of this paper to discuss whether that commitment to space as a military tool was a wise one, endowing the United States with decisive military advantages. But certainly space capabilities are central to what has been described in the United States as a "revolution in military affairs."[3]

It is instructive to observe that countries pursuing rapid social and economic development—China and India are probably the best examples—are investing significant amounts of their scarce financial and human resources in space development. They seem convinced that space capabilities can have fundamental impacts on their future history.

I conclude, then, that by its contributions to the various ways in which everything from international conflicts to day-to-day life unfolds, space development has indeed been a significant influence in recent human history, though one whose specific contributions are difficult to separate out. Comparing a world today without the capabilities provided by space systems to one in which those systems are fully integrated would, I believe, support the validity of this judgment.

FORTY YEARS OF FRUSTRATION

McDougall senses a feeling of "melancholy" because space development has not moved beyond what was predicted for it more than a half century ago. I would substitute the word "frustration" for "melancholy." Both visionaries such as Arthur C. Clarke and hard-nosed analysts at the Rand Corporation by the early 1950s had indeed spelled out most of the various domains in which space capabilities, once they were technologically and financially achievable, could contribute to human life in important ways. What happened in that decade is interesting to remember. First of all, these space visions became part of popular culture well before the first satellites were launched. Those raised in

---

3. See, for example, Steven Lambakis, *On the Edge of Earth: The Future of American Space Power* (Lexington, KY: University Press of Kentucky, 2001) for a discussion of the link between space capabilities and military power.

the 1950s (I was among them) had available in print, in film, and on the then-new medium of television multiple images of a future transformed by space activity. The 1952 *Collier's* cover declaring "Man Will Conquer Space Soon" was typical of the message we were receiving.[4]

At the same time, the leaders of the two Cold War superpowers decided that developing the technologies needed to operate in space were linked to their countries' core national interests. More quickly than anyone could have anticipated at the start of the decade, the U.S. and Soviet governments provided the funds needed to develop a broad array of space capabilities, primarily, as McDougall notes, on the basis of national security considerations. But to those steeped in the space visions of the decade, it seemed that the predictions of Clarke, von Braun, and their colleagues might soon become reality. We did not sense the contingent character of government commitment to space, which linked space to broader geopolitical interests.

The acme of this linkage was, of course, Project Apollo. As I wrote in 1970, by his decision to use American trips to the Moon as a way of symbolizing U.S. power vis-à-vis the Soviet Union, President John F. Kennedy "linked the dreams of centuries to the politics of the moment."[5] By backing up his decision to go to the Moon with a war-like mobilization of human and financial resources to achieve the lunar landing goal, Kennedy created a sense that what was in fact a crash program aimed at a specific political goal was instead a U.S. national commitment to achieve on an accelerated schedule the various elements of the 1950s space vision. This sense was reinforced by NASA Administrator James Webb's argument to Kennedy that the real goal was "preeminence"—a clearly leading position in all areas of space activity. Not only human spaceflight, but all areas of space science and applications, grew rapidly in the 1960s.

Thus it is not surprising that the space community in 1969, as the Apollo goal was achieved, proposed to take the next steps, including large space stations, a lunar base, human missions to Mars, and increasingly ambitious robotic missions. Their expectations were quickly dashed, as President Richard Nixon in March 1970 announced that "We must think of [space activities] as part of a continuing process . . . and not as a series of separate leaps, each requiring a massive concentration of energy." The president added "Space expenditures must take their proper place within a rigorous system of national priorities. . . . What we do in space from here on in must become a normal and regular part of our national life and must therefore be planned in conjunction with all of the other undertakings which are important to us."[6]

---

4. Excerpts from the *Collier's* series on space can be found in John M. Logsdon et al., eds. *Exploring the Unknown: Selected Documents in the History of the U.S. Civil Space Program*, Vol. I, Organizing for Exploration, NASA SP-4407 (Washington, DC: Government Printing Office, 1995), pp. 176-200.

5. John M. Logsdon, *The Decision to Go to the Moon: Project Apollo and the National Interest* (Cambridge, MA: MIT Press, 1970), p. 7.

6. President Nixon's statement can be found at *http://www.presidency.ucsb.edu/ws/index.php?pid=2903&st=&st1=* (accessed April 6, 2008).

This perspective was bound to frustrate those who, in the immediate aftermath of the lunar landings, thought that the government commitment to space that had fueled Apollo would continue. What is unfortunate is that this frustration continues today; in the almost four decades since Nixon set forth the policy that has in effect guided civilian space decisions since, the space community has not adjusted its expectations to a much slower-paced but perhaps ultimately more sustainable approach to space development. Apollo created a large government-industrial-scientific complex optimized for carrying out fast-paced development and operation efforts. That complex exists, albeit in a diminished form, today, and it continues to be frustrated that its aspirations are not fully supported by the White House, Congress, and ultimately the American public. That the space community still hopes to recapture something approaching the Apollo approach to space is what is "melancholy." As Howard McCurdy has commented

> The reality of space travel depleted much of the vision that originally inspired it. Space-flight engineers have not developed technologies capable of achieving the dream; advocates have not formulated alternative visions capable of maintaining it. At the same time, no alternative vision of sufficient force has appeared to supplant the original dream. Advocates still embrace the original vision of adventure, mystery, and exploration. They continue to dream of expeditions to nearby planets and the discovery of habitable worlds. The dreams continue, while the gap between expectations and reality remains unresolved.[7]

That being said, I think one can look back at what has been accomplished over the past 50 years and agree with the late Sir Arthur C. Clarke's observation: "On the whole, I think we have had remarkable accomplishments during the first 50 years of the Space Age. Some of us might have preferred things to happen in a different style or time frame, but when our dreams and aspirations are adjusted for reality, there is much we can look back on with satisfaction."[8]

WHAT ABOUT SPACE EXPLORATION?

McNeill comments that "Space exploration, as opposed to the totality of space programs, could well be relegated to the status of historical footnote.... [E]xploration programs are another matter: they are especially expensive and they probably won't cure cancer or defeat terrorism, so they are at high risk of being phased out.... If

---

7. Howard E. McCurdy, *Space and the American Imagination* (Washington, DC: Smithsonian Institution Press, 1997), p. 243.

8. Arthur C. Clarke, "Remembering Sputnik" at *http://spectrum.ieee.org/oct07/5584* (accessed March 30, 2008).

so, in time space exploration will be forgotten, a dead end, a historical cul-de-sac."
He adds "On the other hand, it could be that space exploration will thrive, find
new budgetary champions in the corridors of power." McNeill suggests that "Space
exploration may survive on one or another basis, but it still will not loom large in
terms of human history unless something really new and interesting happens." If
that occurs, "then the first 50 years of space exploration will look like the beginning
of something of epic significance." If it does not, "it will look like a small step for
mankind that led nowhere, and did not amount to much in the balance before being
consigned to the dustbin of history." McNeill concludes, and I concur, that "It is
indeed too soon to judge whether the whole enterprise is a gigantic folly diverting
money and talent from more urgent applications, a noble calling consonant with our
deepest nature, or something else altogether."[9]

In the first 50 years of the Space Age, only 27[10] Americans ventured
beyond Earth orbit to begin the exploration of the solar system by voyages to the
Moon. In reality, that sentence is not completely accurate. While many space
advocates saw Project Apollo as the beginning of a long period of human space
exploration, the political leaders who provided the funds for Apollo certainly
did not do so out of a commitment to space exploration. Given the dead-end
character of Apollo and the fact that it was driven by geopolitical considerations,
I do not think there is much that can be said about its historical contributions
as an exploratory undertaking. The history of human space exploration is yet
to be written. Whether it will begin to be written in the next few decades is
today's most pressing space policy question.

McNeill cites one of his colleagues, Felipe Fernandez-Armesto, as
suggesting that space exploration has been a "gigantic folly."[11] He is not alone
in that view. The *Economist* recently commented that "a scandalous amount
of money has been wasted on the conceit that voyaging across the cosmos
is humanity's destiny"[12] Aerospace executive Rick Fleeter in October 2004
criticized advocates of space exploration for taking "as axiomatic that space's
highest and true calling is achieving societal goals of research and exploration
into the unknown." In Fleeter's view, "Hauling this burdensome baggage of
an aristocratic calling, now bankrupt both ideologically and financially, is not
helping space—it is hindering our community from reaching our potential to

---

9. McNeill is talking here about both human and robotic space exploration. It is my view that
    robotic exploratory missions of some character will continue for the foreseeable future, although
    ambitious multi-billion dollar undertakings may be few. To me, the key issue is whether
    governments in the early 21st century will support human exploration beyond Earth orbit.

10. Two people—Eugene Cernan, and John Young— both made a trip to lunar orbit without
    landing and a second trip to the lunar surface. One person—James Lovell—went into lunar orbit
    on the Apollo 8 mission and then looped around the Moon on the ill-fated Apollo 13 mission.

11. Felipe Fernandez-Armesto, *Pathfinders: A Global History of Exploration* (New York, NY:
    Norton, 2006), p. 399.

12. The *Economist*, September 29, 2007: 23.

serve humanity." This is so, he argued, because these "old ideas are rigid and anachronistic, no longer pointing us to a brighter tomorrow, but rather back toward a dead end of technological progress for its own sake."[13]

I suggest that there is no compelling evidence one way or the other to assess the validity of these assertions, since the actual experience of human space exploration is so limited. In addition, the belief that sending humans beyond Earth orbit is the correct next step in space development is gaining political acceptance around the world. Leaders of the United States and, more recently, France have committed their countries to the support of human exploration, beginning with a return to the Moon before 2020 and including eventual voyages to Mars. To me, the issue is whether this round of human exploration will be designed to answer, at least for this century, the question of whether such steps are indeed a "gigantic folly," or part of future human history.

The requirements for sustained human exploration beyond Earth orbit were perceptively stated by Harry Shipman in his 1989 study, *Humans in Space*.[14] Shipman says that the future of human activity beyond Earth orbit depends on the answer to two questions:

1. Can extraterrestrial materials be used to support life in locations other than Earth?

2. Can activities of sustained economic worth be carried out at those locations?

Depending on the answer to those questions, Shipman suggests, the following outcomes are probable:

CAN IN SITU MATERIALS BE USED TO SUPPORT HUMAN LIFE?

		NO	YES
CAN SPACE COMMERCE EMERGE?	**NO**	Space science only	Research and tourism
	YES	Robot mines, factories, and labs	Full space settlement

13. Rick Fleeter, *Space News*, October 18, 2004: 10. Fleeter's remarks were in response to an op-ed essay I had published in the same venue two weeks earlier.

14. Harry Shipman, *Humans in Space: 21st Century Frontiers* (New York, NY: Plenum Press, 1989), p. 17.

Humanity may be at a branch point in future space development, one that could provide the answers to Shipman's questions. There is on the table a bold proposition, put forth by U.S. President George W. Bush in January 2004—that the nations of the world, led by the United States, accept as the guiding purpose of their governments' space programs carrying out "a sustained and affordable human and robotic program to explore the solar system and beyond."[15] It seems as if space leaders in other spacefaring countries, and those eager to become more active in space, are also embracing exploration beyond Earth orbit as an essential element in their future activities. For example, 14 space agencies[16] in May 2007 issued a statement of Global Exploration Strategy that argued "This Global Exploration Strategy will bring significant social, intellectual and economic benefits to people on Earth." The document argued that "space exploration is essential to humanity's future." It added that [Emphasis added by the author.] *"Opportunities like this come rarely. The human migration into space is still in its infancy. For the most part, we have remained just a few kilometers above the Earth's surface—not much more than camping out in the backyard."*[17]

The key words here are "opportunities like this come rarely." I would go even further. Never before has a major government, in this case the United States, committed itself to an open-ended vision of space exploration. The pressing issues are: Will the United States sustain that commitment in coming years? Will other countries join the United States in such a long-term exploratory effort? Or will others follow a different path, developing an exploration program of their own? Finally, will space exploration by humans prove not to be sustainable, and thus will humans focus their space efforts on robotic exploration and space applications that provide direct benefits here on Earth?

These are the key questions for the next period of spaceflight. Only after they are answered can we state with any assurance that space exploration was "a false start that led no where and did not amount to much in the balance before being consigned to the dustbin of history."

Other outcomes are also possible, as space dreamers have reminded us. Looking back 50 years from now, it may be that our evaluation of the historical significance of space exploration can be much more definitive, and much more positive.

---

15. The White House, *A Renewed Spirit of Discovery: The President's Vision for U.S. Space Exploration,* January 2004.

16. NASA; Canadian Space Agency, European Space Agency; CNES; DLR; Italian Space Agency; British National Space Center; Russian Space Agency, Roscosmos; Ukrainian Space Agency; Indian Space Research Organization; Chinese National Space Administration; Korean Aerospace Research Institute; Japanese Aerospace Exploration Agency; JAXA; and Australian Commonwealth Scientific and Industrial Research Organization.

17. Each of the 14 agencies issued the document in some form. See, for example, *www.nasa.gov/ pdf/178109main_ges_framework.pdf.*, p. 3 (accessed April 6, 2008).

CHAPTER 20

HAS THERE BEEN A SPACE AGE?

Sylvia Kraemer

O ur conference opened with the observation by John Logsdon that how one remembers the Space Age depends mightily on who does the remembering. I would add that how we remember the Space Age today is also likely to depend on one's angle of repose, or that point in our shared history at which we have acquired sufficient stability to pause and to reflect on the relative importance of striking features in the cultural and political/economic landscape that surrounds us.

So I will begin with some observations that cause me to question whether U.S. or global space activity since Sputnik warrants its characterization as defining an "age." Whether space has fostered globalization or increased nationalism is part of this question. Then I will comment on the ways in which space activity has nonetheless left an indelible and lasting mark on our world.

When we refer to any development as defining an age of human history, we imply that it has been a singular agent of historical change. The notion that space activity is one such development may appeal to those who equate events that receive extensive media attention with the things that are historically important. And space activity has certainly helped to shape the careers of millions of engineers, scientists, and managers in corporate America and within the federal government and many of our universities. For these individuals space activities have defined a substantial portion of their lives.

But space activity has some strong competition as a claimant to defining our world. First, I would offer the Cold War, in which space was an important salient but not principal provocateur. That role of preeminence is held by ideology—ours as well as that of the Soviet Union. No less important were the post-World War II geopolitical changes wrought by the emergence of the United States as the world's dominant "superpower" and the regional realignments in Europe, the Middle East, and Southeast Asia. As we know only too well, those realignments have challenged our military, economic, and diplomatic independence to an arguably unprecedented degree. I also think a strong case can be made for the emergence, popularization, and ramifications of digital communications and information technologies as the defining phenomenon of the "age" following the end of the Cold War.

The panel was also asked to consider whether space activity fostered a new global identity, or reinforced distinct national identities. Here I think a two-handed response is unavoidable. On the one hand, nations do take pride in being able to demonstrate to everyone that they, too, can launch and sustain space missions, including human missions. Along with this we have the national security implications—not only for the United States, but for everyone else—of being able to deliver catastrophic weapons to adversaries' soil and military assets wherever they might be. The same can be said for nations' ability to spy on each other continuously from space and to use satellites for tactical advantages in the field with space-based surveillance and targeting.

Has the ability to amplify national military capabilities in space, one of the most visible manifestations of national technological capacity, strengthened nationalism? We tend to assume it has, but I think that notion is debatable. We might recall the premise, built into President Eisenhower's space policy and illustrated in the case of the Soviet Union, that international belligerence is less sustainable when nations can accurately assess one another's military capacities. So we can debate whether the enhancement of military capabilities by space weaponry, reconnaissance, and targeting actually fosters nationalism or simply elevates the geopolitical balance of powers to a higher plane.

And now, to the other hand: Our ability to observe Earth from space has unquestionably reinforced our understanding that Earth is a solitary and probably unique traveler through space, its natural plenitude the single greatest treasure bequeathed to humankind, whether by a divine creator or the mysterious "fickle finger of fate." But the indirect contribution to globalization may be more important than this more obvious visual paradigm.

To begin with, Earth imagery from space has brought to fruition the historic process of global discovery that began in earnest during that previous "Age of Reconnaissance" of the 14th and 15th centuries. Secondly, by engaging scientists from around the world in the shared investigation of Earth's dynamic climate and physical geography, as well as the relationship of its dynamic processes to those of the Sun, space activity has reinforced the cosmopolitanism of intellectual life—an essential component of genuine "globalization."

I believe that the contribution of space activity to globalization has been far greater than its contribution to nationalism. Indeed, space travel has been largely a product of nationalism, rather than one of its sources. And I believe that a symbiotic relationship between space activity and globalization will prevail over whatever uses individual nations may wish to make of their ability to operate in space. This is because the nation state is being overtaken by the globally invested corporation as the primary means of aggregating economic and allied political interests. Moreover, thanks to now ubiquitous "outsourcing," the functions of government are increasingly carried out by corporations wielding enough financial power to buy favorable, or at least neutral, government policies.

Space activity has contributed to this process by enabling virtually instant communication of information and wealth across national boundaries. If the more adventurous super-rich like Richard Branscomb have their way, in the future we will move around the globe with a speed comparable to that at which information and money now move around the world. We might even be able to travel around the globe in less time than it takes today to get by air from New York to Boston. If and when that day occurs, the great cities of the world will have more in common with each other than they have with their respective hinterlands, and space travel will have, indeed, reshaped us into one world.

CHAPTER 21

CULTURAL FUNCTIONS OF SPACE EXPLORATION

Linda Billings

> Culture: a "historically transmitted pattern of meanings
> embedded in symbols, a system of inherited conceptions
> expressed in symbolic forms by means of which [people]
> communicate, perpetuate and develop their knowledge
> about and attitudes toward life."[1]

What role has space exploration played in the cultural environment of the
U.S. and the world? What has space exploration meant, or done, for the
vast majority of people on Earth outside the space community? Has this role or
function varied across cultural boundaries (for example, gender or nationality),
time, or space? Where, or what, has space exploration been in public discourse?
Has space exploration had subcultures as well as a dominant culture? In short,
what cultural functions has space exploration performed? How have people
remembered, represented, and made use of space exploration?

All these questions may be addressed from a broad range of perspectives.
The papers in this volume illustrate in a variety of ways that space exploration
means different things to different people at different times and in different
geographical and sociocultural places. Official and dominant cultural narratives
of space exploration are not the only sites where meaning is constructed. The
so-called "public" makes meaning out of space exploration in its own ways. Just
how space exploration has affected aspects of social life such as material culture,
education, aesthetics, values and attitudes, and religion and spirituality is an
interesting question in its own right. In her paper in this volume, University of
California, Irvine, historian Emily Rosenberg documented how the Apollo-
era U.S. space program influenced art and architecture and produced "space
spectaculars" for the newly dominant mass medium of television. "Space was
the star of this historical moment," she said. Ultimately, she concluded, space
exploration might mean many things, or it might mean nothing. National Air
and Space Museum historian Martin Collins noted that the traditional narrative

---

1. Clifford Geertz, *The Interpretation of Cultures* (New York, NY: Basic Books, 1973), pp. 14, 34.

of space exploration as a lone, heroic, and progressive enterprise "still resonates, but in a much diminished way."

In a 1945 letter to President Eisenhower accompanying the now-famous July 1945 report, *Science: The Endless Frontier*, White House Office of Scientific Research and Development Director Vannevar Bush wrote "The pioneer spirit is still vigorous within this nation. Science offers a largely unexplored hinterland for the pioneer who has the tools for this task. The rewards of such exploration both for the nation and the individual are great. Scientific progress is one essential key to our security as a nation, to our better health, to more jobs, to a higher standard of living, and to our cultural progress."[2] *Science: The Endless Frontier* laid out a U.S. scientific research and technology development program for the post–World War II era.

By substituting the words "space exploration" for "science" in this passage, Vannevar Bush's post–World War II rhetoric becomes indistinguishable from the rhetoric of contemporary space exploration advocates. An example of current rhetoric is a so-called "elevator speech" developed by NASA's Office of Strategic Communications Planning in 2007 to offer a rationale for the civilian space program:

> NASA explores for answers that power our future. NASA exploration powers inspiration that engages the public and encourages students to pursue studies in challenging high-tech fields. NASA exploration powers innovation that creates new jobs, new markets, and new technologies that improve and save lives every day in every community. . . . NASA exploration powers discovery that enables us to better understand our Solar System and protect Earth through the study of weather and climate change, monitor the effects of the Sun and detect objects that could collide with Earth. Why explore? . . . Because exploration powers the future through inspiration, innovation, and discovery.[3]

In considering what space exploration has meant, or done, for the vast majority of people who are not a part of the "official" space community, what role do these official narratives play? Do people construct their own narratives and make their own meanings, in consideration of their own, specific cultural boundaries of gender, nationality, time, or space? Media commentaries on the 50th anniversary of the launch of Sputnik and the beginning of the Space

---

2. Vannevar Bush, Director of the Office of Scientific Research and Development, *Science: The Endless Frontier. A Report to the President*, July 1945. United States Government Printing Office, Washington: 1945. (letter of transmittal, n.p.)

3. NASA Message Construct, NASA Office of Communications Planning, June 1, 2007.

Age tended to repeat familiar and official narratives. The *New York Times* reported, for example, "Sputnik changed everything: history, geopolitics, the scientific world. It launched careers, too. . . . Sputnik lifted us into the future." The *Houston Chronicle* asserted, "Today the U.S. reigns over a growing cast of nations . . . on a vast new frontier," framing contemporary space exploration as the geopolitical enterprise it was depicted to be in the 1960s. Writing in the *Los Angeles Times*, Matthew Brzezinski (author of *Red Moon Rising: Sputnik and the Hidden Rivalries that Ignited the Space Age*) characterized space exploration since Sputnik as geopolitics as usual.

In contrast, the *Toronto Globe and Mail* offered a different 50th anniversary perspective on the meaning of space exploration. In an editorial entitled "Venturing into space and finding Earth," the paper made the claim that "the most significant achievement of the space age is a better understanding of the vulnerability of our own home planet."

University of California, Santa Barbara, cultural studies scholar Constance Penley is one of a small number of researchers who have explored alternative or subordinate narratives of space exploration. To young people and others for whom the official narrative of space exploration may not have been meaningful, she noted during comments at this meeting, the makers of *Star Trek* offered an alternative narrative, "a sustainable and inclusive vision" of a human future in space.[4] *Star Trek* producers have done a better job than NASA has of articulating a widely appealing vision. Today young people "are not interested in space unless they can participate in some way," Penley said, and while NASA "lives and dies by popular culture," they have just barely begun to engage with 18–35-year-olds via now-dominant social networks such as MySpace that provide broad opportunities for participation. Penley mentioned NASA Ames Research Center's creation of a meeting and working space on the social networking site Second Life and Ames's hosting of a public "Yuri's Night" party in 2007 as first steps toward a more participatory space program. She also mentioned private-sector initiatives to expand public participation in space exploration, such as the Google-sponsored Lunar X Prize competition to land a robotic explorer on the Moon.

During the meeting, Yale University historian, Bettyann Kevles showed how artists working in a range of media, from science fiction to dance to music, have interpreted and remembered the Space Age, making space exploration meaningful in ways not typically considered outside the space community. Kevles played an excerpt of a jazz suite composed and performed by saxophonist Jane Ira Bloom under commission by NASA's space art program. It is not clear what interest space exploration holds for contemporary artists.

---

4. Constance Penley, *NASA/TREK: Popular Science and Sex in America.* (New York, NY: Verso, 1997).

Margaret Weitekamp, the Curator at the National Air and Space Museum in charge of the museum's Social and Cultural Dimensions of Spaceflight collection, offered her views on how social and cultural products of the Space Age tell a story of space exploration that may converge with and diverge from the official narrative that tends to be embodied in the space hardware and technology that people typically think of as artifacts of the Space Age.

Finally, Alan Ladwig contributed a unique perspective to the discussion on what space exploration means to different sorts of people. As a NASA official in the 1980s and 1990s, Ladwig managed a variety of programs including the space agency's Teacher in Space and Journalist in Space programs and the Shuttle Student Involvement Program. These programs were intended to give people outside the traditional aerospace community a chance to engage directly in the experience of spaceflight. The space agency was not enthusiastic about implementing these programs, and in fact NASA did not proceed with the Journalist in Space program. Ladwig advocated organizing public events to engender public discussion about what space exploration means to different sectors of society. Precedent has been set: In the 1970s, the Committee for the Future held a series of syn-cons (synthetic convergences) to find out what space exploration means to different sectors of society[5]; the National Commission on Space, appointed by President Reagan in 1985 to develop a long-term plan for space exploration, held public forums around the country in 1985 and 1986 to solicit public opinion about the human future in space; and in 1992, NASA Administrator Daniel S. Goldin presided over a nationwide series of town meetings designed for the same purpose.

With China's efforts in space exploration typically framed in public discourse as a "race" with the West, it is clear that what we call the Space Age has not yet fostered a new global identity. Will 21st century space exploration achieve this goal? Here at the beginning of the new century, it is clear that the enterprise of space exploration has gone global. Will a new global identity emerge?

The 21st century cultural environment for space exploration is radically different from the cultural environment that nurtured the U.S. space program through its first 50 years. It remains to be seen whether NASA can, or will, respond to shifting public interests and concerns and give the people the kind of space program they want. The first step in reconfiguring the space program to survive and thrive in the 21st century is to involve citizens in the process, to ask what sorts of visions they have for a human future in space.

5. Barbara Marx Hubbard interview by David S. Cohen for the Light Connection (accessed December 21, 2007), *http://208.131.157.96/fce/content/node/30*.

ABOUT THE AUTHORS

Linda Billings is coordinator of communications for NASA's Astrobiology Program in the Science Mission Directorate on assignment to NASA under an Intergovernmental Personnel Agreement with the SETI Institute of Mountain View, CA. As astrobiology communications coordinator, Dr. Billings is responsible for reviewing, assessing, and coordinating communications, education, and public outreach activities sponsored by the Astrobiology Program. From September 2002 through December 2006, Dr. Billings conducted science and risk communication research for NASA's Planetary Protection Office. From September 1999 through August 2002, she was Director of Communications for SPACEHAB, Inc., a builder of space habitats. Dr. Billings has three decades of experience in Washington, DC, as a researcher, journalist, freelance writer, communication specialist, and consultant to the government. She was the founding editor of *Space Business News* (1983-1985) and the first senior editor for space at *Air & Space/Smithsonian Magazine* (1985-1988). She also was a contributing author for *First Contact: The Search for Extraterrestrial Intelligence* (New American Library, 1990). Dr. Billings was a member of the staff for the National Commission on Space (1985-1986). Her freelance articles have been published in outlets such as the *Chicago Tribune, Washington Post Magazine*, and *Space News*. Dr. Billings's expertise is in mass communication, science communication, risk communication, rhetorical analysis, journalism studies, and social studies of science. Her research has focused on the role that journalists play in constructing the cultural authority of scientists and the rhetorical strategies that scientists and journalists employ in communicating about science. She earned her B.A. in social sciences from the State University of New York at Binghamton, her M.A. in international transactions from George Mason University, and her Ph.D. in mass communication from Indiana University's School of Journalism.

Andrew J. Butrica earned his Ph.D. in history of technology and science from Iowa State University in 1986. He then worked at the Thomas A. Edison Papers Project and the Center for Research in the History of Science and Technology in Paris. Subsequently, as a contract historian, he has researched and written *Out of Thin Air*, a history of Air Products and Chemicals, Inc., a Fortune 500 firm; *Beyond the Ionosphere*, a history of satellite communications; *To See the Unseen*, a history of planetary radar astronomy that won the Leopold Prize of the Organization of American Historians; and *Single Stage To Orbit: Politics, Space*

Technology, and the Quest for Reusable Rocketry, which was recently published by Johns Hopkins University Press and won the 2005 Michael C. Robinson Prize of the National Council on Public History. In addition, he was the historian on NASA's X-33 Program and currently is a historian on the Defense Acquisition History Project responsible for researching and writing the fourth volume of *From the Reagan Buildup to the End of the Cold War, 1981–1990*.

Martin J. Collins is a curator in the Smithsonian National Air and Space Museum. He received his Ph.D. from the University of Maryland in history of science and technology. Book publications include *Cold War Laboratory: RAND, the Air Force, and the American State* (2002); *Showcasing Space*, edited with Douglas Millard, Volume 6, Artifacts series: *Studies in the History of Science and Technology* (2005); and *After Sputnik* (2007), editor. He is currently working on a history of satellite telephony and globalism in the 1990s.

Jonathan Coopersmith is an associate professor of history at Texas A&M University, where he has taught the history of technology since 1988. He received his D.Phil. in 1985 in Modern History at Oxford University. His books include *The Electrification of Russia, 1880-1926* and, with coeditor Roger Launius, *Taking Off: A Century of Manned Flight*. Currently, his main areas of research are the history of the fax machine from the 1840s to the present and the close links between communication technologies and pornography.

Steven J. Dick is the Chief Historian for NASA. He obtained his B.S. in astrophysics (1971) and his M.A. and Ph.D. (1977) in history and philosophy of science from Indiana University. He worked as an astronomer and historian of science at the U.S. Naval Observatory in Washington, DC, for 24 years before coming to NASA Headquarters in 2003. Among his books are *Plurality of Worlds: The Origins of the Extraterrestrial Life Debate* from *Democritus to Kant* (1982), *The Biological Universe: The Twentieth Century Extraterrestrial Life Debate and the Limits of Science* (1996), and *Life on Other Worlds* (1998). The latter has been translated into Chinese, Italian, Czech, Polish and Greek. His most recent books are *The Living Universe: NASA and the Development of Astrobiology* (2004) and a comprehensive history of the U.S. Naval Observatory, *Sky and Ocean Joined: The U.S. Naval Observatory, 1830–2000* (2003). The latter received the Pendleton Prize of the Society for History in the federal government. He is editor of *Many Worlds: The New Universe, Extraterrestrial Life and the Theological Implications* (2000), editor (with Keith Cowing) of the proceedings of the NASA Administrator's symposium *Risk and Exploration: Earth, Sea and the Stars* (2005), and (with Roger Launius) of *Critical Issues in the History of Spaceflight* (2006) and *Societal Impact of Spaceflight* (2007). He is the recipient of the Navy Meritorious Civilian Service Medal. He received the NASA Group Achievement Award for his role in NASA's multidisciplinary program in astrobiology. He has served as chairman of the Historical Astronomy

Division of the American Astronomical Society as president of the History of Astronomy Commission of the International Astronomical Union, and he as president of the Philosophical Society of Washington. He is a member of the International Academy of Astronautics.

Slava Gerovitch is a lecturer in the science, technology and society program at the Massachusetts Institute of Technology. His research interests include the history of Soviet science and technology during the Cold War, especially cosmonautics, cybernetics, and computing. His book *From Newspeak to Cyberspeak: A History of Soviet Cybernetics* (2002) received an honorable mention for the Wayne S. Vucinich Book Prize for an outstanding monograph in Russian, Eurasian, or East European studies from the American Association for the Advancement of Slavic Studies. His articles appeared in the journals *Technology and Culture, Social Studies of Science, Science in Context,* and *The Russian Review* and in the collections *Science and Ideology, Cultures of Control,* and *Universities and Empire.* His most recent article on human–machine issues in the Soviet space program was published in the collection *Critical Issues in the History of Spaceflight* (2006). He is the recipient of the 2007-2008 Fellowship in Aerospace History from the American Historical Association. He received two doctorates in the history and social study of science and technology from the Russian Academy of Sciences (1992) and from MIT (1999).

James R. Hansen is professor of history in the department of history at Auburn University in Alabama, where he teaches courses on the history of flight, history of science and technology, space history, and the history of technological failure. He has published nine books and three dozen articles on a wide variety of technological topics including the early days of aviation, the first nuclear fusion reactors, the Moon landings, and the environmental history of golf course development. His books include *First Man: The Life of Neil A. Armstrong* (2005); *The Bird is on the Wing: Aerodynamics and the Progress of the Airplane in America* (2003); *The Wind and Beyond: A Documentary Journey through the History of Aerodynamics in America* (Vol. 1, 2002), *Spaceflight Revolution* (1995), *From the Ground Up* (1988), and *Engineer in Charge* (1987). Hansen earned a B.A. degree with high honors from Indiana University (1974) and an M.A. (1976) and Ph.D. (1981) from The Ohio State University. He served as historian for NASA Langley Research in Hampton, Virginia, from 1981 to 1984, and as a professor at the University of Maine in 1984-85. Professor Hansen has received a number of citations for his scholarship, including the National Space Club's Robert H. Goddard Award, Air Force Historical Foundation's Distinction of Excellence, American Astronautical Society's Eugene Emme Prize in Astronautical Literature (twice), American Institute of Aeronautics and Astronautics's History Book Award; and AIAA Distinguished Lecturer. He has served on a number of important advisory boards and panels, including the Research Advisory Board of the National Air and Space Museum, the Editorial Advisory Board of the Smithsonian Institution Press, the Advisory Board for the Archives of Aerospace Exploration at Virginia Polytechnic Institute and State University, the Museum Advisory Board of the U.S. Space and Rocket Center in Huntsville, Alabama, and the board of directors of the Space Restoration Society. He is a past vice-president of the board of directors of the Virginia Air and Space Museum and Hampton Roads History Center in Hampton, Virginia.

Robert G. Kennedy, III, PE, took Heinlein's advice about a liberal education to heart. He is a registered professional mechanical engineer (robotics specialty for military, nuclear, and industrial applications) in Tennessee and California. He minored in Soviet studies and holds a special master of arts in national security studies, and speaks, reads, and writes Russian, Latin and, to a lesser degree, Arabic and Classical Greek. In 1994, he was selected as the American Society of Mechanical Engineers Congressional Fellow for that year, working for the Subcommittee on Space in the United States House of Representatives. Also in 1994, he was invited to present *Robert A Heinlein: The Competent Man* at the Library of Congress. In 1997, he published, manufactured and distributed the Russian CD-ROM *40th Anniversary of Sputnik: Russians in Space*. Also in 1997-1999, he served on the *Where To?* panel in the inaugural issues of The Heinlein Journal published by The Heinlein Society. He is an amateur military historian, published artist and writer on strategic affairs in the *Journal of the British Interplanetary Society* among other venues, and past chair of the American Society of Mechanical Engineer's (ASME) Technology & Society Division. He currently resides with his spouse and numerous cats under his own vine and fig tree in the Manhattan Project city of Oak Ridge, Tennessee.

Bettyann Kevles is a lecturer in the history department at Yale University. She writes about science, technology, and popular culture. Her most recent book, *Almost Heaven: The Story of Women in Space*, (paperback revised edition MIT Press 2006), is a cross-cultural exploration of the lives and ambitions of women who have traveled into orbit. During 2000, she held the Charles A. Lindbergh Chair in Space History at the Smithsonian's National Air and Space Museum. *Almost Heaven* was selected as the best science book of 2003 by the American Library Association, and in 2005 she was awarded the Educator's Award for the book by Women in Aerospace. Before joining the faculty at Yale, she lived in Pasadena, California, where she wrote a regular science column and science book reviews for the *Los Angeles Times*, taught at the Art Center College of Design, and became an active member of the Planetary Society.

Sylvia K. Kraemer served as NASA Chief Historian from 1983-1990 and as NASA's Director of Policy Development, Director of Interagency Relations, and Chair of Inventions and Contributions Board from 1990-2004. She obtained her Ph.D. in history from The Johns Hopkins University. Among many other publications, she is the author of *NASA Engineers and the Age of Apollo* (1992).

John Krige is an historian who specializes on the place of science and technology in the postwar reconstruction of Europe. He received his Ph.D. in physical chemistry at Pretoria in 1965 and his Ph.D. in philosophy at Sussex in 1979. He was a member of a multinational team that wrote the history of CERN and the leader of the project that produced a two-volume history of the European Space Agency. Krige is the editor of the journal *History and Technology* and has served on the editorial board of the several other journals. His current research deals with the use of science and technology as instruments of U.S. foreign policy during the Cold War, notably in its relations with Western Europe. His new book, *American Hegemony and Postwar Reconstruction of Science in Europe*, was published by the MIT Press in 2006. He also edited (with Kai Henrik Barth, Georgetown University), Vol. 21 of OSIRIS entitled *Global Power Knowledge. Science and Technology in International Affairs* (University of Chicago Press, 2006). In 2005, Krige was the Charles A. Lindbergh Professor of Aerospace History at the National Air and Space Museum in Washington, DC. In May 2005, he was awarded the Henry W. Dickinson medal by the (British) Newcomen Society for the Study of Technology and Society. In Fall 2006, Krige was a Visiting Fellow at the Shelby Cullom Davis Center for Historical Studies at Princeton University.

Monique Laney is a Ph.D. candidate in the American studies program at the University of Kansas. Her dissertation explores the immigration and integration process of German families associated with *Project Paperclip* and Wernher Von Braun who moved to Huntsville, Alabama, in the 1950s. For this investigation, she uses oral histories, archival material, newspaper clippings, literature, and film. Currently Ms. Laney is finishing a pre-doctoral Smithsonian fellowship at the National Air and Space Museum in Washington, DC. She holds an M.A. degree in American studies from the Johann Wolfgang Goethe-Universität, Frankfurt, Germany, and her research interests include national identity, U.S. immigration history, history and memory, German Americans after WWII, and the cultural and social impact of the U.S. on Germany and vice versa. Her first publication, "The New York Times and Frankfurter Allgemeine Zeitung: Two Perspectives on the War in Iraq," appeared in *Safeguarding German-American Relations in the New Century: Understanding and Accepting Mutual Differences*, eds. Kurthen, Hermann; Antonio V. Menendez-Alarcon; and Stefan Immerfall (Lanham, Maryland: Lexington Books-Rowman & Littlefield, 2006).

Roger D. Launius is a member of the division of space history at the Smithsonian Institution's National Air and Space Museum in Washington, DC. Between 1990 and 2002, he served as Chief Historian of NASA. A graduate of Graceland College in Lamoni, Iowa, he received his Ph.D. from Louisiana

State University, Baton Rouge, in 1982. He has written or edited more than 20 books on aerospace history, including *Critical Issues in the History of Spaceflight* (NASA SP-2006-4702, 2006); *Space Stations: Base Camps to the Stars* (Smithsonian Books, 2003), which received the AIAA's history manuscript prize; *Reconsidering a Century of Flight* (University of North Carolina Press, 2003); *To Reach the High Frontier: A History of U.S. Launch Vehicles* (University Press of Kentucky, 2002); *Imagining Space: Achievements, Possibilities, Projections, 1950-2050* (Chronicle Books, 2001); *Reconsidering Sputnik: Forty Years Since the Soviet Satellite* (Harwood Academic, 2000); *Innovation and the Development of Flight* (Texas A&M University Press, 1999); *Frontiers of Space Exploration* (Greenwood Press, 1998, rev. ed. 2004); *Spaceflight and the Myth of Presidential Leadership* (University of Illinois Press, 1997); and *NASA: A History of the U.S. Civil Space Program* (Krieger Publishing Co., 1994, rev. ed. 2001). He served as a consultant to the Columbia Accident Investigation Board in 2003 and presented the Harmon Memorial Lecture on the history of national security space policy at the United States Air Force Academy in 2006. He is frequently consulted by the electronic and print media for his views on space issues, and has been a guest commentator on National Public Radio and all the major television networks.

Cathleen Lewis is curator of International Space programs at the Smithsonian Institution's National Air and Space Museum, specializing in Soviet and Russian programs. Her current research is on the history of the public and popular culture of the early years of human spaceflight in the Soviet Union. She has completed degrees in Russian and East European Studies at Yale University and is currently writing her dissertation, *The Red Stuff: A History of the Public and Material Culture of Early Human Spaceflight in the U.S.S.R., 1959-1968* in the History Department at George Washington University.

John M. Logsdon is director of the Space Policy Institute at George Washington University's Elliott School of International Affairs, where he is also Research Professor and Professor Emeritus of Political Science and of International Affairs. From 1983-2001, he was the director of GW's Center for International Science and Technology Policy. A faculty member at GW since 1970, he also taught at the Catholic University of America from 1966-1970. He holds a B.S. in physics from Xavier University (1960) and a Ph.D. in political science from New York University (1970). Dr. Logsdon's research interests focus on the policy and historical aspects of U.S. and international space activities.

Dr. Logsdon is the author of *The Decision to Go to the Moon: Project Apollo and the National Interest* and is general editor of the eight-volume series *Exploring the Unknown: Selected Documents in the History of the U.S. Civil Space Program*. He has written numerous articles and reports on space policy and history and authored the basic article on the topic of "space exploration" for the most recent edition of Encyclopedia Britannica. Dr. Logsdon has lectured and spoken to a wide variety of audiences at professional meetings, colleges and universities, international conferences, and other settings, and has testified before Congress on several occasions. He has served as a consultant to many public and private organizations and is frequently consulted by the electronic and print media for his views on space issues. Dr. Logsdon is a member of the NASA Advisory Council and of the Commercial Space Transportation Advisory Committee of the Department of Transportation. In 2003, he served as a member of the Columbia Accident Investigation Board. He is a recipient of the Distinguished Public Service and Public Service Medals from NASA, the 2005 John F. Kennedy Astronautics Award from the American Astronautical Society, and the 2006 Barry Goldwater Space Educator Award from the American Institute of Aeronautics and Astronautics. He is a Fellow of the American Institute of Aeronautics and Astronautics and the American Association for the Advancement of Science, and a member of the International Academy of Astronautics and former Chair of its Commission on Space Policy, Law, and Economics. He is a former member of the Board of Directors of the Planetary Society and member of the Society's Advisory Council. He is faculty member of the International Space University and former member of its Board of Trustees. He is on the editorial board of the international journal *Space Policy* and was its North American editor from 1985-2000. He is also on the editorial board of the journal *Astropolitics*.

Dr. Logsdon has served as a member of a blue-ribbon international committee evaluating Japan's National Space Development Agency and of the Committee on Human Space Exploration of the Space Studies Board, National Research Council. He has also served on the Vice President's Space Policy Advisory Board, the Aeronautics and Space Engineering Board of the National Research Council, NASA's Space and Earth Sciences Advisory Committee, and the Research Advisory Committee of the National Air and Space Museum. He has served as the Director of the District of Columbia Space Grant Consortium. He is former Chairman of the Committee on Science and Public Policy of the American Association for the Advancement of Science and of the Education Committee of the International Astronautical Federation. He has twice been a Fellow at the Woodrow Wilson International Center for Scholars and was the first holder of the Chair in Space History of the National Air and Space Museum.

Robert MacGregor is currently a graduate student in the history of science program at Princeton University. Before coming to Princeton he studied at Rice University in Houston, Texas, where he received a B.S. in chemical physics and a B.A. in history. Robert has also studied at Moscow State University in Russia where he studied Russian language, history, and culture. His current work focuses on the processes in the U.S. government that lead to the formation of NASA between the launch of Sputnik in October 1957 and the signing into law of the National Air and Space Act in July 1958. In the future, he plans to delve into the history of the Soviet space program and the early amateur rocket societies in Germany, the United States, and the Soviet Union.

Hans Mark became Deputy Administrator of NASA in July 1981. He had previously served as Secretary of the Air Force from July 1979 until February 1981 and as Under Secretary of the Air Force since 1977. Dr. Mark was born in Mannheim, Germany, June 17, 1929. He came to the United States in 1940 and became a citizen in 1945. He received his bachelor's degree in physics from the University of California, Berkeley, in 1951 and his doctorate in physics from the Massachusetts Institute of Technology in 1954. In February 1969, Mark became director of NASA's Ames Research Center in Mountain View, California, where he managed the Center's research and applications efforts in aeronautics, space science, life science, and space technology. He has taught undergraduate and graduate courses in physics and engineering at Boston University, Massachusetts Institute of Technology, and the University of California (Berkeley and Davis campuses). Following completion of graduate studies, Dr. Mark remained at MIT as a research associate and acting head of the Neutron Physics Group, Laboratory for Nuclear Science until 1955. He then returned to the University of California, Berkeley, as a research physicist at the University's Lawrence Radiation Laboratory in Livermore until 1958. He subsequently served as an assistant professor of physics at MIT before returning to the University of California's Livermore Radiation Laboratory's Experimental Physics Division from 1960 until 1964. He then became chairman of the University's Department of Nuclear Engineering and administrator of the Berkeley Research Reactor before joining the NASA team. Dr. Mark has served as a consultant to government, industry, and business, including the Institute for Defense Analyses and the President's Advisory Group on Science and Technology. He has authored many articles for professional and technical journals. He also coauthored the books *Experiments in Modern Physics* and *Power and Security*, and coedited *The Properties of Matter under Unusual Conditions*. He also published *The Space Station: A Personal Journey* (Duke University Press, 1987), and *The Management of Research Institutions* (NASA SP-481, 1984). When Dr. Mark left NASA in 1984, he became Chancellor of the University of Texas system, a post he held until 1992. He then became a senior professor of aerospace engineering at the University of Texas at Austin. In July 1998, he took a job at the Pentagon as the director, defense research and engineering. In January 2001, he returned to the department of aerospace engineering and engineering mechanics and the University of Texas at Austin.

Walter A. McDougall is the Alloy-Ansin professor of international relations and history and the University of Pennsylvania. His honors include the Pulitzer Prize for history, election to the Society of American Historians, and appointment to the Library of Congress Council of Scholars. McDougall graduated from New Trier High School in Illinois in 1964 and Amherst College, Massachusetts in 1968. After serving in the U.S. Army artillery in Vietnam, he took a Ph.D. under world historian William H. McNeill at the University of Chicago in 1974. The following year he was hired by the University of California, Berkeley, and taught there until 1988, when he was offered the chair at Penn. McDougall is also a Senior Fellow at Philadelphia's Foreign Policy Research Institute where he edited its journal *Orbis* and now codirects its History Academy for secondary school teachers. His articles and columns have appeared in the *New York Times, Wall Street Journal, Los Angeles Times, Commentary*, and other national publications. An unabashed generalist, his books include *France's Rhineland Diplomacy 1914-1924: The Last Bid for a Balance of Power in Europe* (1978), . . . *the Heavens and the Earth: A Political History of the Space Age* (1985), *Let the Sea Make a Noise: A History of the North Pacific From Magellan to MacArthur* (1992), *Promised Land, Crusader State: The American Encounter with the World Since 1776* (1997), and *Freedom Just Around the Corner: A New American History 1585-1828*. His current project, *Throes of Democracy: The American Civil War Era 1829-1877*, will appear early in 2008. A lover of books, music from Bach to Bob Dylan, chess, sports, and politics, McDougall lives with his wife and two teenagers in suburban Philadelphia.

J. R. McNeill was born in Chicago on October 6, 1954. He studied at Swarthmore College and Duke University, where he completed a Ph.D. in 1981. Since 1985 he has taught some 2,500 students at Georgetown University in the history department and school of foreign service, where he held the Cinco Hermanos chair in Environmental and International Affairs before becoming University professor in 2006. His research interests lie in the environmental history of the Mediterranean world, the tropical Atlantic world, and Pacific islands. He has held two Fulbright awards, a Guggenheim fellowship, a MacArthur grant, and a fellowship at the Woodrow Wilson Center. He has published more than 40 scholarly articles in professional and scientific journals. His books are *The Atlantic Empires of France and Spain, 1700-1765* (Chapel Hill: University of North Carolina Press, 1985); *Atlantic American Societies from Columbus through Abolition* (coedited, London: Routledge, 1992); *The Mountains of the Mediterranean World* (New York: Cambridge University Press); *The Environmental History of the Pacific World* (edited, London: Variorum, 2001); *Something New Under the Sun: An Environmental History of the Twentieth-century World* (New York: Norton, 2000), co-winner of the World History Association book prize, the Forest History Society book prize, and runner-up for the BP Natural World book prize, and translated into six languages; and most recently *The Human Web: A Bird's-eye View of World History* (New York: Norton, 2003), coauthored with his father William H. McNeill. He also edited or coedited five more books, including the *Encyclopedia of World Environmental*

History (New York: Routledge, 2003). He is currently working on a history of yellow fever in the Americas from the 17th through the 20th centuries.

Amy Nelson received her Ph.D. from the University of Michigan and is currently an associate professor of history at Virginia Tech. A specialist in Russian and Soviet Culture, her current research focuses on the significance of non-human animals in Russian-Soviet History. She is writing a collective biography of the Soviet space dogs and, together with Jane Costlow (Bates College), is editing a volume of essays entitled, *The Other Animals: Situating the Non-Human in Russian Culture and History.* Nelson is the author of *Music for the Revolution. Musicians and Power in Early Soviet Russia* (Penn State University Press), which received the Heldt Prize for "The Best Book by a Woman in Any Area of Slavic/East European/Eurasian Studies," from the Association of Women in Slavic Studies in 2005. Her recent publications include, "A Hearth for a Dog: The Paradoxes of Soviet Pet Keeping" in *Borders of Socialism: Private Spheres of Soviet Russia,* ed. Lewis Siegelbaum (New York, 2006) and "Accounting for Taste: Choral Circles in Early Soviet Workers' Clubs" in *Chorus and Community,* ed. Karen Ahlquist, (Chicago, 2006).

Michael J. Neufeld is chair of the Space History Division of the National Air and Space Museum, Smithsonian Institution, Washington, DC. Born in Canada, he received history degrees from the University of Calgary and the University of British Columbia followed by a Ph.D. in modern European history from The Johns Hopkins University in 1984. Before Dr. Neufeld came to the National Air and Space Museum in 1988 as an A. Verville Fellow, he taught at various universities in upstate New York. In 1989-1990 he held Smithsonian and NSF fellowships at NASM. In 1990, he was hired as a Museum Curator in the Aeronautics Division, where he remained until early 1999. After transferring to the Space History Division, he took over the collection of German World War II missiles and, from 2003-2007, the collection of Mercury and Gemini spacecraft and components. In fall 2001, he was a Senior Lecturer at The Johns Hopkins University in Baltimore. He was named Chair of Space History in January 2007. In addition to authoring numerous scholarly articles, Dr. Neufeld has written three books: *The Skilled Metalworkers of Nuremberg: Craft and Class in the Industrial Revolution* (1989), *The Rocket and the Reich: Peenemünde and the Coming of the Ballistic Missile Era* (1995), which won two book prizes, and *Von Braun: Dreamer of Space, Engineer of War*, which is forthcoming in September 2007. He has also edited Yves Béon's memoir *Planet Dora* (1997) and is the coeditor of *The Bombing of Auschwitz: Should the Allies Have Attempted It?* (2000).

Emily S. Rosenberg is professor of history at the University of California, Irvine. Two of her books, *Spreading the American Dream: American Economic and Cultural Expansion, 1890-1945* and *Financial Missionaries to the World: The Politics* *and Culture of Dollar Diplomacy, 1900-1930*, deal with the intersections of culture and economics in U.S. international relations. Her most recent book, *A Date Which Will Live: Pearl Harbor in American Memory* (also translated into Japanese), examines the issue of collective historical memory in a media age. She is a coauthor of *Liberty, Equality, Power: A History of the American People* (5th ed., 2007). She has served as president of the Society for Historians of American Foreign Relations (SHAFR); an editor of the *Oxford Companion to United States History*; a board member of the Organization of American Historians; and coedits the *American Encounters, Global Interactions* book series for Duke University Press.

Asif A. Siddiqi is assistant professor of history at Fordham University in New York. He specializes in the social and cultural history of technology and modern Russian history. His forthcoming book, *The Rockets' Red Glare: Spaceflight and the Russian Imagination, 1857-1957* (Cambridge University Press, 2008) is the first archive-based study on the social, cultural, and technological forces that made Sputnik possible.

Michael Soluri is a New York City-based photographer. His work has been published in editorial magazines like *Wired, Time, Discover, BBC Horizons,* and *GEO*, as well as in corporate, institutional, and nonprofit multimedia communications. He is a contributing editor and photographer for *Discover, Space.Com* and *Ad Astra*. Profiled in *Photo District News* and on *Space.Com* for his expertise in the photography and editing of human and robotic space exploration, he has lectured at the Smithsonian Institute and at the National Science Foundation. In an 18-month photographic documentation of the last service mission to the Hubble Space Telescope, Soluri secured exclusive access to the integration of flight hardware, EVA tools, engineering personnel, and the crew of SM4 that resulted in the first creatively controlled portrait session of an astronaut crew in more than 25 years. He was also invited by the crew to present a photo seminar on making more communicative, insightful photographs during their historic mission to the Hubble Space Telescope. In addition, since 2005, Soluri has been following and documenting the project scientists and technicians with NASA's New Horizons mission to Pluto and the Kuiper Belt. Currently published in eight languages, Soluri is coauthor and picture editor of *What's Out There—Images from Here to the Edge of the Universe* and *Cosmos—Images from Here to the Edge of the Universe,* for which he secured Stephen Hawking to write these books' forewords. He was a contributing editor for *The History of Space Travel*, a special edition of *Discover* commemorating 50 years of spaceflight. A former professor of photographic studies at the Rochester Institute of Technology, Soluri is currently adjunct faculty at Pratt Institute in New York City. He holds an MFA in photography from the Rochester Institute of Technology.

Margaret A. Weitekamp is a Curator in the Division of Space History at the National Air and Space Museum, Smithsonian Institution, in Washington, DC. As curator of the Social and Cultural Dimensions of Spaceflight collection, she oversees over 4,000 individual pieces of space memorabilia and space science fiction objects. These social and cultural products of the Space Age—including toys and games, clothing, stamps, medals and awards, buttons and pins, comics and trading cards—round out the story of spaceflight told by the museum's collection of space hardware and technologies.

Her book *Right Stuff, Wrong Sex: America's First Women In Space Program* (published by the Johns Hopkins University Press) won the Eugene M. Emme Award for Astronautical Literature given by the American Astronautical Society. The book reconstructs the history of a privately funded project that tested female pilots for astronaut fitness at the beginning of the Space Age. In addition, Weitekamp has also contributed to the anthology *Impossible to Hold: Women and Culture in the 1960s*, ed. Avital Bloch and Lauri Umansky (New York University Press, 2005). Weitekamp won the Smithsonian Institution's National Air and Space Museum Aviation/Space Writers Award in 2002 and served as an interviewer for *The Infinite Journey: Eyewitness Accounts of NASA and the Age of Space* (Discovery Channel Publishing, 2000). She spent the academic year 1997-1998 in residence at the National Aeronautics and Space Administration Headquarters History Division in Washington, DC, as the American Historical Association/NASA Aerospace History Fellow. She is a 1993 Mellon Fellow in the humanities. Weitekamp received her B.A. summa cum laude from the University of Pittsburgh and her Ph.D. in history at Cornell University in 2001. Before joining the Smithsonian Institution, Weitekamp taught for three years as an assistant professor in the women's studies program at Hobart and William Smith Colleges in Geneva, New York.

Acronyms and Abbreviations

AAAS	American Academy of Arts and Sciences
ABM	Antiballistic Missile
ABMA	Army Ballistic Missile Agency
AEC	Atomic Energy Commission
ARPA	Advanced Research Projects Agency
ASAT	Anti–Satellite
CaLV	Cargo Launch Vehicle
CERN	European Organization for Nuclear Research
CIA	Central Intelligence Agency
CLV	Crew Launch Vehicle
CNES	French National Space Agency (Centre Nationale des Études Spatiales)
CNSA	China National Space Administration
CNTA	China's National Tourism Administration
COPUOS	United Nations Committee on the Peaceful Uses of Outer Space
COSPAR	Committee on Space Research
DOD	Department of Defense
DOT	Department of Transportation
EEO	Equal Employment Opportunity
ELDO	European Launcher Development Organization .
ELV	Expendable Launch Vehicle
ESA	European Space Agency
ESRO	European Space Research Organization
EVA	Extra Vehicular Activity
FSA	Farm Securities Administration
GLONASS	GLObal Navigation Satellite System
GPS	Global Positioning System

ICBM	Intercontinental Ballistic Missile
IGY	International Geophysical Year
INKhUK	Moscow Institute of Artistic Culture
ISPM	International Solar Polar Mission
ISS	International Space Station
ITAR	International Traffic in Arms Regulations
JSC	Johnson Space Center
LEM	Lunar Excursion Module
MAD	Mutual Assured Destruction
MESA	Modularized Equipment Stowage Assembly
MLS	Manana Literary Society
MMU	Manned Maneuvering Unit
MNC	Multinational Corporation
MRI	Midwestern Research Institute
MRO	Mars Reconnaissance Orbiter
NACA	National Advisory Committee for Aeronautics
NAMC	Navy Air Materials Center
NASM	National Air and Space Museum
NASP	National Aero-Space Plane
NDEA	National Defense Education Act
NEAR	Near Earth Asteroid Rendezvous
NEP	New Economic Policy
NGO	Non-Governmental Organization
NKVD	People's Commissariat Internal Affairs
NSC	National Security Council
NSDD	National Security Decision Directive
NST	Nuclear and Space Talks
OCST	Office of Commercial Space Transportation
ODM	Office of Defense Mobilization
OPF	Orbiter Processing Facilities
OSI	Office of Special Investigations
OTA	Office of Technology Assessment

PAO	Public Affairs Office
PSAC	President's Science Advisory Committee
R&D	Research and Development
SAC	Strategic Air Command
SALT	Strategic Arms Limitation Talks
SAO	Smithsonian Astrophysical Observatory
SAR	Synthetic Aperture Radar
SDI	Strategic Defense Initiative
SEI	Space Exploration Initiative
SIG (Space)	Senior Interagency Group for Space
SLBM	Submarine-Launched Ballistic Missile
SOHO	Solar and Heliosperic Observatory
SRTM	Shuttle Radar Topography Mission
START	Strategic Arms Reduction Treaty
TCP	Technological Capabilities Panel
TVA	Tennessee Valley Authority
UAH	University of Alabama in Huntsville
UNESCO	United Nations Educational Scientific and Cultural Organization
USAAF	United States Army Air Force
USAF	U.S. Air Force
VAB	Vehicle Assembly Building
VDNKh	Exhibition of Achievements of the National Economy
VSE	Vision for Space Exploration
WPA	Work Projects Administration

THE NASA HISTORY SERIES

Reference Works, NASA SP-4000:

Grimwood, James M. *Project Mercury: A Chronology*. NASA SP-4001, 1963.

Grimwood, James M., and Barton C. Hacker, with Peter J. Vorzimmer. *Project Gemini Technology and Operations: A Chronology*. NASA SP-4002, 1969.

Link, Mae Mills. *Space Medicine in Project Mercury*. NASA SP-4003, 1965.

Astronautics and Aeronautics, 1963: Chronology of Science, Technology, and Policy. NASA SP-4004, 1964.

Astronautics and Aeronautics, 1964: Chronology of Science, Technology, and Policy. NASA SP-4005, 1965.

Astronautics and Aeronautics, 1965: Chronology of Science, Technology, and Policy. NASA SP-4006, 1966.

Astronautics and Aeronautics, 1966: Chronology of Science, Technology, and Policy. NASA SP-4007, 1967.

Astronautics and Aeronautics, 1967: Chronology of Science, Technology, and Policy. NASA SP-4008, 1968.

Ertel, Ivan D., and Mary Louise Morse. *The Apollo Spacecraft: A Chronology, Volume I, Through November 7, 1962*. NASA SP-4009, 1969.

Morse, Mary Louise, and Jean Kernahan Bays. *The Apollo Spacecraft: A Chronology, Volume II, November 8, 1962–September 30, 1964*. NASA SP-4009, 1973.

Brooks, Courtney G., and Ivan D. Ertel. *The Apollo Spacecraft: A Chronology, Volume III, October 1, 1964–January 20, 1966*. NASA SP-4009, 1973.

Ertel, Ivan D., and Roland W. Newkirk, with Courtney G. Brooks. *The Apollo Spacecraft: A Chronology, Volume IV, January 21, 1966–July 13, 1974*. NASA SP-4009, 1978.

Astronautics and Aeronautics, 1968: Chronology of Science, Technology, and Policy. NASA SP-4010, 1969.

Newkirk, Roland W., and Ivan D. Ertel, with Courtney G. Brooks. *Skylab: A Chronology*. NASA SP-4011, 1977.

Van Nimmen, Jane, and Leonard C. Bruno, with Robert L. Rosholt. *NASA Historical Data Book, Vol. I: NASA Resources, 1958–1968*. NASA SP-4012, 1976, rep. ed. 1988.

Ezell, Linda Neuman. *NASA Historical Data Book, Vol. II: Programs and Projects, 1958–1968*. NASA SP-4012, 1988.

Ezell, Linda Neuman. *NASA Historical Data Book, Vol. III: Programs and Projects, 1969–1978*. NASA SP-4012, 1988.

Gawdiak, Ihor, with Helen Fedor. *NASA Historical Data Book, Vol. IV: NASA Resources, 1969–1978*. NASA SP-4012, 1994.

Rumerman, Judy A. *NASA Historical Data Book, Vol. V: NASA Launch Systems, Space Transportation, Human Spaceflight, and Space Science, 1979–1988*. NASA SP-4012, 1999.

Rumerman, Judy A. *NASA Historical Data Book, Vol. VI: NASA Space Applications, Aeronautics and Space Research and Technology, Tracking and Data Acquisition/Support Operations, Commercial Programs, and Resources, 1979–1988*. NASA SP-4012, 1999.

Astronautics and Aeronautics, 1969: Chronology of Science, Technology, and Policy. NASA SP-4014, 1970.

Astronautics and Aeronautics, 1970: Chronology of Science, Technology, and Policy. NASA SP-4015, 1972.

Astronautics and Aeronautics, 1971: Chronology of Science, Technology, and Policy. NASA SP-4016, 1972.

Astronautics and Aeronautics, 1972: Chronology of Science, Technology, and Policy. NASA SP-4017, 1974.

Astronautics and Aeronautics, 1973: Chronology of Science, Technology, and Policy. NASA SP-4018, 1975.

Astronautics and Aeronautics, 1974: Chronology of Science, Technology, and Policy. NASA SP-4019, 1977.

Astronautics and Aeronautics, 1975: Chronology of Science, Technology, and Policy. NASA SP-4020, 1979.

Astronautics and Aeronautics, 1976: Chronology of Science, Technology, and Policy. NASA SP-4021, 1984.

Astronautics and Aeronautics, 1977: Chronology of Science, Technology, and Policy. NASA SP-4022, 1986.

Astronautics and Aeronautics, 1978: Chronology of Science, Technology, and Policy. NASA SP-4023, 1986.

Astronautics and Aeronautics, 1979–1984: Chronology of Science, Technology, and Policy. NASA SP-4024, 1988.

Astronautics and Aeronautics, 1985: Chronology of Science, Technology, and Policy. NASA SP-4025, 1990.

Noordung, Hermann. *The Problem of Space Travel: The Rocket Motor.* Edited by Ernst Stuhlinger and J.D. Hunley, with Jennifer Garland. NASA SP-4026, 1995.

Astronautics and Aeronautics, 1986–1990: A Chronology. NASA SP-4027, 1997.

Astronautics and Aeronautics, 1991–1995: A Chronology. NASA SP-2000-4028, 2000.

Orloff, Richard W. *Apollo by the Numbers: A Statistical Reference.* NASA SP-2000-4029, 2000.

Management Histories, NASA SP-4100:

Rosholt, Robert L. *An Administrative History of NASA, 1958–1963.* NASA SP-4101, 1966.

Levine, Arnold S. *Managing NASA in the Apollo Era.* NASA SP-4102, 1982.

Roland, Alex. *Model Research: The National Advisory Committee for Aeronautics, 1915–1958.* NASA SP-4103, 1985.

Fries, Sylvia D. *NASA Engineers and the Age of Apollo.* NASA SP-4104, 1992.

Glennan, T. Keith. *The Birth of NASA: The Diary of T. Keith Glennan.* Edited by J.D. Hunley. NASA SP-4105, 1993.

Seamans, Robert C. *Aiming at Targets: The Autobiography of Robert C. Seamans.* NASA SP-4106, 1996.

Garber, Stephen J., editor. *Looking Backward, Looking Forward: Forty Years of Human Spaceflight Symposium.* NASA SP-2002-4107, 2002.

Mallick, Donald L. with Peter W. Merlin. *The Smell of Kerosene: A Test Pilot's Odyssey.* NASA SP-4108, 2003.

Iliff, Kenneth W. and Curtis L. Peebles. *From Runway to Orbit: Reflections of a NASA Engineer.* NASA SP-2004-4109, 2004.

Chertok, Boris. *Rockets and People, Volume 1.* NASA SP-2005-4110, 2005.

Chertok, Boris. *Rockets and People: Creating a Rocket Industry, Volume II.* NASA SP-2006-4110, 2006.

Laufer, Alexander, Todd Post, and Edward Hoffman. *Shared Voyage: Learning and Unlearning from Remarkable Projects.* NASA SP-2005-4111, 2005.

Dawson, Virginia P., and Mark D. Bowles. *Realizing the Dream of Flight: Biographical Essays in Honor of the Centennial of Flight, 1903–2003.* NASA SP-2005-4112, 2005.

Mudgway, Douglas J. *William H. Pickering: America's Deep Space Pioneer.* NASA SP-2008-4113,

Project Histories, NASA SP-4200:

Swenson, Loyd S., Jr., James M. Grimwood, and Charles C. Alexander. *This New Ocean: A History of Project Mercury*. NASA SP-4201, 1966; reprinted 1999.

Green, Constance McLaughlin, and Milton Lomask. *Vanguard: A History*. NASA SP-4202, 1970; rep. ed. Smithsonian Institution Press, 1971.

Hacker, Barton C., and James M. Grimwood. *On Shoulders of Titans: A History of Project Gemini*. NASA SP-4203, 1977, reprinted 2002.

Benson, Charles D., and William Barnaby Faherty. *Moonport: A History of Apollo Launch Facilities and Operations*. NASA SP-4204, 1978.

Brooks, Courtney G., James M. Grimwood, and Loyd S. Swenson, Jr. *Chariots for Apollo: A History of Manned Lunar Spacecraft*. NASA SP-4205, 1979.

Bilstein, Roger E. *Stages to Saturn: A Technological History of the Apollo/ Saturn Launch Vehicles*. NASA SP-4206, 1980 and 1996.

Compton, W. David, and Charles D. Benson. *Living and Working in Space: A History of Skylab*. NASA SP-4208, 1983.

Ezell, Edward Clinton, and Linda Neuman Ezell. *The Partnership: A History of the Apollo-Soyuz Test Project*. NASA SP-4209, 1978.

Hall, R. Cargill. *Lunar Impact: A History of Project Ranger*. NASA SP-4210, 1977.

Newell, Homer E. *Beyond the Atmosphere: Early Years of Space Science*. NASA SP-4211, 1980.

Ezell, Edward Clinton, and Linda Neuman Ezell. *On Mars: Exploration of the Red Planet, 1958–1978*. NASA SP-4212, 1984.

Pitts, John A. *The Human Factor: Biomedicine in the Manned Space Program to 1980*. NASA SP-4213, 1985.

Compton, W. David. *Where No Man Has Gone Before: A History of Apollo Lunar Exploration Missions*. NASA SP-4214, 1989.

Naugle, John E. *First Among Equals: The Selection of NASA Space Science Experiments*. NASA SP-4215, 1991.

Wallace, Lane E. *Airborne Trailblazer: Two Decades with NASA Langley's 737 Flying Laboratory*. NASA SP-4216, 1994.

Butrica, Andrew J., ed. *Beyond the Ionosphere: Fifty Years of Satellite Communications*. NASA SP-4217, 1997.

Butrica, Andrew J. *To See the Unseen: A History of Planetary Radar Astronomy*. NASA SP-4218, 1996.

Mack, Pamela E., ed. *From Engineering Science to Big Science: The NACA and NASA Collier Trophy Research Project Winners*. NASA SP-4219, 1998.

Reed, R. Dale. *Wingless Flight: The Lifting Body Story*. NASA SP-4220, 1998.

Heppenheimer, T. A. *The Space Shuttle Decision: NASA's Search for a Reusable Space Vehicle*. NASA SP-4221, 1999.

Hunley, J. D., ed. *Toward Mach 2: The Douglas D-558 Program*. NASA SP-4222, 1999.

Swanson, Glen E., ed. *"Before This Decade is Out . . ." Personal Reflections on the Apollo Program*. NASA SP-4223, 1999.

Tomayko, James E. *Computers Take Flight: A History of NASA's Pioneering Digital Fly-By-Wire Project*. NASA SP-4224, 2000.

Morgan, Clay. *Shuttle-Mir: The United States and Russia Share History's Highest Stage*. NASA SP-2001-4225.

Leary, William M. *We Freeze to Please: A History of NASA's Icing Research Tunnel and the Quest for Safety*. NASA SP-2002-4226, 2002.

Mudgway, Douglas J. *Uplink-Downlink: A History of the Deep Space Network, 1957–1997*. NASA SP-2001-4227.

Dawson, Virginia P., and Mark D. Bowles. *Taming Liquid Hydrogen: The Centaur Upper Stage Rocket, 1958–2002*. NASA SP-2004-4230.

Meltzer, Michael. *Mission to Jupiter: A History of the Galileo Project*. NASA SP-2007-4231.

Heppenheimer, T. A. *Facing the Heat Barrier: A History of Hypersonics*. NASA SP-2007-4232.

Tsiao, Sunny. *"Read You Loud and Clear!" The Story of NASA's Spaceflight Tracking and Data Network*. NASA SP-2007-4233.

Center Histories, NASA SP-4300:

Rosenthal, Alfred. *Venture into Space: Early Years of Goddard Space Flight Center*. NASA SP-4301, 1985.

Hartman, Edwin, P. *Adventures in Research: A History of Ames Research Center, 1940–1965*. NASA SP-4302, 1970.

Hallion, Richard P. *On the Frontier: Flight Research at Dryden, 1946–1981*. NASA SP-4303, 1984.

Muenger, Elizabeth A. *Searching the Horizon: A History of Ames Research Center, 1940–1976*. NASA SP-4304, 1985.

Hansen, James R. *Engineer in Charge: A History of the Langley Aeronautical Laboratory, 1917–1958*. NASA SP-4305, 1987.

Dawson, Virginia P. *Engines and Innovation: Lewis Laboratory and American Propulsion Technology*. NASA SP-4306, 1991.

Dethloff, Henry C. *"Suddenly Tomorrow Came . . .": A History of the Johnson Space Center, 1957–1990.* NASA SP-4307, 1993.

Hansen, James R. *Spaceflight Revolution: NASA Langley Research Center from Sputnik to Apollo.* NASA SP-4308, 1995.

Wallace, Lane E. *Flights of Discovery: An Illustrated History of the Dryden Flight Research Center.* NASA SP-4309, 1996.

Herring, Mack R. *Way Station to Space: A History of the John C. Stennis Space Center.* NASA SP-4310, 1997.

Wallace, Harold D., Jr. *Wallops Station and the Creation of an American Space Program.* NASA SP-4311, 1997.

Wallace, Lane E. *Dreams, Hopes, Realities. NASA's Goddard Space Flight Center: The First Forty Years.* NASA SP-4312, 1999.

Dunar, Andrew J., and Stephen P. Waring. *Power to Explore: A History of Marshall Space Flight Center, 1960–1990.* NASA SP-4313, 1999.

Bugos, Glenn E. *Atmosphere of Freedom: Sixty Years at the NASA Ames Research Center.* NASA SP-2000-4314, 2000.

Schultz, James. *Crafting Flight: Aircraft Pioneers and the Contributions of the Men and Women of NASA Langley Research Center.* NASA SP-2003-4316, 2003.

Bowles, Mark D. *Science in Flux: NASA's Nuclear Program at Plum Brook Station, 1955–2005.* NASA SP-2006-4317.

Wallace, Lane E. *Flights of Discovery: An Illustrated History of the Dryden Flight Research Center.* NASA SP-4318, 2007. Revised version of SP-4309.

General Histories, NASA SP-4400:

Corliss, William R. *NASA Sounding Rockets, 1958–1968: A Historical Summary.* NASA SP-4401, 1971.

Wells, Helen T., Susan H. Whiteley, and Carrie Karegeannes. *Origins of NASA Names.* NASA SP-4402, 1976.

Anderson, Frank W., Jr. *Orders of Magnitude: A History of NACA and NASA, 1915–1980.* NASA SP-4403, 1981.

Sloop, John L. *Liquid Hydrogen as a Propulsion Fuel, 1945–1959.* NASA SP-4404, 1978.

Roland, Alex. *A Spacefaring People: Perspectives on Early Spaceflight.* NASA SP-4405, 1985.

Bilstein, Roger E. *Orders of Magnitude: A History of the NACA and NASA, 1915–1990.* NASA SP-4406, 1989.

Logsdon, John M., ed., with Linda J. Lear, Jannelle Warren Findley, Ray A. Williamson, and Dwayne A. Day. *Exploring the Unknown: Selected Documents in the History of the U.S. Civil Space Program, Volume I, Organizing for Exploration.* NASA SP-4407, 1995.

Logsdon, John M., ed, with Dwayne A. Day, and Roger D. Launius. *Exploring the Unknown: Selected Documents in the History of the U.S. Civil Space Program, Volume II, External Relationships.* NASA SP-4407, 1996.

Logsdon, John M., ed., with Roger D. Launius, David H. Onkst, and Stephen J. Garber. *Exploring the Unknown: Selected Documents in the History of the U.S. Civil Space Program, Volume III, Using Space.* NASA SP-4407,1998.

Logsdon, John M., ed., with Ray A. Williamson, Roger D. Launius, Russell J. Acker, Stephen J. Garber, and Jonathan L. Friedman. *Exploring the Unknown: Selected Documents in the History of the U.S. Civil Space Program, Volume IV, Accessing Space.* NASA SP-4407, 1999.

Logsdon, John M., ed., with Amy Paige Snyder, Roger D. Launius, Stephen J. Garber, and Regan Anne Newport. *Exploring the Unknown: Selected Documents in the History of the U.S. Civil Space Program, Volume V, Exploring the Cosmos.* NASA SP-4407, 2001.

Logsdon, John M., ed., with Stephen J. Garber, Roger D. Launius, and Ray A. Williamson. *Exploring the Unknown: Selected Documents in the History of the U.S. Civil Space Program, Volume VI: Space and Earth Science.* NASA SP-2004-4407, 2004.

Logsdon, John M., ed., with Roger D. Launius. *Exploring the Unknown: Selected Documents in the History of the U.S. Civil Space Program, Volume VII: Human Spaceflight: Projects Mercury, Gemini, and Apollo.* NASA SP-2008-4407, 2008.

Siddiqi, Asif A., *Challenge to Apollo: The Soviet Union and the Space Race, 1945–1974.* NASA SP-2000-4408, 2000.

Hansen, James R., ed. *The Wind and Beyond: Journey into the History of Aerodynamics in America, Volume 1, The Ascent of the Airplane.* NASA SP-2003-4409, 2003.

Hansen, James R., ed. *The Wind and Beyond: Journey into the History of Aerodynamics in America, Volume 2, Reinventing the Airplane.* NASA SP-2007-4409, 2007.

Hogan, Thor. *Mars Wars: The Rise and Fall of the Space Exploration Initiative.* NASA SP-2007-4410, 2007.

Monographs in Aerospace History (SP-4500 Series):

Launius, Roger D., and Aaron K. Gillette, comps. *Toward a History of the Space Shuttle: An Annotated Bibliography.* Monograph in Aerospace History, No. 1, 1992.

Launius, Roger D., and J. D. Hunley, comps. *An Annotated Bibliography of the Apollo Program.* Monograph in Aerospace History No. 2, 1994.

Launius, Roger D. *Apollo: A Retrospective Analysis.* Monograph in Aerospace History, No. 3, 1994.

Hansen, James R. *Enchanted Rendezvous: John C. Houbolt and the Genesis of the Lunar-Orbit Rendezvous Concept.* Monograph in Aerospace History, No. 4, 1995.

Gorn, Michael H. *Hugh L. Dryden's Career in Aviation and Space.* Monograph in Aerospace History, No. 5, 1996.

Powers, Sheryll Goecke. *Women in Flight Research at NASA Dryden Flight Research Center from 1946 to 1995.* Monograph in Aerospace History, No. 6, 1997.

Portree, David S. F., and Robert C. Trevino. *Walking to Olympus: An EVA Chronology.* Monograph in Aerospace History, No. 7, 1997.

Logsdon, John M., moderator. *Legislative Origins of the National Aeronautics and Space Act of 1958: Proceedings of an Oral History Workshop.* Monograph in Aerospace History, No. 8, 1998.

Rumerman, Judy A., comp. *U.S. Human Spaceflight, A Record of Achievement 1961–1998.* Monograph in Aerospace History, No. 9, 1998.

Portree, David S. F. *NASA's Origins and the Dawn of the Space Age.* Monograph in Aerospace History, No. 10, 1998.

Logsdon, John M. *Together in Orbit: The Origins of International Cooperation in the Space Station.* Monograph in Aerospace History, No. 11, 1998.

Phillips, W. Hewitt. *Journey in Aeronautical Research: A Career at NASA Langley Research Center.* Monograph in Aerospace History, No. 12, 1998.

Braslow, Albert L. *A History of Suction-Type Laminar-Flow Control with Emphasis on Flight Research.* Monograph in Aerospace History, No. 13, 1999.

Logsdon, John M., moderator. *Managing the Moon Program: Lessons Learned From Apollo.* Monograph in Aerospace History, No. 14, 1999.

Perminov, V. G. *The Difficult Road to Mars: A Brief History of Mars Exploration in the Soviet Union.* Monograph in Aerospace History, No. 15, 1999.

Tucker, Tom. *Touchdown: The Development of Propulsion Controlled Aircraft at NASA Dryden.* Monograph in Aerospace History, No. 16, 1999.

Maisel, Martin, Demo J. Giulanetti, and Daniel C. Dugan. *The History of the XV-15 Tilt Rotor Research Aircraft: From Concept to Flight.* Monograph in Aerospace History, No. 17, 2000. NASA SP-2000-4517.

Jenkins, Dennis R. *Hypersonics Before the Shuttle: A Concise History of the X-15 Research Airplane.* Monograph in Aerospace History, No. 18, 2000. NASA SP-2000-4518.

Chambers, Joseph R. *Partners in Freedom: Contributions of the Langley Research Center to U.S. Military Aircraft of the 1990s.* Monograph in Aerospace History, No. 19, 2000. NASA SP-2000-4519.

Waltman, Gene L. *Black Magic and Gremlins: Analog Flight Simulations at NASA's Flight Research Center.* Monograph in Aerospace History, No. 20, 2000. NASA SP-2000-4520.

Portree, David S. F. *Humans to Mars: Fifty Years of Mission Planning, 1950–2000.* Monograph in Aerospace History, No. 21, 2001. NASA SP-2001-4521.

Thompson, Milton O., with J. D. Hunley. *Flight Research: Problems Encountered and What they Should Teach Us.* Monograph in Aerospace History, No. 22, 2001. NASA SP-2001-4522.

Tucker, Tom. *The Eclipse Project.* Monograph in Aerospace History, No. 23, 2001. NASA SP-2001-4523.

Siddiqi, Asif A. *Deep Space Chronicle: A Chronology of Deep Space and Planetary Probes 1958–2000.* Monograph in Aerospace History, No. 24, 2002. NASA SP-2002-4524.

Merlin, Peter W. *Mach 3+: NASA/USAF YF-12 Flight Research, 1969–1979.* Monograph in Aerospace History, No. 25, 2001. NASA SP-2001-4525.

Anderson, Seth B. *Memoirs of an Aeronautical Engineer: Flight Tests at Ames Research Center: 1940–1970.* Monograph in Aerospace History, No. 26, 2002. NASA SP-2002-4526.

Renstrom, Arthur G. *Wilbur and Orville Wright: A Bibliography Commemorating the One-Hundredth Anniversary of the First. Powered Flight on December 17, 1903.* Monograph in Aerospace History, No. 27, 2002. NASA SP-2002-4527.

No monograph 28.

Chambers, Joseph R. *Concept to Reality: Contributions of the NASA Langley Research Center to U.S. Civil Aircraft of the 1990s.* Monograph in Aerospace History, No. 29, 2003. SP-2003-4529.

Peebles, Curtis, editor. *The Spoken Word: Recollections of Dryden History, The Early Years.* Monograph in Aerospace History, No. 30, 2003. SP-2003-4530.

Jenkins, Dennis R., Tony Landis, and Jay Miller. *American X-Vehicles: An Inventory- X-1 to X-50.* Monograph in Aerospace History, No. 31, 2003. SP-2003-4531.

Renstrom, Arthur G. *Wilbur and Orville Wright: A Chronology Commemorating the One-Hundredth Anniversary of the First Powered Flight on December 17, 1903.* Monograph in Aerospace History, No. 32, 2003. NASA SP-2003-4532.

Bowles, Mark D., and Robert S. Arrighi. *NASA's Nuclear Frontier: The Plum Brook Research Reactor.* Monograph in Aerospace History, No. 33, 2004. (SP-2004-4533).

Matranga, Gene J., C. Wayne Ottinger, Calvin R. Jarvis, and D. Christian Gelzer. *Unconventional, Contrary, and Ugly: The Lunar Landing Research Vehicle.* Monograph in Aerospace History, No. 35, 2006. NASA SP-2004-4535.

McCurdy, Howard E. *Low Cost Innovation in Spaceflight: The History of the Near Earth Asteroid Rendezvous (NEAR) Mission.* Monograph in Aerospace History, No. 36, 2005. NASA SP-2005-4536.

Seamans, Robert C., Jr. *Project Apollo: The Tough Decisions.* Monograph in Aerospace History, No. 37, 2005. NASA SP-2005-4537.

Lambright, W. Henry. *NASA and the Environment: The Case of Ozone Depletion.* Monograph in Aerospace History, No. 38, 2005. NASA SP-2005-4538.

Chambers, Joseph R. *Innovation in Flight: Research of the NASA Langley Research Center on Revolutionary Advanced Concepts for Aeronautics.* Monograph in Aerospace History, No. 39, 2005. NASA SP-2005-4539.

Phillips, W. Hewitt. *Journey Into Space Research: Continuation of a Career at NASA Langley Research Center.* Monograph in Aerospace History, No. 40, 2005. NASA SP-2005-4540.

Rumerman, Judy A., Chris Gamble, and Gabriel Okolski, compilers. *U.S. Human Spaceflight: A Record of Achievement, 1961–2006.* Monograph in Aerospace History No. 41, 2007. NASA SP-2007-4541.

Dryden Historical Studies

Tomayko, James E., author, and Christian Gelzer, editor. *The Story of Self-Repairing Flight Control Systems.* Dryden Historical Study #1.

Electronic Media (SP-4600 Series)

Remembering Apollo 11: The 30th Anniversary Data Archive CD-ROM. NASA SP-4601, 1999.

Remembering Apollo 11: The 35th Anniversary Data Archive CD-ROM. NASA SP-2004-4601, 2004. This is an update of the 1999 edition.

The Mission Transcript Collection: U.S. Human Spaceflight Missions from Mercury Redstone 3 to Apollo 17. SP-2000-4602, 2001. Now available commerically from CG Publishing.

Shuttle-Mir: the United States and Russia Share History's Highest Stage. NASA SP-2001-4603, 2002. This CD-ROM is available from NASA CORE.

U.S. Centennial of Flight Commission presents Born of Dreams ~ Inspired by Freedom. NASA SP-2004-4604, 2004.

Of Ashes and Atoms: A Documentary on the NASA Plum Brook Reactor Facility. NASA SP-2005-4605.

Taming Liquid Hydrogen: The Centaur Upper Stage Rocket Interactive CD-ROM. NASA SP-2004-4606, 2004.

Fueling Space Exploration: The History of NASA's Rocket Engine Test Facility DVD. NASA SP-2005-4607.

Conference Proceedings (SP-4700 Series)

Dick, Steven J., and Keith Cowing, ed. *Risk and Exploration: Earth, Sea and the Stars.* NASA SP-2005-4701.

Dick, Steven J., and Roger D. Launius. *Critical Issues in the History of Spaceflight.* NASA SP-2006-4702.

Societal Impact (SP-4800 Series)

Dick, Steven J., and Roger D. Launius. *Societal Impact of Spaceflight.* NASA SP-2007-4801.

INDEX

Numbers in **bold** indicate pages with illustrations